T0250246

Data Mining in Biomedical Imaging, Signaling, and Systems

Edited by
Sumeet Dua and Rajendra Acharya U.

CRC Press
Taylor & Francis Group
Boca Raton London New York

CRC Press is an imprint of the
Taylor & Francis Group, an **Informa** business

AN AUERBACH BOOK

Auerbach Publications
Taylor & Francis Group
6000 Broken Sound Parkway NW, Suite 300
Boca Raton, FL 33487-2742

© 2011 by Taylor and Francis Group, LLC
Auerbach Publications is an imprint of Taylor & Francis Group, an Informa business

No claim to original U.S. Government works

International Standard Book Number: 978-1-4398-3938-6 (Hardback)

Library of Congress Cataloging-in-Publication Data

Data mining in biomedical imaging, signaling, and systems / Editors, Sumeet Dua and Rajendra Acharya U.
 p. ; cm.
 Includes bibliographical references and index.
 ISBN 978-1-4398-3938-6 (hardcover : alk. paper)
 1. Data mining. 2. Medical informatics. 3. Bioinformatics. I. Dua, Sumeet, editor. II. Acharya U, Rajendra, editor.
 [DNLM: 1. Data Mining. 2. Decision Support Systems, Clinical. 3. Image Interpretation, Computer-Assisted--methods. 4. Signal Processing, Computer-Assisted. W 26.55.I4]

 R859.7.D3538 2011
 610.285--dc22   2010051040

Visit the Taylor & Francis Web site at
http://www.taylorandfrancis.com

and the Auerbach Web site at
http://www.auerbach-publications.com

Contents

Preface

Biosignals, medical images, and biosystems can effectively and efficiently characterize the health of the subject and can assist clinicians in making refined diagnoses for treatment. Because of the immense volume, heterogeneity, and complexity of data involved in these techniques, data mining has become an important amalgamation of indispensible and evolving tools for the accurate and fast implementation of biomedical applications. Novel data mining algorithms have aided in discovering and clarifying hidden information in medical data and have helped in accurately classifying pathological data from normal data. Data mining applied to medical data can help automated decision making, visualization, and extraction of hidden complex features from different patient groups and disease states. This book has 16 chapters that describe the applications of data mining in biosignals, medical imaging, and biosystems, and they explain how the applications aid in clinical diagnostic discovery. These chapters contain several examples of heterogeneous data modalities that are frequently encountered in biomedical and clinical applications, as well as detailing the applicability of fundamental data mining approaches and paradigms employed to address the computational challenges in analyzing this data for knowledge discovery.

The analysis of biosignals and medical images is significant in health care. Chapter 1 contains a description of the fundamental feature extraction methods employed in biomedical signaling and imaging. The chapter introduces feature extraction techniques for biomedical signaling, including techniques in frequency and statistical domains and information-theoretic methods. Such techniques are discussed in significant detail. A number of critical feature descriptors in biomedical signaling and imaging are also covered in this chapter.

Machine learning has been evaluated and employed in many biomedical applications for the diagnosis of diseases. The fundamentals of supervised learning and unsupervised learning are outlined in Chapter 2. Supervised and unsupervised learning methods, their performance evaluation measures, and the challenges that may have to be faced during their implementation for biosignals and medical images are also discussed in this chapter.

Depression is the world's fourth most serious health threat, and the incidence of this disease is expected to rise steadily with the aging of world population. Speech patterns of depressed speakers have often been characterized as dull, monotone, monoloud, lifeless, and metallic. These perceptual qualities have been associated with fluctuations involving fundamental frequency, formant structure, and power spectral distribution. Identifying whether individuals are affected with depression is often one of the most important judgments that physicians must make. Chapter 3 contains an analysis of depressed speech using acoustic properties, such as vocal jitter, formant frequencies, formant bandwidths, and power distribution. These parameters are extracted from dyadic wavelet transform (D_yWT) for the detection of normal and depressed speech. The results indicate the use of speech-processing tools for the detection and analysis of depressed subjects.

Biological systems are highly complex, and knowledge about them is fragmented. Thus, the creation and evaluation of clinical prediction rules are time consuming and expensive. However, with growing accessibility to electronically stored medical data and advances in data mining methods, the creation, evaluation, and utilization of predictive models can be supported by automated processes, which can induce models and patterns for large collections of patient medical records and medical data sets used in clinical trials. Chapter 4 contains an explanation of the concept of typicality from a broad perspective of cognitive psychology and of the intracategory and intercategory typicality measures; it presents a fuzzy set representation of typicality and a discussion on the applications of typicality measures used in visual classifications of objects.

Arrhythmia and ischemia are two of the most prominent life-threatening heart abnormalities. The automated screening of a large population of patients to identify those with arrhythmia and ischemia to aid the cardiologist in diagnosing these diseases is one of the biggest challenges faced by the scientific community. Chapter 5 contains a discussion of an integrated system that can identify cardiac diseases using the Gaussian mixture model (GMM) and electrocardiogram (ECG) signals. This chapter explains feature extraction, classification using GMM, and classifier performance evaluation based on error bounds on classification error. The performance of GMM is compared with k-means and fuzzy c-means clustering.

Epilepsy, one of the most common neurological disorders in the world, affects more than 60 million people (approximately 1% of the world population). Nonlinear analysis has been widely used to characterize the dynamics of transitions between states that precede the onsets of seizures. Most studies focus on finding the earliest possible time at which significant changes in system dynamics may indicate an impending seizure. Reported prediction time can range from seconds to a few minutes or even a few hours, depending on the methods used and the recording locations. Chapter 6 contains an explanation of the development of feature extraction and supervised learning related to identifying seizure-related patterns.

Chapter 7 contains a method for the automatic recognition and classification of cardiac arrhythmia. Eight types of ECG signals, including normal beat data and seven types of arrhythmic data that were extracted from the Massachusetts Institute of Technology-Beth Israel Hospital (MIT-BIH) arrhythmia database, have been chosen for this analysis. Preprocessing of the signals is performed using Pan–Tompkins algorithm. The fourth-order autoregressive model coefficients and spectral entropy of the samples around the QRS complex of the signals are used as features for classification. Comparisons of the performance indices of the classifiers reveal that the probabilistic neural network in arrhythmia classification outperforms the conventional multilayered feedforward neural networks. Thus, the chapter contains an investigation of these techniques, such as signal processing, feature extraction, and pattern recognition, to develop computer programs that automatically classify ECG signals.

A migraine is a complex neurological disorder, and migraine sufferers may present alterations in some hematological variables, have an increased risk of cerebrovascular diseases, show an impaired cerebral carbon dioxide autoregulation, and have genetic mutations. Genetic mutations of the *MTHFR* gene are correlated with migraines and with the increased risk of artery pathologies. Chapter 8 combines supervised and unsupervised metabonomic techniques to the classification of 677-*MTHFR* mutations in migraineurs. In this chapter, it is shown that metabonomics can be effectively applied in clinical practice. Results indicate that the overall correlation structure of complex systems in migraine pathology reaches a classification accuracy of roughly 90%. The results of this study confirm the importance of transcranial Doppler sonography in the metabolic profiling and follow-up studies of migraine patients.

Depression grading is a serious problem due to the subjectivity of the data/information. This is performed manually using various rating tools and, hence, is always prone to risks of human errors and bias. Often, the grading is performed as ranges, for example, mild-to-moderate or moderate-to-severe. Such ranges are often confusing for the doctors who are using this information to decide on management options. These ranges can be especially difficult to analyze when a doctor is attempting to choose an option that includes drug dosage. This problem is addressed in Chapter 9 using a backpropagation neural network for making a more accurate decision on the grade of a set of real-life depression data.

Traditional clustering methods based on conventional similarity measures are not suitable for time-series clustering. Clustering, or grouping of similar entities based on the similarity of their temporal profiles, involving a large data set is a daunting task in an experiment. Various clustering methods that take the temporal dimension of the data into account have been proposed for time-series gene expression data. Chapter 10 provides a review of alignment-based clustering approaches for time-series profiles. The method also employs the temporal relationships between and within the time-series profiles. This chapter contains an explanation of the performances of these alignment methods on many data sets, a comparison of recently proposed methods, and a discussion on their strengths and weaknesses.

A unique segmentation method for mining a three-dimensional (3-D) imaging biomarker to evaluate osteoarthritis (OA) is presented in Chapter 11. The proposed method addresses knee cartilage segmentation by constructing a 3-D smoothing B-spline active surface. An adaptive combination of edge-based and balloon parameters that enforces the capture range of external forces in the case of noises and occlusions caused by tissues is also introduced. A comparison between the results of the experiments using this method and previous 3-D validated snake segmentation show the accuracy and robustness of this new method. The resulting 3-D B-spline surface can also be extended to mining other imaging biomarkers.

Most mammogram-classification techniques can be based on either density or abnormality. Density-based classification techniques categorize mammograms into tissue density classes like fatty, glandular, and dense or into the breast imaging-reporting and data system (BIRADS) I–IV categories. Abnormality-based classification techniques perform categorization based on tissue abnormality into normal and abnormal; or normal, benign cancerous, and malignant cancerous. Chapter 12 contains an explanation of the density-based classification paradigms and abnormality-based classification.

Chapter 13 is a review of various techniques for the segmentation, analysis, and quantification of biofilm images. A combination of techniques for the segmentation of biofilm images through optimal multilevel thresholding algorithms and a set of clustering validity indices, including the determination of the best number of thresholds used for the segmentation process, is presented in this chapter. Clustering validity indices are used to find the correct number of thresholds. The results are validated through the Rand index and a quantification process performed in a laboratory. The automatic segmentation and quantification results shown in this chapter are comparable to those performed by an on-site expert.

Chapter 14 contains an analysis of applications of text mining in the medical field. Applications such as text summarization, text categorization, document retrieval, and information extraction are highlighted. Knowledge extraction from medical reports of the upper gastrointestinal (GI) tract endoscopy is discussed. The upper GI tract covers three organs: (1) esophagus, (2) stomach, and (3) duodenum. Endoscopy reports highlight the observations of these organs in text form. Text mining techniques are applied to discover the relationship between the observations of the organs in the upper GI tract. Text-classification applications as applied to medical reports are discussed in this chapter.

Information communication technology (ICT) applications in global health care are poised to render effective health care once some theoretical debates are solved. Today, mental health has gained immense attention worldwide through various organizational initiatives. As a consequence, telepsychiatry, a form of ICT used in mental health, is practiced in several international environments. Chapter 15 delves beyond the boundary of traditional telemedicine use and focuses on various research methodologies and computational scopes in mental health that are used for the screening, prediction, and management of diseases. This chapter uses knowledge engineering and management techniques to focus especially on mental health and cases of suicide. In a nutshell, the chapter swings from applications of electronic health (e-health) through telepsychiatry service to medical decision support with the help of evidence-based computational research. In the chapter, a new term, "mental health informatics," is introduced. Mental health informatics is a cutting-edge research field that is still primarily unexplored due to various technical issues and challenges, which are discussed in Chapter 15.

In recent years, there has been rapid development and widespread deployment of biomedical systems, which have progressed from single-purpose island systems such as traditional X-ray machines to massively networked health-care systems. In Chapter 16, the complexity of these systems is discussed along with the systems' engineering design methodologies. These discussions include a proposal of complex biomedical systems, which work reliably, on time, and within budget. The second half of the chapter contains a sample design for an automated mental state detection system that follows system-engineering principles.

We have made an effort in this book to provide sufficient information and methodologies for data mining that is applied to biosignals, medical images, and biosystems for the benefit of researchers, professionals, practitioners, and educators of biomedical science and engineering.

MATLAB®is a registered trademark of The MathWorks, Inc. For product information, please contact:

The MathWorks, Inc.
3 Apple Hill Drive
Natick, MA 01760-2098 USA
Tel: 508 647 7000
Fax: 508 647 7001
E-mail: info@mathworks.com
Web: www.mathworks.com

Editors

Sumeet Dua, PhD, is currently an Upchurch endowed associate professor and the coordinator of IT research at Louisiana Tech University, Ruston, Louisiana. He also serves as adjunct faculty to the School of Medicine, Louisiana State University, Health Sciences Center in New Orleans, Louisiana. He received his PhD in computer science from Louisiana State University, Baton Rouge, Louisiana.

His areas of expertise include data mining, image processing and computational decision support, pattern recognition, data warehousing, biomedical informatics, and heterogeneous distributed data integration. The National Science Foundation (NSF), the National Institutes of Health (NIH), the Air Force Research Laboratory (AFRL), the Air Force Office of Sponsored Research (AFOSR), and the Louisiana Board of Regents (LA-BoR) have funded his research with over US$3 million. He frequently serves as a study section member (expert panelist) for the NIH's Center for Scientific Review and has served as a panelist for the NSF/Computing in Science and Engineering (CISE) Directorate. Dr. Dua has chaired several conference sessions in the area of data mining and bioinformatics, and is the program chair for the 5th International Conference on Information Systems, Technology, and Management (ICISTM-2011). He has given over 26 invited talks on data mining and bioinformatics at international academic and industry arenas, advised over 25 graduate theses, and currently advises several graduate students in this field. Dr. Dua is a coinventor of two issued U.S. patents, has (co)authored over 50 publications and book chapters, and is an author/editor of 4 books in data mining and bioinformatics. Dr. Dua received the Engineering and Science Foundation Award for Faculty Excellence (2006) and the Faculty Research Recognition Award (2007); he has been recognized as a Distinguished Researcher (2004–2010) by the Louisiana Biomedical Research Network (sponsored by NIH) and has won the Outstanding Poster Award at the NIH/National Cancer Institute (NCI) caBIG–NCRI Informatics Joint Conference; Biomedical

Informatics without Borders: From Collaboration to Implementation. Dr. Dua is a senior member of the IEEE Computer Society, a senior member of the Association for Computing Machinery (ACM), and a member of SPIE and the American Association for Advancement of Science.

 **Rajendra Acharya U., PhD, DEng**, is a visiting faculty at Ngee Ann Polytechnic, Singapore; an associate faculty member at Singapore Institute of Management (SIM) University, Singapore; and an adjunct faculty at the Manipal Institute of Technology, Manipal, India. Dr. Acharya received his PhD from the National Institute of Technology, Karnataka, Surathkal, India, and DEng from the Chiba University, Japan. He has published more than 120 papers in refereed international journals (94), international conference proceedings (30), and books (9 including those that were in press at the time of this book's publication), and has an H-index of 12. Currently, Dr. Acharya is on the editorial board of the following journals: *Journal of Medical Imaging and Health Informatics*, *Open Access Medical Informatics Journal*, *Open Access Artificial Intelligence Journal*, *Journal of Biomedical Nanotechnology*, *Journal of Mechanics in Medicine and Biology*, and the *Biomedical Imaging and Intervention Journal*. He has served as the guest editor for many journals, and is a senior IEEE member and a fellow of IETE.

Contributors

Rajendra Acharya U.
Department of Electronics and
  Communication Engineering
Ngee Ann Polytechnic and SIM
  University
Singapore

Ananthakrishna T.
Department of Electronics and
  Communication Engineering
Manipal Institute of Technology
Manipal, India

Najib T. Ayas
Faculty of Medicine
University of British Columbia
Vancouver, British Columbia, Canada

Venkataraya P. Bhandary
Department of Psychology
KMC
Manipal, India

Conrad Burden
Centre for Bioinformation Science
Mathematical Sciences Institute
  and John Curtin School of
  Medical Research
The Australian National University
Canberra, Australia

Gerardo Carcamo
Center for Biotechnology and Faculty
  of Biological Sciences
University of Concepción
Concepción, Chile

Chandan Chakraborty
School of Medical Science and
  Technology
Indian Institute of Technology
Kharagpur, India

Subhagata Chattopadhyay
Department of Computer Science and
  Engineering
National Institute of Science and
  Technology
Berhampur, India

Matilde Chessa
Sardinian Mediterranean Imaging
  Research Group
Sassari, Italy

Alan Chiu
Department of Biomedical Engineering
Louisiana Tech University
Ruston, Louisiana

Nicola Culeddu
National Council of Research
Institute for Biomolecular Chemistry
Sassari, Italy

Xian Du
Computer Science Program
Louisiana Tech University
Ruston, Louisiana

Sumeet Dua
Computer Science Program
Louisiana Tech University
Ruston, Louisiana

Oliver Faust
School of Engineering
Ngee Ann Polytechnic
Singapore

Pierangela Giustetto
Department of Neuroscience
Gradenigo Hospital
Torino, Italy

Santosh D. Hajare
Department of Gastroenterology
K.L.E. Prabhakar Kore Hospital and
    Research Center
Belgaum, India

Haseena H.
Department of Electrical and
    Electronics Engineering
MES College of Engineering
Kuttippuram, India

K. Paul Joseph
Department of Electrical Engineering
National Institute of Technology
Calicut, India

Preetisha Kaur
Department of Biomedical
    Engineering
Manipal Institute of Technology
Manipal, India

Krzysztof Kielan
Department of Psychiatry
Diana, Princess of Wales Hospital
Grimsby, United Kingdom

Kumar K. B.
Department of Psychology
KMC
Manipal, India

Mila Kwiatkowska
Department of Computing Science
Thompson Rivers University
Kamloops, British Columbia,
    Canada

William Liboni
Department of Neuroscience
Gradenigo Hospital
Torino, Italy

Teik-Cheng Lim
School of Science and Technology
SIM University
Singapore

Cesare Manetti
Department of Chemistry
"La Sapienza" University
Rome, Italy

Roshan Joy Martis
School of Medical Science and
    Technology
Indian Institute of Technology
Kharagpur, India

Abraham T. Mathew
Department of Electrical
    Engineering
National Institute of Technology
Calicut, India

Filippo Molinari
Department of Electronics
Politecnico di Torino
Torino, Italy

Alioune Ngom
School of Computer Science
University of Windsor
Windsor, Ontario, Canada

Niranjan U. C.
Department of Electronics and
    Communication Engineering
Manipal Institute of Technology
Manipal, India

Fethi Rabhi
School of Computer Science and
    Engineering
The University of New South Wales
Sydney, Australia

Ajoy Kumar Ray
Department of Electronics and
    Electrical Communications
Indian Institute of Technology
Kharagpur, India

Dario Rojas
Department of Computer Science
University of Atacama
Copiapo, Chile

Luis Rueda
School of Computer Science
University of Windsor
Windsor, Ontario, Canada

C. Frank Ryan
Faculty of Medicine
University of British Columbia
Vancouver, British Columbia,
    Canada

S. S. Saraf
Research Center, Department of
    Electronics and Communications
    Engineering
Gogte Institute of Technology
Belgaum, India

Chong Wee Seong
School of Science and Technology
SIM University
Singapore

Kumara Shama
Department of Electronics and
    Communication Engineering
Manipal Institute of Technology
Manipal, India

Harpreet Singh
Computer Science Program
Louisiana Tech University
Ruston, Louisiana

Numanul Subhani
School of Computer Science
University of Windsor
Windsor, Ontario, Canada

G. R. Udupi
Vishwanathrao Deshpande Rural
    Institute of Technology
Haliyal, India

Homero Urrutia
Center for Biotechnology and Faculty
    of Biological Sciences
University of Concepción
Concepción, Chile

Maria Cristina Valerio
Department of Chemistry
"La Sapienza" University
Rome, Italy

Chapter 1

Feature Extraction Methods in Biomedical Signaling and Imaging

Xian Du and Sumeet Dua

Contents

1.1 Introduction to Biomedical Signaling and Imaging

Biomedical signaling addresses certain types of signals that are obtained from in vivo organisms. Each living organism consists of a variety of systems. Each system consists of organs, tissues, and cells that fulfill certain physiological functions. For example, a cardiovascular system pumps blood to deliver nutrients to the body. Each physiological function corresponds to a biological process, which has a group of characteristic signals. According to their origin and physics, these signals can be categorized into biochemical signals (e.g., hormones and neurotransmitters), electrical signals (e.g., potentials and currents), and mechanical signals (e.g., pressure and temperature). A normal biological process typically presents a group of stable signals with tolerable variations. The abnormal deviation of these signals can indicate pathology of diseases. The retrieval of normal and abnormal features from biomedical signals can help pathological analysis and decision making. For example, a high body temperature may indicate the flu, arrhythmias in an electrocardiogram (ECG) may indicate cardiovascular disorders, and peaks or valleys in an electroencephalogram (EEG) may help to diagnose sleep disorders.

Biomedical signals can be obtained quantitatively or qualitatively from a variety of biomedical detection instruments. These signals can be analog, digital, or data stream. Among biomedical signals, imaging is employed the most because it is noninvasive and observable for operations. As shown in Figure 1.1, biomedical imaging

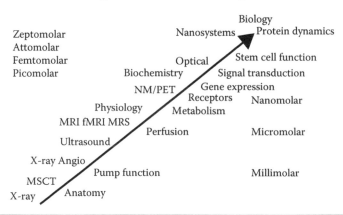

Figure 1.1 Relative sensitivity of imaging technologies. (Reprinted from *Mol Oncol* 2, Fass, L., Imaging and cancer: A review, 115–52, Copyright 2008, with permission from Elsevier.)

technologies are ubiquitous in disease measurement. The measured data are used for biomedical diagnosis and analysis after signal and image processing.

In this chapter, we first introduce biomedical signaling techniques. Second, we focus on biomedical imaging techniques that compose principal components in biomedical signaling.

1.2 Biomedical Signaling

The obtained biomedical data are of two parts: (1) signals and (2) noise. In general, a biomedical signal is defined as the useful information of the in vivo system, and noise refers to the undesired information involved in data sources. Biomedical noises originate from two sources: (1) measurement process and (2) biological system. Noises from the measurement process are caused by the limited capacity of measurement devices or software, for example, nonuniform lighting conditions cause microscopy images to have nonuniform distribution of intensity in the background. Noises from the biological system are caused by the occlusion of other uninteresting tissues in the region of interest (ROI), for example, lesions and liquids generate similar intensities with cartilage and bone in magnetic resonance imaging (MRI) images. Noises are deleted through spectral analysis or various filters.

Furthermore, we can partition the remaining signal into two parts: (1) irrelevant information and (2) information of interest. Which part constitutes the information of interest depends on specific application purposes. For example, cell images from a microscope contain cell information and information about other tissues and light. Clinical researchers are interested in cell shape for cell growth analysis, and in such studies information about light is considered as noise. In biomedical signal processing, researchers attempt to remove noises and irrelevant and redundant information in biomedical signals to facilitate the extraction of biomarkers and other characteristics of pathology for clinical or research decision making.

In accordance with different roles of signal processing, the workflow of biomedical signaling can be partitioned into five stages, as shown in Figure 1.2. In this chapter, we assume signals have been captured and preprocessed (e.g., by denoising filters) for specific feature extraction and pattern recognition. Depending on specific clinical applications and purposes, various biomedical features are selected and extracted from the preprocessed signals. In terms of applications, feature extraction in biomedical signaling is application specific, for example, minimum thickness in cartilage image segmentation. In terms of signal processing, we can understand

Figure 1.2   The workflow of biomedical signaling and imaging.

feature extraction topologically and statistically. For example, we can use high frequencies and low frequencies to represent noise and signals. We can use statistical information contained in signals, such as mean and standard deviation, and frequencies of certain patterns to characterize signals of interest. In this chapter, we concentrate on the feature extraction process. Pattern learning is used here to associate the extracted feature values with class labels, for example, normal and cancerous. Using the labeled sample signals, the learning model is trained until the classification result satisfies an adaptable accuracy. The obtained model can be applied to upcoming signals and can help clinicians and scientists to make diagnostic decisions.

1.3 Feature Extraction in Biomedical Signaling

In biomedical signaling, feature extraction can be conducted in two domains: frequency domain and statistical domain. In Section 1.3.1, we discuss frequency-feature extraction methods, which include Fourier transform (FT) and discrete wavelet transform (DWT).

1.3.1 Feature Extraction in the Frequency Domain

Biomedical signaling addresses two types of features, the overall shape of the spectrum and the local parameters, by using Fourier transform and discrete wavelet transform.

Let $x(t)$ denote a biomedical signal function of time t. In continuous FT, the continuous signal $x(t)$, $t \in (-\infty, +\infty)$, is decomposed into a series of sine waves with respect to the series of frequencies u, $u \in (-\infty, +\infty)$. Conversely, the original signal $x(t)$ can be obtained by using inverse FT. The pair of FT and inverse FT can be expressed as follows:

$$F(u) = \int_{-\infty}^{+\infty} x(t)e^{-j2\pi ut}\,dt \leftrightarrow x(t) = \int_{-\infty}^{+\infty} F(u)e^{j2\pi ut}\,du \qquad (1.1)$$

In Equation 1.1, $j = \sqrt{-1}$ and $u \in (-\infty, +\infty)$. In the discrete form, the FT and inverse FT of the discrete variable $x(n)$, $n \in (0, 1, 2, \ldots, N-1)$, can be expressed as follows:

$$F(u) = \frac{1}{N}\sum_{n=0}^{N-1} x(n)e^{-j2\pi un/N} \leftrightarrow x(n) = \frac{1}{N}\sum_{n=0}^{N-1} F(u)e^{j2\pi un/N} \qquad (1.2)$$

In Equation 1.2, n denotes the index of frequency members in the FT series, and N is the total number of such members. The magnitude of Fourier coefficients decreases with the increase of frequencies. The first few Fourier coefficients have more significant impacts on the signal. Hence, these coefficients are generally extracted from signals as principal features for signal analysis in the frequency domain. The FT provides no signal information in the time domain.

Wavelet transform (WT) is different from FT in that WT has a localization feature in the time domain and wavelet localization in frequency. The WT allows multiresolution analysis. For example, WT allows image description in terms of frequencies at a local position, whereas FT transforms the spatial image to a frequency domain across the image. In continuous WT, the continuous signal $x(t)$, $t \in (-\infty, +\infty)$, is discomposed into a series of wavelets in different times as follows:

$$x_w(a, b) = \frac{1}{\sqrt{a}} \int_{-\infty}^{+\infty} x(t) \varphi\left(\frac{t-b}{a}\right) dt \tag{1.3}$$

In Equation 1.3, function $\varphi\left(\frac{t-b}{a}\right)$ denotes the mother wavelet, and it is continuous in both the time and frequency domains.

In the discrete form, the WT of sequence variables $x(n)$, $n \in (0, 1, 2, \ldots, N-1)$, can be expressed in the following orthonormal form:

$$x[n] = \sum_k \langle \varphi_k[l], x[l] \rangle \varphi_k[n] = \sum_k X[k] \varphi_k[n] \tag{1.4}$$

In Equation 1.4, $X[k]$ is the transform of $X[n]$ and $X[k] = \langle \varphi_k[l], x[l] \rangle = \sum_l \varphi_k \times [n]x[l]$.

Function $\varphi_k[n]$ satisfies the following orthonormal constraint: $\langle \varphi_k[n], \varphi_l[n] \rangle = \delta[k-1]$.

Both FT and WT are broadly employed in time-series signal analysis and image processing in biomedical engineering, such as ECG and multiresolution analysis of biomedical image processing. The Gabor transform (GT) is among the popular WT approaches in image processing. One-dimensional (1-D) GT is given by

$$G_w(t) = e^{-ju_0 t} \, e^{-\left(\frac{t-t_0}{a}\right)^2} \tag{1.5}$$

Equation 1.5 can be regarded as the convolution of a Gaussian kernel function and sine waves. Thus, the most significant difference between GT and FT is that FT has no Gaussian kernel function. By setting constant a to be large or u_0 to be zero, we can obtain signal information using the frequency or spatial domain, respectively. To apply GT in a two-dimensional (2-D) image, we need to extend Equation 1.5 as follows:

$$G_w(x, y) = \frac{1}{\sigma\sqrt{\pi}} e^{-j2\pi u_0((x-x_0)\cos(\theta)+(y-y_0)\sin(\theta))} \, e^{-\left(\frac{(x-x_0)^2+(y-y_0)^2}{2\sigma^2}\right)} \tag{1.6}$$

In Equation 1.6, x_0 and y_0 denote the center of the Gaussian ellipse; θ denotes orientation of the 2-D Gaussian ellipse; u_0 denotes the frequency variance in radial direction, and σ denotes spatial variance of the wavelet orientation in radians.

1.3.2 Feature Extraction in the Statistical Domain

In Section 1.3.2, we discuss three statistical feature extraction methods, which include principle component analysis (PCA), independent component analysis (ICA), and information-theoretic feature extraction.

1.3.2.1 Principal Component Analysis

PCA represents raw data in a lower dimensional and combined feature space to describe the variances of the data along the extracted dimensions, expressed by eigenvectors. In PCA, principal feature components, called eigenvalues, are extracted along the new dimensions. Given a data set $\{x_1,...,x_n\}$ in a d-dimensional feature space, we put these data points in a matrix X in which each row represents a data point and each column represents an attribute. Now we have the matrix $X = [x_1,...,x_n]^T$, where T denotes transpose. We adjust the data points to be centered around zero by $X - \bar{X}$, where $\bar{X}$ denotes the matrix in space $\mathbb{R}^{n \times d}$, with each row representing the mean of all rows in matrix X. Such an operation ensures that the PCA result will not be skewed due to the difference between features. Then, an empirical covariance matrix of $X - \bar{X}$ can be obtained by $C = 1/d \sum (X - \bar{X})(X - \bar{X})^T$. Next, we obtain a matrix V, $V = [v_1,...,v_d]$, of eigenvectors in space $\mathbb{R}^d$, which consists of a set of d principal components in d dimensions. Each eigenvector v_i, $i = 1,...,m$, in matrix V corresponds to an eigenvalue λ_i in the diagonal matrix D, where $D = V^{-1}CV$ and $D_{ij} = \lambda_i$, $i = j$; otherwise, $D_{ij} = 0$. Finally, we rank eigenvalues and reorganize the corresponding eigenvectors such that we can find the significance of variances along eigenvectors. Mathematically, we can represent the ith principal component and eigenvector v_i as follows:

$$v_i = \arg\max_{|v|=1} \left\| \left((X - \bar{X}) - \sum_{k=1}^{i-1} (X - \bar{X}) v_k v_k^T \right) v \right\| \tag{1.7}$$

In Equation 1.7, $(X - \bar{X})v_j$ captures λ_i, the amount of variance projected along v_j. When $i = 1$, $v_1 = \arg\max_{|v|=1} \|(X - \bar{X})v\|$. The PCA algorithm includes the following four steps:

Step 1: Subtract the data mean in all dimensions to produce a data set with zero mean.
Step 2: Calculate the covariance matrix.
Step 3: Calculate the eigenvectors and eigenvalues of the covariance matrix.
Step 4: List the eigenvectors according to the ranks of the eigenvalues from the highest to the lowest.

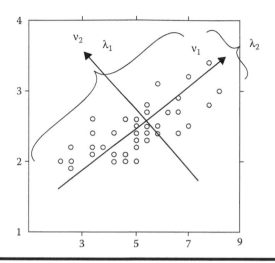

Figure 1.3 Example of principal component analysis application in a two-dimensional Gaussian mixture data set.

As shown in Figure 1.3, given a 2-D Gaussian mixture data set, v_1 and v_2 are the first and second principal components obtained by PCA. The values λ_1 and λ_2 are the corresponding first and second eigenvalues. The principal components are orthogonal in the feature space, whereas v_1 represents the most significant component of original variance in the data set, and v_2 explains the second most significant component of the remaining variance.

1.3.2.2 Independent Component Analysis

ICA attempts to discover the independent source signals from a set of observations. To simplify the expression, we reuse the data set X described in Section 1.3.2.1. Assume that k signal sources generate k signals $\{s_1,\ldots,s_k\}$ independently and that each signal has n dimensions. We obtain a $k \times n$ matrix $S = [s_1,\ldots,s_k]^T$. Given an $d \times k$ mixture matrix A, we obtain the following equation:

$$X^T = AS \tag{1.8}$$

Approximating the inverse mixture matrix A by $W = \hat{A}^{-1}$, we obtain

$$S = WX^T \tag{1.9}$$

In ICA algorithms, a function $f(s_1,\ldots,s_k)$ is proposed to measure the independence between signal sources. The maximization of this function solves the independence analysis among signals. For example, the following joint probability function is commonly employed to measure the independence between observed signals:

$$P(S) = \Pi p(s_i) \tag{1.10}$$

In Equation 1.10, $p(s_i)$ is the probability intensity of signal source s_i, $i = 1,...,k$. Using the maximum likelihood estimation (MLE), we can find the optimum statistic parameters to describe Equation 1.10.

1.3.3 Information-Theoretic Feature Extraction

Entropy is defined as follows:

$$H(x) = -\sum_{x \in \chi} p(x) \log_2 p(x) \tag{1.11}$$

In Equation 1.11, $p(x)$ is the probability of finding the feature x in the feature space χ given some end points.

Mutual information (MI) is well-known for representing the dependence between features, and it can be used to represent correlation in clinical practice. Minimum MI equals zero when no correlation exists between independent features. Higher values of MI occur when all features are strongly dependent in a group. Pairwise MI, the correlation between features X and Y, is represented as follows:

$$MI(X,Y) = H(X) + H(Y) - H(X,Y) \tag{1.12}$$

1.4 Biomedical Imaging

Biomedical imaging is a type of noninvasive signal processing. It is ubiquitously employed in almost all phases of disease diagnosis, such as cancer biopsy guidance for detection, staging, prognosis, therapy planning, therapy guidance, therapy response, recurrence, and palliation (Fass 2008).

Biomedical imaging has the same workflow as biomedical signaling, as shown in Figure 1.2. Most biomedical image capturing techniques depend on interactive signals generated by electromagnetic radiation (e.g., MRI) with or without reflection (e.g., ultrasound) on in vivo organisms. As shown in Figure 1.4, such imaging technologies play an important role in generating various frequencies of electromagnetic radiation. The underlying physics and properties of imaging technologies cause various sensitivities, including temporal and spatial resolution sensitivities, for example, positron emission tomography (PET) and MRI have 1 nmol/kg and 10 umol/kg sensitivities, respectively. For different purposes of clinical diagnosis and analysis, various imaging technologies of various sensitivities are properly employed in biomedical engineering, such as MRI for knee osteoarthritis (OA),

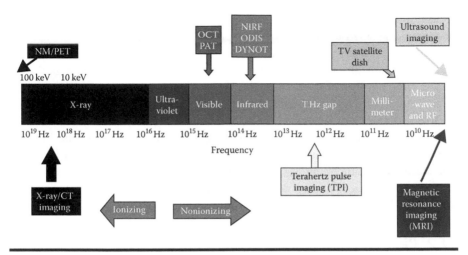

Figure 1.4 Frequency spectrum of electromagnetic radiation imaging technologies. (Reprinted from *Mol Oncol* 2, Fass, L., Imaging and cancer: A review, 115–52, Copyright 2008, with permission from Elsevier.)

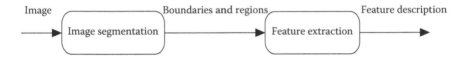

Figure 1.5 Image feature extraction.

computed tomography (CT) for mammogram, and ECG for arrhythmias. Image preprocessing consists of denoising, image enhancement, image transformation, image smoothing, and image sharpening. Image preprocessing is aimed at providing easily recognizable ROIs for image segmentation. Image segmentation refers to the process of segmenting ROIs in a preprocessed image. It constitutes the first step in feature extraction and faces the most challenges in practice.

1.4.1 Image Feature Extraction

Image feature extraction consists of two procedures: (1) image segmentation and (2) feature description. As shown in Figure 1.5, image segmentation algorithms result in boundaries and ROIs. Representative features need to be selected for machine learning. For example, boundaries generally denote the discontinuity of image intensities or abrupt changes of intensity in neighboring pixels/voxels. Each region consists of the pixels/voxels that satisfy a similarity criterion. The segmentation result of biomedical images depends on imaging modalities, intensity resolution, and anatomy quality, as these factors are the constraints of image processing techniques. For example, MRI technology can produce three-dimensional (3-D) images with around 10-umol/kg resolution. For application-specific purposes, biomedical features can be extracted and represented from the segmented boundaries or regions. For example, geometric features of cell shape are extracted for cell growth analysis and pathological analysis of cancerous cells. Biomedical image segmentation and feature extraction varies according to medical applications and related organisms. Thus, a large number of research studies in the literature report the progress of image feature extraction in specific domains, for example, MRI image segmentation (Clarke et al. 1995; Wells et al. 1996) and brain image segmentation (Ashton et al. 1995; Zhang, Brady, and Smith 2001). A higher-level brief of biomedical imaging techniques over these application-specific topics can help readers in learning biomedical image processing horizontally in depth. Following this idea, we summarize the image segmentation methods and feature descriptions that are the most employed in biomedical engineering in Sections 1.4.2.1 through 1.4.2.5.

1.4.2 Biomedical Image Segmentation

In Section 1.4.2, we categorize biomedical image-segmentation methods into five groups, which include intensity-discontinuity-based segmentation, regional-intensity-based segmentation, hybrid algorithms combining edge-based and region-based segmentation, deformable-model-based, and pattern-classification-based segmentation.

1.4.2.1 Segmentation Based on Intensity Discontinuity

This category of segmentation has a premise that the boundary of region (structure or object) of interest shows discontinuous intensity transitions. The discontinuity from the ROI to its surroundings can be measured by the first and second derivatives of profiles across the boundaries of the ROI. The first derivatives are obtained by approximating the gradient of an image, and the second derivatives can be represented by Laplacian operators.

Let us consider a 2-D image $I(x, y)$, and (x, y) denotes the coordinates of each pixel in image $I(x, y)$. The first derivatives of the image $I(x, y)$ can be obtained by

$$\left[\frac{\partial I(x, y)}{\partial x}, \frac{\partial I(x, y)}{\partial y} \right] \tag{1.13}$$

To implement Equation 1.13 at image pixels, various masks are designed for each pixel $I(x, y)$, which is located at the center of the masks. These masks can approximate the gradient of an image by using Roberts cross-gradient, Prewitt, Sobel, or Canny operators. For example, pixel $I(x, y)$ and its neighborhood are shown in Figure 1.6, and various 3×3 masks (see Figure 1.7) can be applied on pixel $I(x, y)$. The operation of masks can be expressed as follows:

$$G(x, y) = \sum_{j=-1,0,1} \sum_{i=-1,0,1} I(x+i, y+j)w(x+i, y+j) \tag{1.14}$$

$I(x-1, y-1)$	$I(x-1, y)$	$I(x-1, y+1)$
$I(x, y-1)$	$I(x, y)$	$I(x, y+1)$
$I(x+1, y-1)$	$I(x+1, y)$	$I(x+1, y+1)$

Figure 1.6   Pixel $I(x, y)$ and its 3 × 3 neighborhood.

$w(x-1, y-1)$	$w(x-1, y)$	$w(x-1, y+1)$
$w(x, y-1)$	$w(x, y)$	$w(x, y+1)$
$w(x+1, y-1)$	$w(x+1, y)$	$w(x+1, y+1)$

Figure 1.7   The 3 × 3 mask of pixel $I(x, y)$.

−1	−1	−1
−1	8	−1
−1	−1	−1

Figure 1.8   Template of Laplacian.

The Laplacian operator of image $I(x, y)$ is defined as

$$\nabla^2 I(x, y) = \left[\frac{\partial I^2(x, y)}{\partial x^2}, \frac{\partial I^2(x, y)}{\partial y^2} \right] \quad (1.15)$$

To implement the Laplacian in Equation 1.15 at image pixels, approximation masks can be obtained, for example, the template in Figure 1.8 (Gonzalez and Woods 2002). This form of the Laplacian generates problems of unacceptable sensitivity to noise and double edges. In practice, the Laplacian is employed after convolving the following Gaussian function with an image $I(x, y)$:

$$G(x, y) = -\exp(-0.5 \times (x^2 + y^2)/\sigma^2) \quad (1.16)$$

The convolution can be further obtained by

$$\nabla^2(G(x, y) \times I(x, y)) = \nabla^2 G(x, y) \quad (1.17)$$

In Equation 1.17, $\nabla^2 G(x, y)$ denotes the Laplacian of a Gaussian (LoG), which can be obtained by

$$\nabla^2 G(x, y) = -((x^2 + y^2)/\sigma^4 - 2/\sigma^2)\exp(-0.5 \times (x^2 + y^2)/\sigma^2 \quad (1.18)$$

The other popular segmentation techniques include edge linking by Hough transform (Kittler and Illingworth 1988), graph-based techniques (Malik and Jitendra 2000), and Canny's approach (Canny 1986).

1.4.2.2 Segmentation Based on Regional Intensity Similarity

This category of segmentation has a premise that the region (structure or object) of interest has the same intensity distribution inside, although different intensity distributions exist across regions. Thresholding has a premise that the segmented objects distribute distinctively in histograms. Threshold segmentation is a method that separates an image into a number of meaningful regions by selecting threshold values, such as intensity values. If the image is a gray image, thresholds are integers in the range $[0, L-1]$, where $L-1$ is the maximum intensity value. Normally, this method is used to segment an image into two regions, background and object, with one threshold. The following is the equation for threshold segmentation:

$$I_B(x,y) = \begin{cases} 1, & \text{if } I(x,y) > T \\ 0, & \text{if } I(x,y) \le T \end{cases} \tag{1.19}$$

In this equation, I_B is the segmentation resultant. The most famous threshold method was proposed by Otsu (1979). Otsu's method finds the optimal threshold T among all the intensity values from 0 to $L-1$ and chooses the value that produces the minimum within-class variance σ^2_{within} as the optimal threshold value. Consequently, the optimal value of T, T_{opt}, is obtained by the following optimal computation:

$$\sigma^2_{within}(T_{opt}) = \min_{0 \le T \le L-1} \left[\sigma^2_{within}(T) \right] \tag{1.20}$$

In the whole image, variances σ^2 are composed of two parts: $\sigma^2 = \sigma^2_{within}(T) + \sigma^2_{between}(T)$. Otsu shows that $\min_{0 \le T \le L-1} \left[\sigma^2_{within}(T) \right]$ is the same as $\max_{0 \le T \le L-1} \left[\sigma^2_{between}(T) \right]$. Therefore, the optimal value of T can also be obtained using the following alternative optimization process:

$$\sigma^2_{between}(T_{opt}) = \max_{0 \le T \le L-1} \left[\sigma^2_{between}(T) \right] \tag{1.21}$$

Equation 1.21 is often used to find the optimal threshold value for simple calculations. Theoretically, $\sigma^2_{between}(T)$ is expressed in the following equation:

$$\sigma^2_{between}(T_{opt}) = \omega_1(T)\omega_2(T)(\mu_1(T) - \mu_2(T))^2 \tag{1.22}$$

Using an intensity histogram, the optimal threshold T is exhaustively searched among $[0, L-1]$ to meet the objective according to Equation 1.22.

1.4.2.3 Hybrid Algorithms Combining Edge-Based and Region-Based Segmentation Methods

Both edge-based and region-based segmentation methods focus on a part of image information: the boundaries of ROIs or the interior features of ROIs. To improve segmentation accuracy, these two parts of information can be combined to incorporate a more complete feature set for segmentation. The new segmentation methods are normally hybrid edge-based segmentation and region-based segmentation algorithms. Related methods include region growing, region splitting and merging, and graph cutting. We focus on watershed algorithms, and readers can extend their understanding to other methods easily.

In watershed algorithms, image pixels consist of three groups (as shown in Figure 1.9): The first group includes the pixels that denote a local minimum in the ROI. This group of pixels can be regarded as located at the bottom of barriers. The second group of pixels is located at the catchment basins of those minimums. These pixels tend to fall to a single local minimum. The third group of pixels is located at the watershed lines and has equal likelihood of falling to two or more minimums. In watershed algorithms, the objective is to detect the location of watershed lines and use those lines as boundaries to segment the ROIs.

Watershed algorithms normally start with all the pixels that have local minimum values. The algorithms use these pixels as the basis for initial watersheds. Assuming an image has an intensity level set $1 \leq k \leq N$, for example, $N = 255$, these algorithms iterate in the following steps to converge: For the pixels in intensity level k,

If the pixels are nearest to only one minimum, assign the pixels to the region of that minimum.
Else if the pixels are nearest to more than one minimum, label the pixels as boundary.
Else, assign the pixels to a new region.
Iterate until pixels in all levels are labeled.

Normaly, watershed algorithms use the gradient image as input.

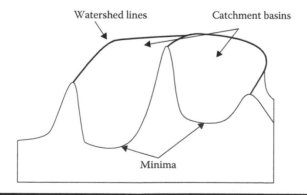

Figure 1.9   Groups of pixels/voxels in watershed algorithms.

1.4.2.4 Segmentation Based on Deformable Models

Deformable models attempt to define an evolving contour in equations such that the contour is activated toward the boundary of the ROI. According to the different representations of the contour, deformable models are categorized into two groups: (1) parametric models and (2) level-set methods. Parametric models include the active contour model (ACM; Kass, Witkin, and Terzoploulos 1988), gradient vector flow (GVF; Xu and Prince 1998b), balloon model (Cohen 1991), and active shape model (Cootes et al. 1995). Level-set methods include the level-set model by Osher and Sethian (1988), model of active contours without edges (ACWE) by Chan and Vese (2001), and Mumford and Shah's (MS) function (Jayant and Shah 1989).

1.4.2.4.1 Parametric Models

In this section, we present three popular parametric models: (1) ACM, (2) active shape model (ASM), and (3) GVF. The key difference between ACM and ASM is that "ASM can only deform to fit the data in ways consistent with the training set" (Coots et al. 1995, p. 38).

In ACM, contours can be represented as explicit or implicit parametric mathematical formulas. The contours can be approximated using polynomial expressions such as spline equations. The objective of ACM is to find the best contour, or the contour having a minimal energy. The energy of a contour is composed of two parts: (1) internal energy and (2) external energy. Kass et al. described internal energy with respect to tension (first derivative) and bending (second derivative) in the contour. The external energy is represented by the high intensity gradient. The formulation is replaced with the corresponding force balance, a vector-valued partial differential Euler–Lagrange equation. Using the gradient descent method, the desired contour can be obtained.

Traditional ACMs suffer from a number of problems: First, the initialization should be close enough to the real contour or it may lead to an unexpected result (snake has limited searching ability). The proposed solutions include multiresolution and pressure forces. Second, the internal energy term of ACMs generates a shrinking force, which implements regularization (internal continuity and smoothing) on the contour. Such a force may cause contraction of the contour, and the active contour faces difficulties in converging to a concave boundary. It is also difficult to choose a pair of proper internal parameters. The proposed solutions include balloon force (Cohen 1991) and robust active contours (Xu, Segawa, and Tsuji 1994). Third, the external forces have limited capture range. The magnitude of an external force disappears from the boundary quickly because this force is generated from the boundary. In a study by Xu and Prince (1998b), GVF has been introduced as an external force using the internal information.

Cohen (1991) introduced the balloon model to deal with the constant problem implicated in the traditional "snake" evolution: If the initialization is not close enough to the real contour, it cannot be attracted. The balloon force is introduced into the external force as follows:

$$F = k_1 \vec{n}(s) - k \frac{\nabla I}{\|\nabla I\|} \tag{1.23}$$

where $\vec{n}(s)$ is the normal unitary vector to the curve at the point $v(s)$ and k_1 is the amplitude of the balloon force. Both k and k_1 are smaller than a pixel size. Further, k is slightly bigger than k_1 so that the edge force can stop inflation of the balloon. The numeric parameters are chosen in the same order of magnitude for elasticity and rigidity. The balloon model poses another problem in implementation. Its use raises the question of how to choose the numeric parameters. Most of the parameters have to be adjusted based on experience.

Xu et al. (1994) introduced GVF to address the problems of initialization and poor convergence to concave boundaries. The GVF is defined as a vector field $\vec{v}(x, y) = (u(x, y), v(x, y))$ that minimizes the following function:

$$\varepsilon = \iint \mu(\nabla I)\left(u_x^2 + u_y^2 + v_x^2 + v_y^2\right) + |\nabla I|^2 |\vec{v} - \nabla I|^2 \, dx \, dy \tag{1.24}$$

In Equation 1.24, $\mu(\nabla I)$ is a weighting function that is implemented to adjust the smoothing effect according to the distance between vector field $\vec{v}(x, y)$ and image gradient $\nabla I(x, y)$ (Xu and Prince 1998b).

The GVF is found by using the calculus of variations by solving the Euler equations. The external force is replaced with this GVF in the traditional snake force equation. Xu et al. (1994) applied the following weighting function in Equation 1.24:

$$u(|\nabla I|) = e^{-(|\nabla I|/K)} \tag{1.25}$$

In this equation, K is a nonnegative smoothing parameter for the force field (u, v). This weighting function is a decreasing function of edge-force magnitude, and smoothes the force field only when the edge strength is low.

In ASM, contours should incorporate knowledge about the desired object in a general model. Shapes are represented by groups of landmarks. Shape models involve information of the types of objects to model, locations of the landmarks, and collection of the object images. This approach can be summarized in the following two steps: (1) training and (2) application (Cootes et al. 1995).

In the training step, we attempt to describe the available training data (clouds) using PCA and the point distribution model (PDM). Training data are chosen (dimension of the cloud or hyperellipsoid) manually with respect to the global and local characteristics in the images. The allowable shape is deduced by the mean shape and linear combination of eigenvectors.

The moving step size of each landmark is decided by the maximum edge strength (suggested or gradient) along the normal to boundary. The shape parameters (including translation, rotation, scaling, and residual adjustments, which are adjustments of shape parameters) are adjusted according to the movements.

The training result for ASM is expressed in the form of PDM as follows: $X = \bar{X} + Pb$, where X is a shape model that consists of a number of landmarks for a 2-D image. In this equation, $\bar{X}$ is the mean shape, in the dimension of 2 × the number of landmarks for the 2-D image; P is the matrix of the first several principal eigenvectors, in the dimension of $2 \times t$ for the 2-D image; b is a vector of weights, in the dimension of t; and Pb is the weighted sum of different shape-variation modes.

Here, we describe how the search method works (see Figure 1.10). It includes three steps:

Step 1: For a given point, consider a sample of k pixels along a profile on either sides of the model point (see Figure 1.10a). Put the derivative of those $2k + 1$ samples in a vector g_i. Normalize the sample by dividing the sum of absolute element values.

Step 2: Assume the sample points are distributed as a multivariate Gaussian. A criterion for evaluating the fitting of a new sample to the model is represented by Mahalanobis distance (see Figure 1.10b) as follows:

$$f(g_i) = (g_s - \hat{g})^T S_g (g_s - \hat{g}) \tag{1.26}$$

where g_s is the set of normalized samples for a given model point, $\hat{g}$ is the mean of the set of sample points, and S_g is the covariance of the set of sample points.

Step 3: Sample more pixels along the profiles ($m > k$), and choose the point with the lowest value of $f(g_s)$ as the moving destination for the model point (see Figure 1.10c).

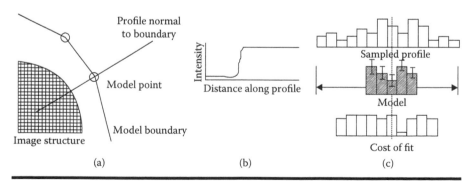

Figure 1.10 Intensity distribution: Parts (a) and (b) show the intensity distribution along a profile normal to the image boundary through a model point. Part (c) shows a statistical model for searching the fitful image boundary. (From Cootes, T. F. and C. J. Taylor, *Proc. SPIE Medical Imaging*, 2001. With permission.)

The difficulty of ASM lies in how to choose the location of training data, which determines the dimension of the ellipsoid. If any important point is missed, the final shape may definitely not be robust.

1.4.2.4.2 Level-Set Methods

Osher and Sethian (1988) introduced the level-set function to active contours in the 1980s. Zero level set represents the active contour by a level-set function $\phi(x, y)$ as follows:

$$C(t) = \{(x, y) \in I \,|\, \phi(x, y, t) = 0\} \tag{1.27}$$

Here, $\phi(x, y, t)$ denotes a Lipschitz continuous function:

$$\begin{cases} \phi(x, y, t) > 0, & \text{if } (x, y) \text{ locates inside contour } C(t) \\ \phi(x, y, t) = 0, & \text{if } (x, y) \text{ locates on contour } C(t) \\ \phi(x, y, t) < 0, & \text{if } (x, y) \text{ locates outside of contour } C(t) \end{cases} \tag{1.28}$$

According to the nonlinear partial directive equation, the evolution of an active contour can be represented as follows:

$$\frac{\partial \phi}{\partial t} = v_N(\phi) \cdot |\nabla \phi| \tag{1.29}$$

Equation 1.29 indicates that contour ϕ evolves with the velocity $v_N(\phi)$ normal to the contour.

Chan and Vese (2001) introduced the ACWE model. The ACWE model addresses the difficulty of applying boundary-based segmentation techniques to objects whose boundary is either smooth or cannot be defined by gradient. Mathematically, image segmentation is defined as a minimization problem in the ACWE model by

$$\min_{\phi, c_1, c_2} E(\phi, c_1, c_2, \lambda) \tag{1.30}$$

Here, $E(\phi, c_1, c_2, \lambda)$ denotes the energy defined by

$$E(\phi, c_1, c_2, \lambda) = \int_\Omega |\nabla H(\phi)| dx + \lambda \int_\Omega (H(\phi)(c_1 - I(x))^2 \\ + H(-\phi)(c_2 - I(x))^2) dx \tag{1.31}$$

where Ω denotes the image domain and H denotes the Heaviside function.

1.4.2.5 Segmentation Based on Pattern Classification Methods

We include the application of pattern classification methods in image segmentation in Chapter 2.

1.4.3 Imaging Feature Description

We categorize imaging features into low-level and high-level features. Low-level features can be extracted by processing images locally without involving any spatial integration. Low-level imaging features include pixel values, gradients, edges, corners, curvatures, histograms, colors, and statistical information such as means and variance. We can easily segment these features in biomedical images by using the intensity distribution in the images. For example, an edge-based segmentation algorithm obtains both local and global edge and gradient features in images. High-level features can be extracted by globally integrating pixel/voxel information in an image and describing regional signatures, such as shape, features, and texture. We illustrate some of the most commonly employed feature descriptors in biomedical imaging in Sections 1.4.3.1 and 1.4.3.2.

1.4.3.1 Shape Features

We categorize shape features into two groups: (1) geometric shape features and (2) topological shape features. Given a closed contour $c(\varphi) = [x(\varphi), y(\varphi)]$, we can describe the shape features of the contour, also called "shape descriptors," using $d(\varphi)$, which is the radial distance from the points on the object boundary to the center of mass of the object (reference point). A shape descriptor is an operator applied to the binary image of a shape, resulting in a numerical quantity. Shape descriptors can be region or boundary based. For example, we can obtain the object "area" A and "perimeter" P as follows:

$$P = \int_0^{2\pi} \sqrt{c(\varphi)^2 + \left(\frac{dc(\varphi)}{d\varphi}\right)^2} \tag{1.32}$$

$$A = \frac{1}{2}\int_0^{2\pi} c^2(\varphi)\,d\varphi \tag{1.33}$$

In this section, we introduce four popular shape features in biomedical imaging. These features have common advantages in measuring shape: the features are dimensionless and invariant to rotation, translation, and reflection. First, we define "circularity" C_r as a number that measures deviations of radial distance in a shape:

$$C_r = \frac{u_R}{\sigma_R}, \quad 0 \leq C_r \tag{1.34}$$

Here, u_R is the mean value of shape radial distance, and σ_R is the standard deviation of those distances. Shape "roughness" R calculates the number of angles in which more than one boundary point is observed, and divides this number by the total number of angles. Low-order moments can also calculate roughness. However, roughness is dependent on a digitalization step for computation. Shape "compactness" C_p describes the diffused nature of a brain tumor compared to a circle, which is defined as follows:

$$C_P = 4\pi \frac{A}{P^2}, \quad 0 \le C \le 1 \tag{1.35}$$

Topological shape features include connectivity, number of holes, Euler number, convex hull, skeleton, and object counting. For further information, we suggest the work by Gonzalez and Woods (2002).

1.4.3.2 Texture Feature Description

Texture features are extracted on a gray-tone spatial-dependence matrix. In this matrix, each component $M(i, j)$ represents the occurrence of paired pixels that have a specified distance d along a specified direction and combined gray levels of i and j. As shown in Figure 1.11, four directions are used in Harrilick's (Harralick, Shanmugam, and Dinstein 1973) texture feature calculation: $0°$, $45°$, $90°$, and $135°$.

We illustrate three representative Harrilick's features in Equations 1.36 through 1.38, to measure homogeneity, contrast, and gray-tone linear dependencies in images. In these equations, c_{nor} denotes the normalization constant, L denotes the gray level in image I, and $\mu_x, \mu_y, \sigma_x,$ and σ_y denote the means and standard deviations of $M(i, j)$ along the coordinates x and y:

$$f_{homogeneity} = \sum_{i=1}^{L} \sum_{j=1}^{L} \left(\frac{M(i, j)}{C_{nor}} \right) \tag{1.36}$$

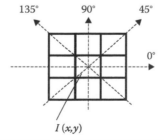

Figure 1.11   Specified direction and neighborhood of an image pixel.

$$f_{contrast} = \sum_{n=0}^{L-1} n^2 \left(\sum_{|i-j|=n}^{L} \left(\frac{M(i,j)}{C_{nor}} \right)^2 \right)$$ (1.37)

$$f_{linearity} = \frac{1}{\sigma_x \sigma_y} \left(\sum_{i=0}^{L} \sum_{j=0}^{L} \left(\frac{ijM(i,j)}{C_{nor}} \right) - \mu_x \mu_y \right)$$ (1.38)

The feature $f_{homogeneity}$ can capture significant dominant gray-tone transitions in an image. A homogeneous image results in few significant values of $f_{homogeneity}$. The feature $f_{contrast}$ can capture local gray-tone variations in an image. The feature $f_{linearity}$ can capture linear dependent and correlative structures of gray tone along different directions in an image. Generally, additive noises result in the reduction of linear correlation between structures in an image. Harralick, Shanmugam, and Dinstein (1973) also suggests other texture features that can be extracted from a gray-tone spatial-dependencies matrix, such as variance, entropy, and information measures of correlation. Correlations or dependencies exist among the texture features when applied in image analysis. Thus, the feature selection procedure is generally employed in feature extraction. Readers are referred to the work by Huang and Thomas (1999) for details regarding the application of feature selection techniques in image retrieval.

1.5 Summary

In this chapter, we introduce the fundamental feature extraction techniques in biomedical signaling and imaging. This chapter is not a survey, and we do not review all the literature in this domain.

We categorize feature extraction techniques in biomedical signaling into frequency-based, statistics-based, and informatics-based techniques. These techniques can also be applied to biomedical imaging. Frequency-based feature extraction is used to decompose time-series signals or images into frequency components for further processing and analysis. Statistics-based feature extraction can extract principal or independent components according to a variance distribution or a dependency estimate between the source and the signals. Informatics-based feature extraction uses entropy, MI, and other related techniques to describe the significance of a feature or a subset of features.

We categorized feature extraction techniques for biomedical imaging into image segmentation and image feature descriptors. We classified biomedical image segmentation techniques into five groups: (1) boundary-based, (2) region-based, (3) integration of boundary-based and region-based, (4) deformable model–based, and (5) pattern recognition–based groups. Boundary-based segmentation attempts to find discontinuous intensity pixels for the ROI, and the related techniques most commonly use gradient to find edges and boundaries. Region-based segmentation methods search for regions with more similar pixel intensities inside the ROI than pixel intensities across the ROI. Integration of the aforementioned two technologies can regularize the weights of both boundary-based and region-based techniques to combat difficult segmentation when the ROI is contaminated highly by noises, artifacts, and other tissues. Deformable model–based techniques are popular in biomedical imaging because these techniques are adaptive to biomedical images, especially for soft organisms. The "active" nature of these models highlights the extraction of complex and evolving features in biomedical images. For a discussion of pattern recognition–based image segmentation, readers are referred to Chapter 3. We briefly introduce a number of geometric shape feature descriptors and texture feature descriptors. Readers can formulate specific feature descriptors for various application purposes after extracting the ROIs or boundaries.

In biomedical signaling and imaging, features are sensitive and application specific. We do not include all biomedical features in this chapter. However, all the features and feature extraction techniques discussed in this chapter are applicable to any types of biomedical signals and images. The features extracted are applied in machine learning methods for clinical pattern recognition and decision making.

References

Ashton, E. A., M. J. Berg, K. J. Parker, J. Weisberg, C. W. Chen, and L. Ketonen. 1995. Segmentation and feature extraction techniques, with applications to MRI head studies. *Magn Reson Med* 33:670–7.

Canny, J. 1986. A computational approach to edge detection. *IEEE Trans Pattern Anal Mach Intell PAMI* 8:679–98.

Chan, T., and L. Vese. 2001. Active contours without edges. *IEEE Trans Image Process* 10:266–77.

Clarke, L. P., R. P. Velthuizen, M. A. Camacho, J. J. Heine, M. Vaidyanathan, L. O. Hall, R. W. Thatcher, and M. L. Silbiger. 1995. MRI segmentation: Methods and applications. *Magn Reson Imaging* 13:343–68.

Cootes, T. F., C. J. Taylor, D. H. Cooper, and J. Graham. 1995. Active shape models: Their training and application. *Comput Vis Image Underst* 61:38–59.

Cohen, L. 1991. On active contour models and balloons. *CVGIP Image Underst* 53:211–8.

Fass, L. 2008. Imaging and cancer: A review. *Mol Oncol* 2:115–52.

Gonzalez, R. C., and R. E. Woods. 2002. *Digital Image Processing*. 2nd ed. Upper Saddle River, NJ: Prentice Hall.

Harralick, R.-M., K. Shanmugam, and I. Dinstein. 1973. Texture features for image classification. *IEEE Trans Syst Man Cybern* SMC-3:610–21.

Huang, Y. R., and S. Thomas. 1999. Image retrieval: Current techniques, promising directions, and open issues. *J Vis Commun Image Represent* 10:39–62.

Jayant, M., and D. Shah. 1989. Optimal approximations by piecewise smooth functions and associated variational problems. *Commun Pure Appl Math* 42:577–685.

Kass, M., A. Witkin, and D. Terzoploulos. 1988. Snakes: Active contour models. *Int J Comput Vis* 1:321–31.

Kittler, J., and J. Illingworth. 1988. A survey of the hough transform. *Comput Vis Graph Image Process* 44:87–116.

Malik, J., and S. Jitendra. 2000. Normalized cuts and image segmentation. *IEEE Trans Pattern Anal Mach Intell* 22:888–905.

Otsu, N. 1979. A threshold selection method from gray-level histogram. *IEEE Trans Syst Man Cybern* SMC-9 1:62–6.

Osher, S., and J. Sethian. 1988. Fronts propagating with curvature dependent speed: Algorithms based on Hamilton–Jacobi formulations. *J Comput Phys* 79:12–49.

Wells, W. M., W. L. Grimson, R. Kininis, and F. A. Jolesz. 1996. Adaptive segmentation of MRI data. *IEEE Trans Med Imaging* 15:429–42.

Xu, C., and J. L. Prince. 1998a. Generalized gradient vector flow external forces for active contours. *Signal Process* 71:131–9.

Xu, C., and J. L. Prince. 1998b. Snakes, shapes, and gradient vector flow. *IEEE Trans Image Process* 7:359–69.

Xu, G., E. Segawa, and S. Tsuji. 1994. Robust active contours with insensitive parameters. *Pattern Recognit* 27:879–84.

Zhang, Y., M. Brady, and S. Smith. 2001. Segmentation of brain MR images through a hidden Markov random field model and the expectation-maximization algorithm. *IEEE Trans Med Imaging* 20:45–57.

Chapter 2

Supervised and Unsupervised Learning Methods in Biomedical Signaling and Imaging

Xian Du and Sumeet Dua

Contents

2.1 Introduction

As we describe in Chapter 1, the objective of biomedical signaling and imaging is to detect the significant features in data sets and help clinicians or scientists recognize pathological evolution in order to better track the development of diseases and further make decisions about how to treat those diseases. As shown in Chapter 1, Figure 1.2, pattern learning methods are designed to partition the given signals and images into classes, such as those indicating healthy patients and those indicating cancerous patients. The feature extraction methods presented in Chapter 1 can provide features to compose input values of pattern learning models. Using pattern learning procedures, we attempt to discover associative features and to associate patterns of feature subsets with class labels. This learning procedure consists of training, testing, and performance evaluation (as shown in Figure 2.1). Given a pattern learning model that satisfies some performance criteria, we can apply the trained model to upcoming data from the same space and generate labels for decision making. In other words, we employ machine learning methods in biomedical signaling and imaging to facilitate the analysis, diagnosis, prognosis, and treatment of a disease or pathology, such as biomedical image segmentation and content-based retrieval of information from biomedical imaging. In biomedical signaling and imaging, machine learning performs signal and image analysis, interpretation of extracted features using semantic concepts in specific applications, and decision making on the disease or pathology. For example, given microscopy cell imaging

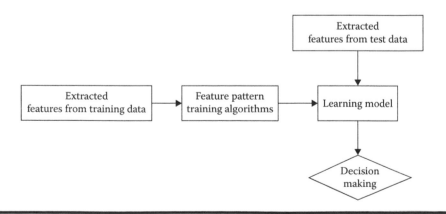

Figure 2.1 The workflow of pattern learning procedure.

and its segmentation results (as shown in Figure 2.2), we can extract cell shape features from normal cells and cancerous cells, and use these features as input to train a machine learning model. The trained model can be employed for the classification and prediction of healthy persons and cancerous patients.

As machine learning has been evaluated and employed in a large number of specific biomedical applications, readers can find related literature reviews on various topics, for example, machine learning for biomedical image segmentation (Bezdek, Hall, and Clarke 1993; Kapetanovic, Rosenfeld, and Izmirlian 2004) and for detection and diagnosis of diseases (Sajda 2006). We focus on the introduction and explanation of fundamental machine learning methods so that readers understand why and how these techniques can be employed in specific biomedical signaling and imaging domains.

In this chapter, we first introduce the fundamentals of machine learning in Section 2.2, as well as supervised learning and unsupervised learning. Then, we briefly elucidate a variety of classic supervised learning methods and unsupervised learning methods, respectively, in Sections 2.3 and 2.4. Next, we discuss a number of performance evaluation measures for machine learning methods. Finally, we describe the challenges that must be faced in biomedical signaling and imaging when we apply supervised and unsupervised learning methods. Furthermore, we explore potential research directions in the application of machine learning methods to biomedical signaling and imaging.

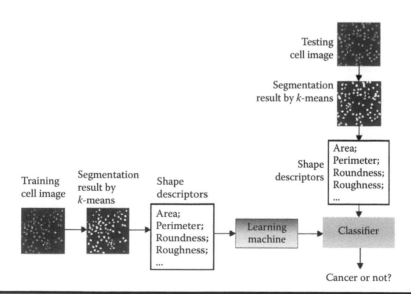

Figure 2.2 Example of machine learning for biomedical signaling and imaging: The *k*-means segmentation result of nucleus images and shape features for classification or prediction are shown. (Adapted from http://www.cellprofiler.org/)

2.2 Machine Learning

Machine learning is a computational process of automatically inferring and generalizing a learning model from sample data. Learning models most often employ statistical functions or rules to describe the dependencies among data and causalities and correlations between input and output. Theoretically, given an observed data set S, a set of parameters θ and variable x, and a learning model $f(x, \theta)$, a machine learning method aims to minimize the learning errors $E(f(x, \theta), X)$ between the learning model $f(x, \theta)$ and the ground truth. Without loss of generalization, we obtain the learning errors using the difference between the predicted output $f(x, \hat{\theta})$ and the observed sample data, where $\hat{\theta}$ is the set of approximated parameters derived from optimization procedures for the minimization of the objective function of learning errors. Machine learning methods differ from each other mainly in the selection of learning model $f(x, \theta)$, parameters θ, and the expression of learning error $E(f(x, \theta), X)$.

There are two types of machine learning methods: (1) supervised learning and (2) unsupervised learning. In supervised learning, data labels are given to training data, whereas in unsupervised learning data labels are not given to training data. In supervised learning, training data are labeled by experts. A supervised learning algorithm uses the labeled samples for training and generalizes a model structure for upcoming data points. The objective of using a supervised learning algorithm is to obtain the highest accuracy of labeling. Typical supervised learning methods include decision trees, Bayesian networks (BNs), artificial neural networks (ANNs), support vector machines (SVMs), and so on. Supervised learning takes advantage of prior knowledge and experience gathered by experts. The appropriate selection of learning structures and accurate labels can always lead to confident labels of upcoming data points that lie in the same data space as the training data. The accuracy of supervised learning can be quantitatively evaluated by various metrics, such as classification accuracy, sensitivity, and specificity. However, supervised learning relies on an expert's accuracy in labeling. No expert can guarantee that he or she is always able to label biomedical signals or images with 100% accuracy. We can employ voting among experts to solve the discrepancy problem; but voting also cannot guarantee 100% accuracy.

In order to reduce sensitivity to prior labels, unsupervised learning algorithms generalize the model structure from unlabeled data, for partitioning new data without human interference. Data are clustered based on similarity measure, density distribution, association or correlation, and other metrics. Typical unsupervised learning methods include k-means clustering, hierarchical clustering, density-based clustering, grid-based clustering, and self-organizing map (SOM) ANN (Han and Kamber 2006).

2.3 Supervised Learning Methods in Medical Signaling and Imaging

In supervised learning, pairs of input and target output are given to train a function, and a learning model is trained such that the value of the function can be predicted at a minimum cost. Based on the structure of learning algorithms according to the different objective functions, we introduce several supervised learning methods including SVM, ANN, decisions trees, Bayesian Network (BN), and Hidden Markov Model (HMM).

2.3.1 Support Vector Machines

Given a data set X in an n-dimensional feature space, SVM separates the data points in X with an $n-1$-dimensional hyperplane. In SVM, the objective is to classify the data points with the hyperplane that has the maximum distance to the nearest data point on each side. Subsequently, such a linear classifier is also called the "maximum margin classifier." As shown in Figure 2.3a, any hyperplane can be written as the set of points X satisfying $w^T x + b = 0$, where the vector w is a normal vector perpendicular to the hyperplane and b is the offset of the hyperplane $w^T x + b = 0$ from the original point along the direction of w. Given the labels of data points X for two classes, class 1 and class 2, we present the labels as $Y = +1$ and $Y = -1$. Meanwhile, for a pair of (w^T, b), we classify data X into class 1 or class 2 according to the sign of the function $f(x) = \text{sign}\,(w^T x + b)$ as shown in Figure 2.3. Thus, the linear separability of the data X in these two classes can be expressed in the combinational equation form as $y \cdot (w^T x + b) \geq 1$. In addition, the distance from a data point to the separator hyperplane $w^T x + b = 0$ can be computed as $r = \dfrac{w^T x + b}{\| w \|}$, and the data

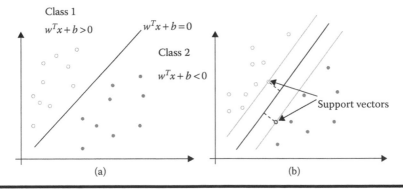

Figure 2.3   Support vector machine classification.

points closest to the hyperplane are called "support vectors." The distance between support vectors is called the "margin of separator" (Figure 2.3b). Linear SVM is solved by formulating the quadratic optimization problem as follows:

$$\arg\min_{w,b}\left(\frac{1}{2}\|w\|^2\right)$$

$$s.t.\, y(w^T x + b) \geq 1 \tag{2.1}$$

Under normal circumstances, nonlinear SVM is formulated in the same problem as linear SVM by mapping the original feature space to some higher-dimensional feature space where the training set is separable using kernel functions. Nonlinear SVM is solved by using soft margins to separate classes or by adding slack variables to the equation.

Compared to ANN, SVM has two advantages in achieving global optimization and controlling the overfitting problem by suitably selecting support vectors for classification. The SVM can solve linear, nonlinear, and complex classification boundaries accurately even with small training sample sizes. The SVM is extensively employed for multitype data by incorporating kernel functions to map data spaces. However, selection of kernel functions and fine-tuning the corresponding parameters by trial and error is time consuming. One can speedily solve the SVM problem when sample size is moderate. Unfortunately, the running time of SVM increases by four times when sample data size is doubled. Moreover, SVM algorithms are rooted in binary classification. To solve multiclass classification problems, multiple binary-class SVMs can be combined by classifying each class or by classifying each pair of classes.

2.3.2 Artificial Neural Network

An ANN transforms inputs into outputs through nonlinear information processing in a connected group of artificial neurons (as shown in Figure 2.4), which make up the

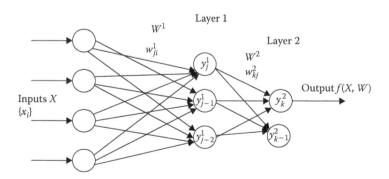

Figure 2.4   Example of a two-layer artificial neural network framework.

layers of "hidden" units. The activity of each hidden unit and output $\hat{Y}$ is determined by the composition of its input X and a set of neuron weights $W : \hat{Y} = f(X, W)$, where W refers to the matrix of weight vectors of hidden layers. For example, Figure 2.4 presents an ANN structure with four inputs, one output, and two hidden layers. The weight vectors for layer 1 and layer 2 are W^1 and W^2, respectively. Layer 1 has three neurons and layer 2 has two neurons.

When ANN is used as a supervised machine learning method, efforts are made to determine a set of weights that minimizes the classification error. One well-known method, which is common to many learning paradigms, is least-mean-square (LMS) convergence. The objective of ANN is to minimize the errors between the ground truth Y and the expected output of ANN, $E(X) = (f(X; W) - Y)^2$. The behavior of an ANN depends on both the weights and the transfer function T_f that are specified for the connections between neurons. For example, in Figure 2.4, the net activation at the jth neuron of layer 1 can be presented as

$$y_j^1 = T_f \left(\sum_i x_i \cdot w_{ji}^1 \right) \tag{2.2}$$

Subsequently, the net activation at the kth neuron of layer 2 can be represented as

$$y_k^2 = T_f \left(\sum_j y_j^1 \cdot w_{kj}^2 \right) \tag{2.3}$$

This transfer function typically falls into one of three categories: (1) linear (or ramp), (2) threshold, or (3) sigmoid. Using a linear function, the output of T_f is proportional to the weighted output. Using a threshold, the output of T_f depends on whether the total input is greater than or less than some threshold value. Using the sigmoid function, the output of T_f varies continuously but not linearly as the input changes. The output of the sigmoid function bears a greater resemblance to real neurons than do linear or threshold units. In any application of these three functions, we must consider rough approximations.

In practice, according to the learning algorithms, ANN is composed of diverse types and several of the most employed ANN algorithms include feedforward backpropagation (BP) networks, radial basis function (RBF) networks, and SOMs.

In feedforward BP ANN, information is transformed from an input layer through hidden layers to an output layer in a straightforward direction without any loops included in the structure of the network (e.g., Figure 2.4). The BP requires that we train the ANN structure as follows: First, input data are fed to the network and, then, the activations for each level of neurons are cascaded forward. By comparing the desired output with the real output, the BP ANN structure, for example, weights in different layers, is updated layer by layer in the direction of BP from the output layer to the input layer.

The RBF ANN has only one hidden layer and uses a linear combination of nonlinear RBFs in the transfer function T_f. For instance, we can express the output of an RBF ANN as follows:

$$f(X,W) = \sum_{i=1}^{n} w_i \cdot T_f \left(\|X - c_i\| \right) \tag{2.4}$$

where w_i and c_i are the weight and center vector for neuron i, and n is the number of neurons in the hidden layer. Typically, the center vectors can be found by using the k-means or k-nearest neighbor (KNN) method. The norm function can be Euclidean distance, and the transfer function T_f can be the Gaussian function.

The clustering method is SOM ANN. It is discussed in Section 2.4.

The ANN methods have advantages in classifying or predicting latent variables that are difficult to measure, in solving nonlinear classification problems, and in solving problems that are insensitive to outliers. The ANN models implicitly define the relationships between input and output, which offers solutions for tedious pattern recognition problems, especially when users have no idea about the relationship between variables. However, compared to the classification methods that assume functional relationships between data points, BN ANN may generate hard-to-interpret classification results. The ANN methods are data dependent such that ANN performance can improve with increasing sample data size.

2.3.3 Decision Trees

A decision tree is a treelike structural model with leaves representing classifications or decisions and branches representing the conjunctions of features that lead to those classifications. A binary decision tree is shown in Figure 2.5, where C is the root node of the tree; A_i, $(i = 1,2)$, are the leaves (terminal nodes) of the tree; and B_j, $(j = 1,2,3,4)$, are branches (decision points) of the tree. Tree classification of an input vector is done by traversing the tree, beginning from the root node and ending at the leaf. Each node of the tree computes an inequality based on a single input variable. Each leaf is assigned to a particular class. Each inequality that is used to split the input space is only based on one input variable. Linear decision trees are

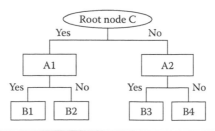

Figure 2.5 Decision tree structure.

similar to binary decision trees, except that the inequality computed at each node takes on the sign of an arbitrary affine combination of multiple variables. With different selections of splitting criteria, classification and regression trees and other tree models are developed.

As shown in Figure 2.5, a decision tree depends on if–then rules but requires no parameters and no metrics. This simple and interpretable structure of decision trees allows them to solve multitype attribute problems. Decision trees also have the ability to deal with missing values or noise data. They cannot guarantee optimal accuracy as the other machine learning methods. Although decision trees are easy to learn and implement, they do not seem to enjoy much popularity in the context of intrusion detection. A possible reason for the lack of popularity is that seeking the smallest decision tree that is consistent with a set of training examples is known to be nondeterministic polynomial-time (NP) hard.

2.3.4 Bayesian Network

The BN, also called the "belief network," uses factored joint probability distribution in a graphical model for making decisions about uncertain variables. A BN classifier is based on the Bayes rule: Given a hypothesis H of classes and data x, we have

$$P(H \mid x) = \frac{P(x \mid H)P(H)}{P(x)} \tag{2.5}$$

In this equation, $P(H)$ denotes a priori probability of each class without information about a variable x. Further, $P(H \mid x)$ denotes a posteriori probability of variable x over the possible classes and $P(x \mid H)$ denotes conditional probability of x given the likelihood H.

As shown in Figure 2.6, BNs are represented with nodes indicating random variables and arcs indicating probabilistic dependencies between variables;

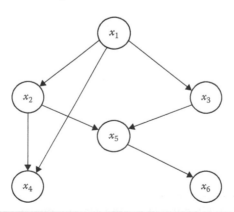

Figure 2.6   Bayes network with sample factored joint distribution.

conditional probabilities encode the strength of dependencies, whereas uncon-nected nodes refer to variables independent of each other. Each node is associ-ated with a probability function corresponding to the node's parent variables. The BN algorithms always need the computation of posterior probabilities given evidence of parents about the selected nodes. For example, in Figure 2.6, the factored joint probability of the network is computed as $p(x_1,x_2,x_3,x_4,x_5,x_6) = p(x_6 \mid x_5) \, p(x_5 \mid x_3,x_2) \, p(x_4 \mid x_2,x_1) \, p(x_3 \mid x_1) p(x_2 \mid x_1) p(x_1)$. In this equation, $p(.)$ denotes the probability of a variable and $p(.|.)$ denotes the conditional probability of variables.

Naive Bayes is a simple form of BN model that assumes all variables are inde-pendent. Using the Bayes rule for Naive Bayes classification, we need to find the best hypothesis, which refers to a class label, for each testing data x. Given observed data x and a group of class labels $C = \{c_j\}$, a Naive Bayes classifier can be solved by the maximum a posteriori (MAP) hypothesis for the data as follows:

$$\underset{c_j \in C}{\arg\max} \, P(x \mid c_j) P(c_j) \tag{2.6}$$

Naive Bayes is efficient for inference tasks. However, Naive Bayes is based on a strong independence assumption. Surprisingly, it gives good results even if the independence assumption is violated, triggering further research in this area.

2.3.5 Hidden Markov Model

In Sections 2.3.1 through 2.3.4, we discuss machine learning methods for data sets that consist of samples independently and identically (iid) from sample space. In some cases, we may have sequential data in which the sequences have correlation. To solve sequential learning problems, a dynamic BN method, hidden Markov model (HMM), was proposed for the supervised learning of sequential patterns, for example, speech recognition (Rabiner 1989).

In HMM, the observed samples y_t, $t = 1, \ldots, T$, have an unobserved state x_t at time t. Figure 2.7 shows the general architecture of an HMM. Each node represents a random variable with the hidden state x_t and observed value y_t at time t. In HMM, it is assumed that state x_t has a probability distribution over the observed samples

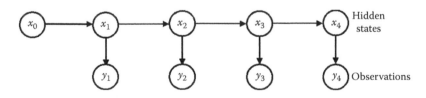

Figure 2.7   General architecture of hidden Markov model.

y_t and the sequence of observed samples embed information about the sequence of states. Statistically, HMM is built on the Markov property that the current true state x_t is conditioned only on the value of the hidden variable x_t-1 and is independent of the past and future states. Similarly, the observation y_t depends only on the hidden state x_t. The most famous solution to HMM is the Baum–Welch algorithm, which derives the maximum likelihood estimate of the parameters of the HMM given a data set of output sequences.

Let us formulate the HMM using the aforementioned notations as follows: Here, Y and X are the fixed observed sample and state sequences of length T defined earlier in this section, with $Y = (y_1, ..., y_T)$ and $X = (x_1, ..., x_T)$; we have the state set S and the observable data set O, with $S = (s_1, ..., s_M)$ and $O = (o_1, ..., o_N)$. Let us define A as the state transition array $[A_{i,j}]$, $i = 1, ..., M, j = 1, ..., M$, where each element $A_{i,j}$ represents the probability of state transformation from s_i to s_j. It can be calculated as follows:

$$A_{i,j} = \text{prob}(x_t = s_j \mid x_{t-1} = s_i) \qquad (2.7)$$

Let us define B as the observation array $[B_{j,k}]$, $j = 1, ..., M, k = 1, ..., N$, where each element $B_{j,k}$ represents the probability that the observation o_k has the state s_j. It can be calculated as follows:

$$B_{j,k} = \text{prob}(y_t = o_k \mid x_t = s_j) \qquad (2.8)$$

Let us define π as the initial probability array $[\pi_i]$, $t = 1, ..., T$, where π_t represents the probability that the observation y_t has the state s_i, $i = 1, ..., I$. It can be expressed as

$$\pi_i = \text{prob}(x_1 = s_i) \qquad (2.9)$$

Then we define an HMM using the aforementioned definitions, as follows:

$$\lambda = (A, B, \pi) \qquad (2.10)$$

The aforementioned analysis is the evaluation of the probability of observations, which can be summarized in the algorithm in four steps as follows:

Step 1: Initialization of state for $t = 1$, according to the initial state distribution π
Step 2: Deduction of the observation value at time t corresponding to Equation 2.8
Step 3: Deduction of the new state at time $t + 1$ according to Equation 2.9
Step 4: Iteration by returning to step 2, until $t = T$

Given the HMM described in Equation 2.10, we are able to predict the probability of observations Y for a specific state sequence X and the probability of the state sequence X as

$$\text{prob}(Y \mid X, \lambda) = \prod_{t=1}^{T} \text{prob}(y_t \mid x_t, \lambda) \tag{2.11}$$

and

$$\text{prob}(X \mid \lambda) = \pi_1 \cdot A_{12} \cdot A_{23} \ldots A_{T-1T} \tag{2.12}$$

Then we obtain the probability of observation sequence Y for state sequence X as follows:

$$\text{prob}(Y \mid \lambda) = \sum_X \text{prob}(Y \mid X, \lambda) \cdot \text{prob}(X \mid \lambda) \tag{2.13}$$

In practice, users are more interested in predicting the hidden state sequence for a given observation sequence. This decoding process has a famous solution known as the "Viterbi algorithm," which uses the maximized probability at each step to obtain the most probable state sequence for the partial observation sequence. Given an HMM model λ, we can find the maximum probability of the state sequence $(x_1, \ldots, x_t)$ for the observation sequence $(y_1, \ldots, y_t)$ at time t as follows:

$$\rho_t(i) = \max_{x_1, \ldots, x_{t-1}} (\text{prob}(x_1, \ldots, x_t = s_i, y_1, \ldots, y_t \mid \lambda)) \tag{2.14}$$

The Viterbi algorithm mainly consists of the following steps:

Step 1: Initialization of the state for $t = 1$, according to the initial state distribution π:

$$\rho_1(i) = \pi_i B_i(y_1), \ 1 \le i \le M, \ \psi_1(i) = 0 \tag{2.15}$$

Step 2: Deduction of the observation value at time t, corresponding to the following equations:

$$\rho_t(j) = \max_i [\rho_t(j) A_{ij}] B_j(y_t), \ 2 \le t \le T, 1 \le j \le M \tag{2.16}$$

$$\psi_t(j) = \arg \max_i [\rho_{t-1}(i) A_{ij}], \ 2 \le t \le T, 1 \le j \le M \tag{2.17}$$

Step 3: Iteration by returning to step 2, until $t = T$

The HMM approach attempts to solve sequential supervised learning problems. It offers an elegant and sound method to classify or predict the hidden state of observed sequences with high accuracy when data fit the Markov property. However, when true relationships between hidden sequential states do not fit the proposed HMM structure, HMM will result in poor classification or prediction. Meanwhile, HMM suffers from large training data sets and complex computation, especially when sequences have long length and many labels. The assumption of independence between the historical states, or future states, and the current states also limits the accuracy of HMM.

2.3.6 Kalman Filter

Different from HMM, the Kalman filter has an assumption that the true state is dependent on and evolved from the previous state. This state transition is expressed as

$$x_t = A_t x_{t-1} + B_t u_t + w_t, w_t \sim N(0, Q_t) \tag{2.18}$$

$$y_t = H_t x_t + v_t, v_t \sim N(0, R_t) \tag{2.19}$$

where A_t is the state transition array between the states x_t and x_{t-1} at time t and $t-1$, B_t refers to the control model for control vector u_t, w_t represents the process noise, H_t is the observation transition array between the hidden state x_t and the observation y_t at time t, v_t denotes the measurement noise in observation, Q_t denotes the variance of noise of hidden state x_t, and R_t denotes the variance of noise of observation y_t. As shown in Equations 2.18 and 2.19, the Kalman model recursively estimates the current systematic state x_t based on the previous state x_{t-1} and the present observation y_t. The Kalman filter estimates the posterior state using a minimum mean-square error estimator. Two phases are included in Kalman filter algorithms: (1) a prior estimate phase, when the current state is estimated from the previous state, and (2) a posterior estimate phase, when the current a priori estimate is combined with current observation information to refine the state estimate. In a prior estimate phase, the model is assumed to be perfect without process noise, and the error covariance of the next state is estimated. In a posterior estimate phase, a gain factor, called "Kalman gain," is computed to correct the state estimation and minimize the error covariance. This is presented in detail in Figure 2.8.

The most commonly employed Kalman filters include the basic Kalman filter, extended Kalman filter, unscented Kalman filter, and Stratonovich–Kalman–Bucy filter (Roweis and Ghahramani 1999). The Kalman filter enables continuous online estimation of state vectors for updating observations. The implication is that the Kalman filter uses all the historical and current information for state prediction, which results in smooth interpretation and estimation of states. However, the Kalman filter relies most on the assumption that noises and initial states have normal distributions. The loss of the normality assumption can result in biased estimators.

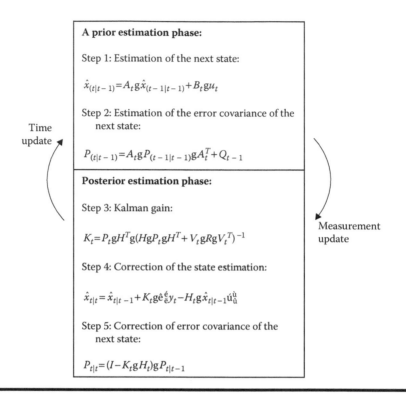

A prior estimation phase:

Step 1: Estimation of the next state:

$$\hat{x}_{(t|t-1)} = A_t g \hat{x}_{(t-1|t-1)} + B_t g u_t$$

Step 2: Estimation of the error covariance of the next state:

$$P_{(t|t-1)} = A_t g P_{(t-1|t-1)} g A_t^T + Q_{t-1}$$

Posterior estimation phase:

Step 3: Kalman gain:

$$K_t = P_t g H^T g (H g P_t g H^T + V_t g R g V_t^T)^{-1}$$

Step 4: Correction of the state estimation:

$$\hat{x}_{t|t} = \hat{x}_{t|t-1} + K_t g ê_e^é y_t - H_t g \hat{x}_{t|t-1} ú_ù^ù$$

Step 5: Correction of error covariance of the next state:

$$P_{t|t} = (I - K_t g H_t) g P_{t|t-1}$$

Time update

Measurement update

Figure 2.8 The workflow of Kalman filter.

2.4 Unsupervised Learning Methods in Medical Signaling and Imaging

Unsupervised learning is also called clustering. Clustering is the assignment of objects into groups (called clusters) so that objects from the same cluster are more similar to each other than objects from different clusters. The sameness of objects is usually determined by the distance between objects over multiple dimensions of the data set. Clustering is widely used in various domains like bioinformatics, text mining, pattern recognition, images segmentation, and content-based image retrieval (CBIR; Chen, Wang, and Krovetz 2005). Clustering is an approach of unsupervised learning in which examples used are unlabeled, that is, they are not preclassified.

2.4.1 k-Means Clustering

The *k*-means clustering algorithm partitions the given data points *S* into *k* clusters, in which each data point is more similar to its cluster mean than to the other cluster means. The k-means clustering algorithm generally consists of the following steps:

Step 1: Select the k initial cluster centroids, $c_1, c_2, c_3, \ldots, c_k$.
Step 2: Assign each instance x in S to the cluster that has a centroid nearest to x.
Step 3: Recompute the centroid of each cluster based on which elements are contained in it.
Step 4: Repeat steps 2 and 3 until convergence is achieved.

Two key issues are involved in the k-means method: (1) the number of clusters k for partitioning and (2) the distance metric. Euclidean distance is the most commonly employed metric in k-means clustering. In practice, unless the cluster number k is known before clustering, no evaluation method can guarantee the selected k is optimal, although researchers have tried to use stability, accuracy, and other metrics in the quality analysis.

2.4.2 Hierarchical Clustering

Hierarchical clustering agglomerates data points into a binary tree structure. At each level of the tree, data points or clusters are grouped into clusters of the next level according to the paired similarities between data points. Hierarchical clustering algorithms mainly consist of the following three steps:

Step 1: Obtain the paired-similarity matrix using a similarity metric for each combination of data points.
Step 2: Group data points or clusters into a hierarchical structure according to the similarity between data points.
Step 3: Select the acceptable clustering part in the hierarchical structure.

In step 1, the similarity metric used can be Euclidean distance, Mahalanobis distance, Minkowski distance, city block distance, or Hamming distance. In step 2, higher-level clusters are formed from the lower-level clusters using a variety of measures, such as nearest neighbor, furthest neighbor, and average distance of all paired distance between clusters. In step 3, the cluster boundaries have to be decided based on the quantitative measure of consistency. Consistency refers to how significantly different is the distance between the data points that are considered to join the two low-level clusters into a high-level cluster and the distance among data points in the low-level clusters.

2.4.3 Grid-Based Clustering

Grid-based clustering partitions each cell at a high level into smaller cells at a low level corresponding to various resolutions. Clustering operations are conducted on cells at a lower level. Typical grid-based clustering methods include the statistical

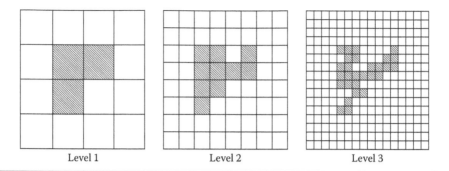

| Level 1 | Level 2 | Level 3 |

Figure 2.9 Illustration of statistical information grid using multiresolution clustering.

information grid (STING), WaveCluster, and clustering in quest (CLIQUE) methods. We present the STING technique to illustrate grid-based clustering.

In STING (Wang, Yang, and Muntz 1997), cells at high levels are split into subcells at low levels in a quadrant manner as shown in Figure 2.9. The statistical information of subcells at a low level, such as means and standard deviations, can be aggregated to obtain the statistical information of cells at a high level. For example, the information "mean" of a cell at a high level can be obtained by taking the average of the means of its subcells. The obtained information is saved in each cell of the structure for querying. For example, STING answers the query for image regions where the variance of gray levels should fall into a desired range. The STING algorithm consists of the following steps:

Step 1: Partition the spatial region of interest into subcells according to hierarchical resolutions.
Step 2: Calculate the statistical information of cells at a preselected level and store it in the corresponding cell.
Step 3: Integrate the statistical information in subcells to a higher level.
Step 4: Follow the up–down method to answer spatial queries.

2.4.4 Density-Based Clustering

Density-based clustering attempts to partition similar density-connected points into clusters of irregular shapes. The most commonly employed algorithms include density-based spatial clustering of applications with noise (DBSCAN), ordering points to identify the clustering structure (OPTICS), density-based clustering (DENCLUE), and CLIQUE. We illustrate this category of clustering by presenting the DBSCAN algorithm as follows:

The DBSCAN algorithm (Ester et al. 1996) groups data points into clusters that have a higher density than a threshold value (MinPts) within a window of a specified size defined by the distance to the data point (Eps). As shown in Figure 2.10,

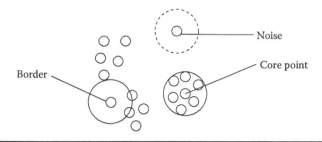

Figure 2.10 Example of the concepts in density-based spatial clustering of applications with noise.

given the specified Nth = 5 and Eps, as well as the radius of the clustering window Eps = 1, data points are classified into three types of clusters according to the local density around them: (1) core points, (2) border points, and (3) noises. A core point has more than MinPts neighboring data points within Eps distance, whereas a border point is located in the neighborhood of a core point but has less neighboring data points within Eps distance. Noises include all the other data points except core points and border points. Given a core point p, any data point q of the other data points within Eps distance from p is within the density range of p. Any data point q is within the density range of core point p if q is within Eps distance from any other data points, which are directly density reachable or density reachable from p. Two data points are density connected if they share at least one common density-reachable data point. The DBSCAN algorithm attempts to group the core points within a specified Eps and MinPts into one cluster, group the border points within a specified neighborhood of a core point in the same cluster, and discard noises.

The DBSCAN algorithm includes the following steps:

Step 1: Find all the data points that are density reachable from a data point of interest p.

Step 2: Group the detected data points in a cluster if p is a core point; if p is a noise, move to another data point of interest and return to step 1.

These steps are continued until all the data points are clustered.

2.4.5 Model-Based Clustering

Model-based clustering assumes that parametric statistical models, such as the Gaussian mixture model (GMM), can describe the distribution of a set of data points. The histogram method is commonly employed to describe the distribution of data points in various dimensions. For example, the histogram of data points can be regarded as an estimate of the probability density function. The parameters of the function can then be estimated by using the histogram. To optimize the estimation of model parameters, the expectation-maximum (EM) algorithm is commonly employed to find the maximum likelihood estimates (machine learning) of the parameters in probabilistic models. Correspondingly, in EM, the expectation (E) and maximization (M) steps are performed iteratively. The E step computes an expectation of the log likelihood with respect to the current estimate of the distribution for latent variables, and the M step computes the parameters that maximize the expected log likelihood found on the E step. These parameters are then used to determine the distribution of latent variables in the next E step. The two steps are as follows:

Step 1—Expectation step: Given sample data x and undiscovered or missed data z, the expected log likelihood function of parameters θ can be estimated by θ^t:

$$f(\theta \mid \theta^t) = E[\log L(\theta; x, z)] \qquad (2.20)$$

Step 2—Maximization step: Using the estimated parameter at step 1, the maximum likelihood function of the parameters can be obtained as follows:

$$\theta^{t+1} = \arg \max_{\theta} (f(\theta \mid \theta^t)) \qquad (2.21)$$

In this equation, the maximum likelihood function is determined by the marginal probability distribution of the observed data $L(\theta; x,z)$.

2.4.6 Self-Organizing Map Artificial Neural Network

Model-based clustering employs the knowledge of a hidden data structure to guide the parametric description of data. With a prior knowledge of data structure, such as a Gaussian distribution, we can efficiently obtain accurate clustering results. The agglomerative input data can improve the accuracy of the estimated parameters in the clustering model. Conversely, a wrong assumption of a data structure can result in a biased clustering model and the absence of meaningful clusters.

The SOM ANN, also known as "Kohonen network," characterizes ANN as visualizing low-dimensional views of high-dimensional data by preserving the neighborhood properties of input data. For example, a two-dimensional SOM consists of lattices. Each lattice corresponds to one neuron. Each lattice contains a vector of weights of the same dimension as the input vectors and no neuron connects with each other. As shown in Figure 2.11, the weight vector $\vec{W}$ of lattices corresponds to the distribution of input vector $\vec{X}$ in the original data space. The objective of SOM ANN is to optimize the area of lattice to resemble the data for the class to which the input vector belongs.

The SOM ANN algorithms consist of the following steps:

Step 1: Initialize neuron weights.
Step 2: Select the vector randomly from training data for the lattice.

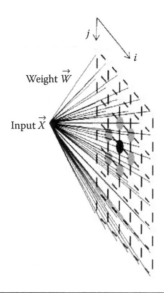

Figure 2.11   Illustration of self-organizing map artificial neural network structure.

Step 3: Find the neuron that has weights most closely matching the input vector, called the "best matching unit" (BMU), for example, the darkest lattice in Figure 2.11.

Step 4: Find the neurons inside the neighborhood of the matched neuron in step 3.

Step 5: Fine-tune the weight of each neighboring neuron obtained in step 4 to increase the similarity of these neurons and the input vector.

Step 6: Iteratively run the process from step 1 to step 5 until convergence.

Figure 2.11 illustrates the mapping result of the SOM ANN algorithm with one cluster: The BMU is the center and the neurons that are furthest from the BMU have more dissimilar weights from the BMU, which are shown in a lighter gray color.

2.4.7 Graph-Based Clustering

Graph-based clustering attempts to group nodes in a graph into clusters such that nodes within clusters have higher similarity than nodes across clusters. The similarity between nodes is measured by edges between nodes. Graph-based clustering includes mainly the following approaches: hierarchical graph clustering, graph cuts, and block models. We illustrate graph-based clustering algorithms by discussing the graph-cut method (Shi and Malik 2000).

When a set of nodes and a set of edges are given, the graph-cut algorithm attempts to partition the set of nodes into two sets such that this cut has minimum cost. The cut cost is calculated by the sum of weights of cut edges (dashed line in Figure 2.12).

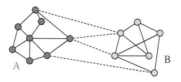

Figure 2.12   Illustration of graph-cut clustering.

Mathematically, the dissimilarity between two clusters C_i and C_j can be computed as similarities between points across those clusters (Shi and Malik 2000):

$$\text{Cut}(C_i, C_j) = \sum_{u \in C_i, v \in C_j} w(u, v) \qquad (2.22)$$

Based on this definition, the disassociation measure cut is defined as follows:

$$\text{Ncut}(C_i, C_j) = \frac{\text{cut}(C_i, C_j)}{\text{assoc}(C_i, V)} + \frac{\text{cut}(C_i, C_j)}{\text{assoc}(C_j, V)} \qquad (2.23)$$

where $V = C_i + C_j$, and $\text{Assoc}(C_{i \text{ or } j}, V) = \sum_{u \in C_{i \text{ or } j}, t \in V} w(u, t)$ is the "tightness weighting" within a cluster.

2.5 Performance Evaluation of Machine Learning Methods

Different biomedical applications have different requirements and algorithms have various features. To ensure a clustering algorithm's suitability for a problem, users need to select proper quantitative and qualitative measures, such as sensitivity, scalability, robustness, and computation complexity. The evaluation measures depend on the confusion matrix as shown in Figure 2.13, in which TP denotes true positive, TN denotes true negative, FP denotes false positive, and FN denotes false negative values.

Traditional classification metrics include classification accuracy and error, defined as follows:

$$\text{Accuracy} = \frac{\#\text{TP} + \#\text{TN}}{|S|} \qquad (2.24)$$

$$\text{Error} = 1 - \text{accuracy} \qquad (2.25)$$

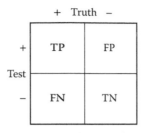

Figure 2.13   Confusion matrix for machine learning performance evaluation.

The metrics are sensitive to changes in the data set and are effective when data are not balanced. For example, we have a data set that has a distribution in which 95% of samples are negative and 5% are positive. If 5 of a given test data set of 100 samples are positive and 95 samples are negative, then even if all the test results are classified as negative the accuracy is 95%. This value is preserved when the number of TN increases while the number of TP decreases by the same amount. When the positive result is more important for researchers, the aforementioned metrics cannot provide the exact information of class labels.

To comprehensively evaluate imbalanced learning, especially for minority classification, other metrics are used including precision, recall, *F*-score, *Q*-score, *G*-mean, receiver operating characteristics (ROCs), areas under receiver operating characteristics (AUCs), precision recall curves (PRCs), and cost curves. These metrics are defined as follows:

$$\text{Precision} = \frac{\#\text{TP}}{\#\text{TP} + \#\text{FP}} \tag{2.26}$$

$$\text{Recall} = \frac{\#\text{TP}}{(\#\text{TP} + \#\text{FN})} \tag{2.27}$$

$$F\text{-score} = \frac{(1+\beta)^2 \cdot \text{recall.precision}}{\beta^2 \cdot \text{recall.precision}} \tag{2.28}$$

$$Q\text{-score} = \frac{(1+\beta)^2 \cdot \#\text{TP}}{(1+\beta)^2 \cdot \#\text{TP} + \beta^2 \cdot \#\text{FN} + \#\text{FP}} \tag{2.29}$$

$$G\text{-mean} = \sqrt{\frac{\#\text{TP}}{\#\text{TP} + \#\text{FN}} \cdot \frac{\#\text{TN}}{\#\text{TN} + \#\text{FP}}} \tag{2.30}$$

Precision, as is clear from Equation 2.26, measures the exactness of positive labeling and the coverage of the correct positive labels among all positively labeled samples. Recall, as is clear from Equation 2.28, measures the completeness of positive labeling and the coverage of correctly labeled positive samples among all positive class samples. Precision is sensitive to data distribution, whereas recall is not (see confusion matrix in Figure 2.13). Recall cannot reflect how many samples are labeled "false" incorrectly, and precision cannot provide any information about how many positive samples are labeled incorrectly. *F*-measure combines these two metrics and assigns the weighted importance on either precision or recall using the coefficient β. Consequently, the *F*-measure provides more insight into the accuracy of a classifier than recall and precision, while remaining sensitive to data distribution. *G*-mean evaluates the inductive bias of the classifier using the ratio of positive to negative accuracy.

The ROC curves provide more insight into the relative balance between the gains (true positive) and costs (false positive) of classification on a given data set. Two evaluation metrics are used in ROC curves as follows:

$$\text{TP}_{\text{rate}} = \frac{\#\,\text{TP}}{|S_{mi}|} \tag{2.31}$$

$$\text{FP}_{\text{rate}} = \frac{\#\,\text{FP}}{|S_{ma}|} \tag{2.32}$$

As shown in Figure 2.14, ROC curves are composed of the combinational values of TP rate and FP rate. Each point on the ROC curve (indicated by the line in Figure 2.14) corresponds to the performance of a classifier on a given data set. Two important points, A and B, are illustrated with their coordinates. Point A is the perfect classification result with no errors. Point B is the worst classification result, in which all positive labels are incorrect. The point located nearer to point A has a better classification result than the point nearer to point B.

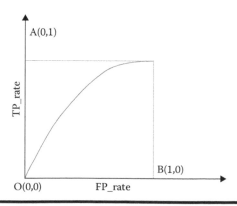

Figure 2.14 Receiver operating characteristic curve representation.

2.6 Research Challenges and Directions in the Future

Although various machine learning methods and tools have been discovered for medical signaling and imaging, a number of challenges must be faced before applying these methods to medical data.

First, the heterogeneity of medical signals and images and other medical data exacerbates the integration of medical information for machine learning. For example, CBIR offers semantic access to a large number of biomedical images for clinicians and scientists (Müller et al. 2004; Datta et al. 2008). Meanwhile, other modal signals or information, such as electrocardiogram, electroencephalogram, and patient medical history, include knowledge of sequences of biomedical records. The integration of multimodal medical knowledge and the result of CBIR can facilitate biomedical queries and clinical diagnosis and prognosis or research.

Second, biomedical signals and images are temporary. Time-varying data need temporary learning, analysis, and visualization tools that can facilitate clinical analysis by detecting temporal patterns within multiple medical data streams (Stacey and McGregor 2007; Klimov, Shahar, and Taieb-Maimon 2009).

Third, machine learning methods face challenges from nontypical data among various applications in biomedical signaling and imaging. Among such challenges, imbalanced data attracts significant attention (He and Garcia 2009). The data imbalance problem arises because it is impossible to collect samples from the minority class for inclusion in some applications, such as the identification of cancer patients. Imbalanced data is often accompanied by a small sample size and a high-dimensional feature space, making the application of standard classification methods in the data difficult (Wasikowski and Chen 2009). A small sample size makes classification methods difficult to generalize statistically, whereas a high-dimensional feature space blocks valuable feature information. Subsequently, an innovative machine learning method is needed to generalize the classification of small samples, especially minority-class samples. To combat the imbalanced learning problem, researchers focus on four groups of methods: (1) modification of sample data, (2) classification methods, (3) classification metrics, and (4) feature selection methods.

Fourth, medical researchers and clinicians, or other end users, need interactive tools to perform analysis, diagnosis, and prognosis of medical data sets. The designing of successful interactive tools depends on the development of CBIR, computer-assisted diagnosis systems, and computer vision systems. The former two types of systems provide integrated information for specialists to visualize.

To combat the aforementioned problems, we have to either design innovative machine learning methods for biomedical signaling and imaging or adapt well-known classic machine learning algorithms to specific problems in biomedical signaling and imaging. Further, as we discuss in this chapter, machine learning methods have been developed successfully in the past for various applications, including some specific domains in biomedical signaling and imaging. However, there exists no machine learning method that generalizes well for any biomedical signaling and imaging application. In such a context, the successful selection or design of machine learning methods relies highly on expert knowledge of used biomedical signals or images. In Chapters 3 through 16, readers will understand how to apply machine learning algorithms to specific biomedical signaling and imaging applications.

References

Bezdek, J., L. Hall, and L. Clarke. 1993. Image segmentation using pattern recognition. *Med Phys* 20:1033–48.

Chen, Y., J. Z. Wang, and R. Krovetz. 2005. CLUE: Cluster-based retrieval of images by unsupervised learning. *IEEE Trans Image Process* 14:1187–201.

Datta, R., D. Joshi, J. Li, and J. Z. Wang. 2008. Image retrieval: Ideas, influences, and trends of the new age. *ACM Comput Surv* 40:1–60.

Ester, M., H.-P. Kriegel, J. Sander, and X. Xu. 1996. A density-based algorithm for discovering clusters in large spatial databases with noise. In Evangelos Simoudis, Jiawei Han, Usama M. Fayyad *Proceedings of 2nd International Conference on Knowledge Discovery and Data Mining (KDD-96)*, AAAI press. pp.226–231.

Han, J., and M. Kamber. 2006. *Data Mining. Concepts and Techniques.* San Francisco, CA: Morgan Kaufmann.

He, B., and E. Garcia. 2009. Learning from imbalanced data. *IEEE Trans Knowl Data Eng* 21:1263–84.

Jones, T. R., A. Carpenter, and P. Golland. 2005. Voronoi-based segmentation of cells on image manifolds. *Lect Notes Comput Sci* 3765:535–43.

Kapetanovic, I. M., S. Rosenfeld, and G. Izmirlian. 2004. Overview of commonly used bioinformatics methods and their applications. *Ann N Y Acad Sci* 1021:10–21.

Klimov, D., Y. Shahar, and M. Taieb-Maimon. 2009. Intelligent interactive visual exploration of temporal associations among multiple time-oriented patient records. *Methods Inf Med* 48:254–62.

Müller, H., N. Michoux, D. Bandon, and A. Geissbuhler. 2004. A review of content-based image retrieval systems in medical applications-clinical benefits and future directions. *Int J Biomed Inform* 73:1–23.

Rabiner, L. R. 1989. A tutorial on hidden Markov models and selected applications in speech recognition 77:257–286.

Roweis, S., and Z. Ghahramani. 1999. A unifying review of linear Gaussian models. *Neural Comput* 11:305–45.

Sajda, P. 2006. Machine learning for detection and diagnosis of disease. *Annu Rev Biomed Eng* 8:537–65.

Shi, J. B., and J. Malik. 2000. Normalize cuts and image segmentation. *IEEE Trans Pattern Anal Mach Learn Intell* 22:888–905.

Stacey, M., and C. McGregor. 2007. Temporal abstraction in intelligent clinical data analysis: A survey. *Artif Intell Med* 39:1–24.

Wang, W., J. Yang, and R. R. Muntz. 1997. STING: A statistical information grid approach to spatial data mining. In Jarke M., M. J. Carey, K. R. Dittrich, F. H. Lochovsky, P. Loucopoulos, M. and A. Jeusfeld: *Proceedings of the 23rd International Conference on Very Large Data Bases*, 186–95. Athens, Greece: Morgan Kaufmann.

Wasikowski, M., and X. Chen. 2009. Combining the small sample class imbalance problem using feature selection. *IEEE Trans Knowl Data Eng* 22:1388–1400.

Chapter 3

Data Mining of Acoustical Properties of Speech as Indicators of Depression

Ananthakrishna T., Kumara Shama, Venkataraya
P. Bhandary, Kumar K. B., and Niranjan U. C.

Contents

3.1 Introduction to Acoustical Properties of Speech

Speech and hearing, a human being's most used means of communication, have been the objects of intense study and research for more than 150 years. Human beings have been gifted with the unique ability to communicate with each other through speech. This amazing ability of human beings to communicate through speech sets them apart from other earthly species. Although textual language also has become important in modern life, speech has dimensions of richness that text cannot approximate. For example, health, sex, and attitude of a person are communicated by that person's speech. Speech production is one of the most complex and delicate operations undertaken by the human body. Acoustic analysis of speech plays an important role in providing health care for emotional disorders as speech signals can carry an extremely high amount of information on the conditions of the body and mind.

Depression is a psychiatric disorder and one of world's most serious health threats. It is an illness that can challenge one's ability to perform routine activities and may even lead people to contemplate or commit suicide (Depression Overview, http://www.emedicinehealth.com). Though depression is more common in the elderly than in young people, it affects people of almost all ages. Depression can be diagnosed and treated effectively in most people. The biggest problem in treating depression is that patients must recognize the signs of depression and seek appropriate treatment. Experienced clinicians use history, clinical interviews, psychological testing, self-reports, and many other methods to assess the presence of depression and the degree of depression in individuals (Gullick and King 1979; Johnson 1974; Pokorny 1983; Popkin and Callies 1987).

Depressed speech has often been characterized as dull, monotone, monoloud, lifeless, and metallic. These perceptual qualities have been associated with fluctuations involving "fundamental frequency," formant structure, and power spectral distribution. Fundamental pitch frequency (F_0) is the most widely studied excitation-based feature in the study of speech and psychopathology. Fundamental frequency range, mean, variance, and kurtosis have been identified as key statistics correlated with depression. It has also been demonstrated that loudness range and mean speaking intensity are correlated with depressed mood. Formant studies, which provide valuable insight into vocal tract behavior during speech production and articulation, show that the central frequencies of an individual's second and third formants are reduced during periods of depression. Further, significant increases in these formant frequencies have been measured on an individual's improvement from a depressive episode. The power spectral distribution of speech has also been determined to exhibit a similar shift before and after treatment for depressive illness. Studies reveal that the speech of severely depressed individuals contains more power at frequencies above 500 Hz. After treatment, power shifts were measured toward lower frequencies.

3.1.1 Anatomy and Physiology of Speech Production

The main components of speech production system are lungs, trachea (windpipe), larynx (organ for voice production), pharyngeal cavity (throat), oral cavity (mouth), and nasal cavity (nose). Figure 3.1 shows the basic anatomy of the human speech production system (Quatieri 2002). In technical terms, pharyngeal and oral cavities are grouped together into a single entity called the "vocal tract." The vocal tract begins at the output of the larynx and terminates at the input to the lips. The finer anatomical features critical to speech production include the vocal folds or vocal cords, velum, tongue, teeth, and lips. These components move in different positions to produce various speech sounds. The lungs, larynx, and vocal tract form the three important organs of speech production.

The anatomy of these organs, as well as the associated physiology and its importance in speech production, is detailed in Sections 3.1.2 through 3.1.4.

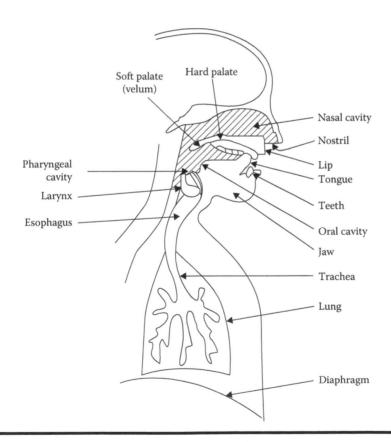

Figure 3.1   Anatomy of human speech production system.

3.1.1.1 Lungs

The lungs are the source of airflow that passes through the larynx and vocal tract before exiting the mouth as pressure variations constituting the speech signal. Situated in the chest or thorax cavity, lungs primarily serve breathing, inspiring, and expiring the tidal volume of air. When we inhale, we enlarge the chest cavity by expanding the rib cage surrounding the lungs and lowering the diaphragm that sits at the bottom of the lungs and separates it from the abdomen. This action lowers the air pressure in the lungs, causing air to rush in through the vocal tract and down the trachea into the lungs. When we exhale, we reduce the volume of the chest cavity by contracting the muscles in the rib cage, thus increasing air pressure in the lung. This increase in pressure causes air to flow through the trachea into the larynx. During speaking, on the other hand, a slightly different functioning is observed. During speaking, we take in short spurts of air and release them steadily by controlling the muscles around the rib cage. We override our rhythmic breathing by making the duration of exhalation roughly equal to the length of a sentence or a phrase. During this timed exhalation, air pressure of the lung is maintained at more or less a constant level, slightly above atmospheric pressure, by the steady and slow contraction of the rib cage, although the air pressure varies around this level due to the time-varying properties of the larynx and the vocal tract.

3.1.1.2 Larynx

The larynx is an important organ involved in speech production. Figure 3.2 shows a cross section of the larynx viewed from the front. The structure of the larynx consists of four cartilages, muscles, and ligaments, which, in the context of speech production, primarily control the vocal folds (Quatieri 2002). The vocal folds are two masses of flesh, ligament, and muscle, which stretch between the front and back of the larynx, as illustrated in Figure 3.3. The folds are about 15-mm long in men and 13-mm long in women. The glottis is the slitlike orifice between the two folds that are fixed at the front of the larynx, where they are attached to the stationary thyroid cartilage. The thyroid cartilage is located at the front of the larynx (or Adam's apple) and sides of the larynx. The folds are free to move at the back and sides of the larynx; they are attached to the two arytenoids cartilages that move in a sliding motion at the back of the larynx along with the cricoid cartilage. The size of the glottis is controlled in part by the arytenoids cartilages and in part by muscles within the folds. Another important property of the vocal folds, in addition to controlling the size of the glottis, is their tension.

The tension of the vocal folds is controlled primarily by muscles within the folds, as well as the cartilage around the folds. The epiglottis, false vocal folds, and true vocal folds provide a triple barrier across the windpipe; they are all closed during swallowing and wide open during breathing. Although the false vocal folds can be closed and they can vibrate, they are likely to be open during speech production.

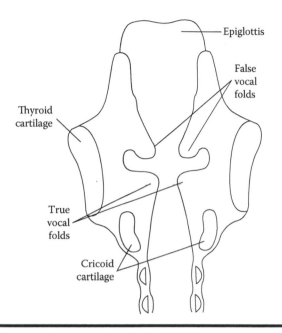

Figure 3.2   Cross section of larynx.

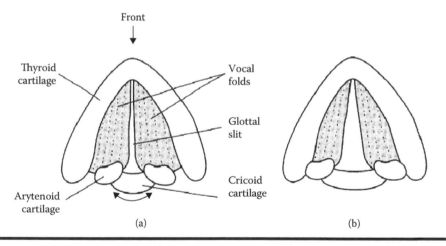

Figure 3.3   Downward-looking view of larynx during (a) voicing and (b) breathing.

There are three primary states for vocal folds: (1) breathing, (2) voicing, and (3) unvoicing. Sketches of the larynx in the downward-looking view during voicing and breathing are shown in Figure 3.3a and b, respectively (Quatieri 2002), with the sketch of the unvoiced state being almost similar to that of the breathing state. In the breathing state, the arytenoids cartilages are held outward, they maintain a wide glottis, and the muscles within the vocal folds are relaxed. In this state, air from the lungs flows freely through the glottis with negligible resistance. In speech production, on the other hand, the folds obstruct airflow.

Voicing is the interesting form of excitation that involves self-sustained oscillations of vocal folds. Voicing is accomplished when the abdominal muscles force the diaphragm up, pushing air out from the lungs into the trachea and then up to the glottis, where the movement of vocal folds periodically interrupts flow of air. Sufficient air pressure below sufficiently adducted vocal folds produces oscillation of vocal folds and, thereby, voiced sounds when the vocal fold tissues are pliable. In fact, the vocal folds open and close regularly, resulting in a quasiperiodic flow of air exiting the vocal tract. The plot of airflow velocity at the glottis during voicing as a function of time is approximately as shown in Figure 3.4 (Quatieri 2002). Typically, with the folds in a closed position, the flow begins slowly, builds up to a maximum, and then quickly decreases to zero when the vocal folds are abruptly shut. The time interval during which the vocal folds are closed and no flow occurs is called the "glottal closed phase," and the time interval over which there is nonzero flow and maximum airflow velocity is called the "glottal open phase." Time interval from the time of airflow maximum to the time of glottal closure is referred to as the "return phase." The time at which the glottis is closed is referred to as the "glottal closing instant" (GCI). The specific flow shape can change with the speaker, speaking style, and specific speech sound. In some cases, the folds do not close completely, so that a closed phase does not exist.

The time duration of one glottal cycle is called the "pitch period," and the reciprocal of the pitch period is called the fundamental frequency, F_0. The fundamental frequency has an average value that varies with the size of speaker's vocal folds. Since neither F_0 nor vocal tract shape stays constant for long, voiced speech is not truly periodic; it is quasiperiodic over short intervals of time. Natural glottal pulses are not truly periodic; they exhibit jitter and shimmer, which are variations in pitch and amplitude, respectively. Normal voices have jitter of 0.5%–1%

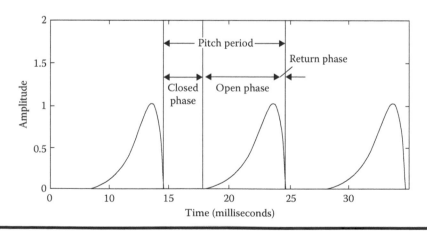

Figure 3.4 **Plot of periodic glottal airflow velocity.**

and shimmer of about 0.04%–0.21 dB (Heiberger and Horii 1982). The jitter and shimmer over successive pitch periods help to give the vowel its naturalness, in contrast to a monotone pitch and fixed amplitude that can result in a machinelike sound. However, the extent and form of jitter and shimmer can contribute to voice character, in addition to naturalness. A high degree of jitter, for example, results in a voice with a hoarse quality, which can be characteristic of a particular speaker or created under specific speaking conditions such as stress or fear.

Like the unvoicing state, the breathing state contains no vocal fold vibration. In the unvoiced state, however, the folds are closer together and more tense than in the breathing state, thus allowing turbulence to be generated at the folds. Turbulence at the vocal folds is called "aspiration." Aspiration occurs in normal speech as with the letter "h" in the word "he." Turbulence is also created in the glottis during whispering. In certain voice types, aspiration occurs normally simultaneously with voicing, resulting in a breathy voice. Aspiration occurs to some extent in all speakers, and the amount of aspiration may serve as a distinguishing feature of depression.

3.1.1.3 The Role of the Vocal Tract

The vocal tract comprises the oral cavity from the larynx to the lips and the nasal passage that is coupled to the oral tract by way of the velum. The vocal tract acts like an organ pipe, producing perceptually distinct speech sounds. In the spectrum of vowel sound, there is evidence of periodic excitation and a well-defined region of emphasis. These resonances are a consequence of articulators having been formed by various acoustical cavities and subcavities out of vocal tract cavities. The locations of these resonances in the frequency domain depend on the shape and physical dimensions of the vocal tract. These resonance frequencies of the vocal tract are, in speech science context, called "formant frequencies" or simply "formants." The shape of the vocal tract varies as a function of time during desired speech production and, therefore, spectral properties of speech vary with time.

In this chapter, we investigate the efficacy of these acoustic features of speech as indicators of depression. In Section 3.2, we explain these parameters and discuss in detail how they are extracted. In Section 3.3, we illustrate the statistical analysis of these parameters. The results of the analysis are presented in Section 3.4, and we conclude the chapter with Sections 3.5 and 3.6.

3.2 Acoustical Measures

In this section we discuss the three important feature sets; viz vocal jitter, formant frequency & bandwidth, and power spectral density.

3.2.1 Vocal Jitter

Jitter is a measure of short fluctuations occurring in the fundamental frequency of voicing. This measure is quantified by the "jitter factor," which is the absolute mean difference between the fundamental frequencies of adjacent glottal cycles divided by the mean fundamental frequency and multiplied by 100. Thus, it is expressed as follows (France et al. 2000):

$$\text{Jitter factor (\%)} = \frac{\dfrac{1}{N-1}\sum_{k=1}^{N-1}|F_k - F_{k+1}|}{\dfrac{1}{N}\sum_{k=1}^{N}F_k} \times 100 \tag{3.1}$$

where F_k and F_{k+1} are the fundamental frequencies for the kth and $(k+1)$th glottal cycles, and N is the total number of glottal cycles considered. During the production of the voiced sound, the vocal folds vibrate and produce what is known as glottal pulse. The fundamental frequency of glottal pulse is the pitch or the fundamental frequency of voicing. Thus, vocal jitter requires the estimation of the glottal cycle duration. Since our speech database consists of continuous, running speech samples that contain voiced, unvoiced, and silence segments, the first step in estimating vocal jitter will be segmenting the speech samples into voiced, unvoiced, and silence segments.

To detect the voiced segments, we use the dyadic wavelet transform (D_yWT)–based algorithm described by Ozdas et al. (2004). The D_yWT of a signal $s(t)$ is given as

$$D_y\text{WT}(b, 2^j) = \frac{1}{2^j}\int_{-\infty}^{\infty} s(t)g^*\left(\frac{t-b}{2^j}\right)dt \tag{3.2}$$

Equation 3.2 computes the wavelet transform using a scale parameter $a = 2^j$ that is discretized along the dyadic sequence. Here, $g^*(t)$ is the complex conjugate of a wavelet function $g(t)$, and

$$g_{2^j}(t) = \frac{1}{2^j}g\left(\frac{t}{2^j}\right) \tag{3.3}$$

From the signal processing point of view, D_yWT can be considered as the output of a bank of constant-Q octave-band band-pass filters, which have impulse responses of $(1/2^j)g(t/2^j)$. The bandwidth and center frequency of each such filter are proportional to $1/2^j$.

We employed D_yWT to detect voice speech segments by comparing the wavelet-transform output energies across scales. The wavelet used was a quadratic spline wavelet with compact support and one vanishing moment. The D_yWT was implemented using filter banks without any subsampling as shown in Figure 3.5.

Here, $L(2^j\omega)$ and $H(2^j\omega)$ represent the Fourier transform of finite-impulse-response (FIR) low-pass and high-pass filters, respectively, that correspond to the quadratic spline wavelet at level $j + 1$, as shown in Equations 3.4 and 3.5:

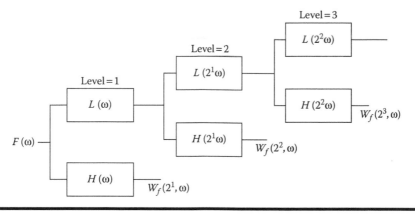

Figure 3.5   Block diagram of the dyadic wavelet transform implementation.

$$L(2^j \omega) = e^{i2^j \omega/2} \left(\cos \frac{2^j \omega}{2} \right)^3 \tag{3.4}$$

$$H(2^j \omega) = 4i\, e^{i2^j 2^j \omega/2} \left(\sin \frac{2^j \omega}{2} \right) \tag{3.5}$$

The output of the D_yWT at each level was calculated as follows:

$$
\begin{aligned}
W_f(2^{j+1}, \omega) &= H(\omega)F(\omega) && j = 0, \text{ level } 1 \\
&= H(2\omega)L(\omega)F(\omega) && j = 1, \text{ level } 2 \qquad (3.6) \\
&= H(2^{j-1}, \omega)L(2^{j-2}, \omega)\ldots L(\omega)F(\omega) && j \geq 2, \text{ level } \geq 3
\end{aligned}
$$

In this equation, $F(\omega)$ and $W_f(2^{j+1}, \omega)$ are the analyzed signal and the D_yWT of the analyzed signal at level $j + 1$ in the frequency domain, respectively. As a result, the D_yWT output of the analyzed signal is the output of the filter bank, where the output of the band-pass filter corresponding to scale 2^{j+1} is equivalent to the resultant of D_yWT at scale 2^{j+1}. The D_yWT values at small scales reflect high-frequency components, whereas at large scales D_yWT reflects low-frequency components of the analyzed signal. Frequency responses and approximate 3-dB bandwidths of the aforementioned filters are given in Figure 3.6 and Table 3.1, respectively.

The algorithm for voiced, unvoiced, and silence classification exploits the fact that unvoiced speech segments are mainly composed of high-frequency components, whereas voiced speech segments are quasiperiodic and are low frequency in nature. In order to classify a speech segment as voiced, unvoiced, or silence, first we computed its D_yWT at scales $a = 2^1$ through $a = 2^5$ and calculated the signal energy at each scale.

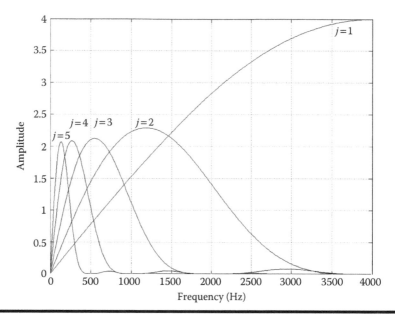

Figure 3.6 **Amplitude responses of the band-pass filters corresponding to different scales.**

Table 3.1   3-dB Bandwidths of the Band-Pass Filters Corresponding to the Output of the Filter Bank at Each Scale

Scale	3-dB Bandwidth (Hz)
2^1	2000–4000
2^2	560–1880
2^3	260–880
2^4	140–440
2^5	70–220

We then employed a decision scheme that is scale–energy based to detect the voice speech segments. Figure 3.7 illustrates the D_yWT energies at different scales among voiced, unvoiced, and silence segments.

Due to the high-frequency nature of unvoiced speech, its D_yWT energy is largest for the scale $a = 2^1$. Therefore, the algorithm first determines the scale at which the D_yWT of the underlying segment has the largest energy. If the segment D_yWT has its maximum energy at scale $a = 2^1$, the segment is classified as unvoiced. Otherwise, it can either be a voiced segment or a silent segment depending on its energy at higher scales. The D_yWT values of both voiced and silent segments

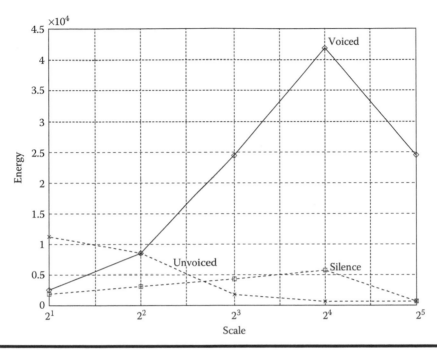

Figure 3.7 Dyadic wavelet transform energies of voiced, unvoiced, and silence speech segments across scales.

exhibit their highest energy content at higher scales. Therefore, in order to separate voiced and silence segments, the D_yWT energy at scale $a = 2^3$ of each segment that is not detected as unvoiced is compared with a predetermined threshold. The median of the energy at scale 2^3 is taken as the threshold. The segments at scale $a = 2^3$ that exceed this threshold are classified as voiced, and those that are below the threshold are classified as silence.

Each speech sample considered in the study is segmented into 512 sample segments and then classified as voiced, unvoiced, and silence. For all those segments that are identified as voiced, the fundamental frequency is estimated. Once the fundamental frequency contours are estimated for each speech sample, the vocal jitter factor is measured according to Equation 3.1.

3.2.2 Formant Frequencies and Bandwidths

Formant frequencies represent the resonance of the vocal tract and are closely related to vocal tract geometry. Different techniques are used for the automatic estimation of formant frequencies (Hogberg 1997; Zolfaghari and Robinson 1996; M. Lee et al. 1999). Here, we use the linear prediction coding (LPC) method (Quatieri 2002), which is the most commonly used method because of its accuracy and computational efficiency. The method assumes a vocal tract response as a linear filter with certain resonances. The vocal tract is, therefore, modeled as a time-invariant all-pole linear system with poles corresponding to vocal tract resonances (formants). The frequency response of the all-pole linear system can be expressed as

$$H(z) = \frac{A_v}{\prod\limits_{k=1}^{N}\left(1-c_k z^{-1}\right)\left(1-C_k^* Z^{-1}\right)} \tag{3.7}$$

where A_v is the gain controlling loudness of sound, and $\left(1-c_k z^{-1}\right)$ and $\left(1-C_k^* z^{-1}\right)$ are complex conjugate poles inside the unit circle with $|c_k| \langle 1.$ Figure 3.8 shows a discrete-time model of the speech production system with periodic excitation.

For the periodic case, the z-transform of the speech output is given by

$$Y(z) = G(z)H(z) \tag{3.8}$$

where $G(z)$ is the z-transform of glottal flow input $g(n)$ over one cycle, which depends on the particular sound (i.e., phoneme), speaker, and speaking style. In Equation 3.8, $R(z)$, the radiation impedance, is lumped into the transfer function $H(z)$.

In order to estimate filter coefficients, an LPC algorithm can be adapted. Equation 3.8 can be rewritten as follows:

$$H(z) = \frac{Y(z)}{G(z)} = \frac{A_v}{1-\sum\limits_{k=1}^{p}a_k z^{-k}} \tag{3.9}$$

or

$$Y(z) - \sum\limits_{k=1}^{p}a_k Y(z)z^{-k} = A_v G(z) \tag{3.10}$$

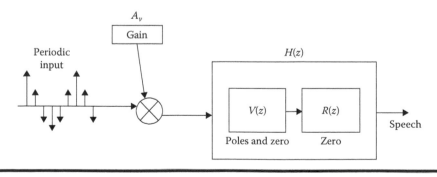

Figure 3.8   Discrete-time speech production model with periodic excitation.

In the time domain, Equation 3.10 becomes

$$y(n) = \sum_{k=1}^{p} a_k y(n-k) + A_v g(n) \qquad (3.11)$$

Equation 3.11 is known as an autoregressive model, because the output is regressive on itself. The coefficients a_k are known as "linear prediction coefficients."

A twelfth-order autoregressive model using the LPC method is used to estimate formants in this study. All the voiced segments corresponding to each speech sample are merged to get the continuous voiced speech signal. The signal is then divided into equal-sized frames of 20 milliseconds. Since the properties of the speech signal vary continuously with time, it is desirable to consider a segment of speech, which is assumed to be stationary. The LPC coefficients were then calculated for each frame. The formant center frequency and bandwidths of each speech frame were determined by computing the roots of the predictor polynomial, or all-pole model of the vocal tract, derived from the LPC coefficients. Finally, the formant frequencies and bandwidths calculated from all the frames were time-averaged to obtain a single formant vector for each class member. The first three formant frequencies denoted as F_1, F_2, and F_3 and the corresponding formant bandwidths FBW_1, FBW_2, and FBW_3 are considered for the analysis.

3.2.3 Power Spectral Density

Power spectral density (PSD) can be defined as the amount of power per unit (density) of frequency (spectral) as a function of the frequency. It shows the strength of the variation (energy) as a function of frequency. The PSD describes how the power (or variance) of a signal is distributed with frequency. Mathematically, PSD is the squared modulus of the Fourier transform of the signal, scaled by a proper constant term. An equivalent definition of PSD is that it is the Fourier transform of the autocorrelation sequence of the signal. In other words, it shows the frequencies at which autocorrelation is strong and those at which it is weak. One can obtain power within a specific frequency range by integrating PSD within that frequency range.

The power distribution of speech signals in the frequency band ranging from 0 to 2 kHz is considered in this study. The PSD is estimated using the Welch (Welch 1967) method of PSD estimation. The spectra were computed using 40-millisecond frames of running speech for each class member with a 100-point nonoverlapping Hamming window and a 1024-point fast Fourier transform (FFT). Four 500-Hz subbands called PSD_1, PSD_2, PSD_3, and PSD_4 are considered. The percentage of total power in each of these four 500-Hz subbands are calculated and used as features.

3.3 Statistical Analysis

To determine any statistically significant differences in the voice parameters, two sample (i.e., control–depressed) statistical analyses are performed. The student *t*-test is performed over all the parameters independently to evaluate differences in class means. In order to evaluate the discriminating power of these parameters among the control and depressed classes, discriminant analysis is employed. The discriminating power of each parameter is tested separately using a linear discriminant classifier. The performance of the classifier for the combined feature set is also considered. A total of 17 normal subjects (control group) and 17 subjects with major depression (depressed group) are considered for the analysis. Since this sample size is small, the leave-one-out method (Fukunaga 1990) is used in all the cases. In the leave-one-out method, one sample is excluded from the database and the classifier is trained with the remaining samples. Then, the excluded sample is used as the test data, and classification accuracy is determined. This process is repeated for all the samples of the database. This procedure is useful for a small database size because it is possible to use the same sample for training and testing rather than using only a part of the samples for training and the remaining for testing.

3.4 Results

The class distributions for vocal jitter, power distribution, formant frequencies, and formant bandwidths are presented in Figures 3.9, 3.10, 3.11, and 3.12, respectively. In these box plots, the lines in the middle of the boxes represent the sample

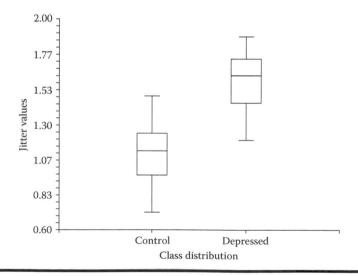

Figure 3.9 Plot of jitter distributions.

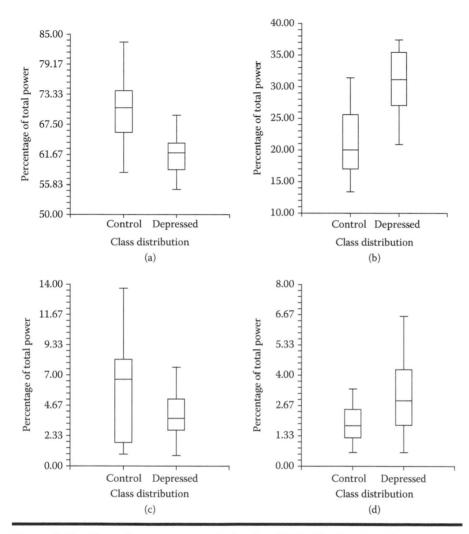

Figure 3.10 Plot of power spectral density (PSD) distributions for (a) PSD₁, (b) PSD₂, (c) PSD₃, and (d) PSD₄.

medians, and the lower and upper lines of the boxes represent the twenty-fifth and the seventy-fifth percentiles of the samples. The lines extending above and below the boxes show the extent of the samples in each group except for the outliers. The solid circle at the top of the box plot indicates the presence of the outlier. Outliers are seen in the case of the control class distributions for F_1, FBW_1, and FBW_2.

The aforementioned patterns need not imply statistically significant differences. The student *t*-test detected the presence of a significant difference between the two classes. The mean, standard deviation, and 95% confidence intervals from the *t*-test calculated for acoustical features are shown in Table 3.2.

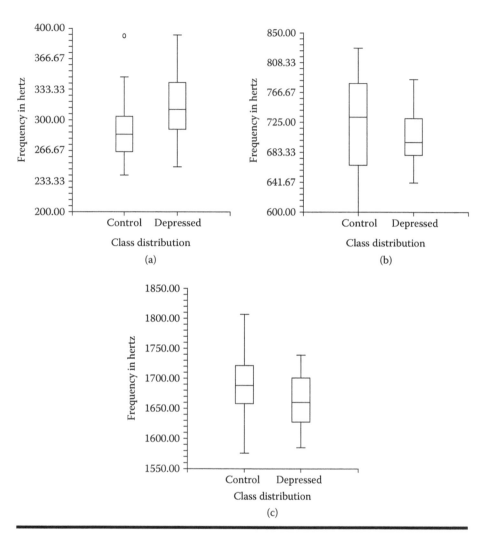

Figure 3.11 Plot of formant frequency (*F*) distributions for (a) F_1, (b) F_2, and (c) F_3.

The results of the study indicate that depression causes statistically significant changes in these parameters. For all the parameters, the class means are found to be significantly different between control and depressed classes with *p*-value less than 0.05.

The discriminating power of each parameter is obtained through a linear discriminant classifier. Table 3.3 shows the results of the classifier for individual parameters. Performance is given in terms of the class sensitivity (SE), specificity (SP), positive prediction (PP), negative prediction (NP), and overall accuracy. For the parameters to be effective in discriminating between classes, it is important

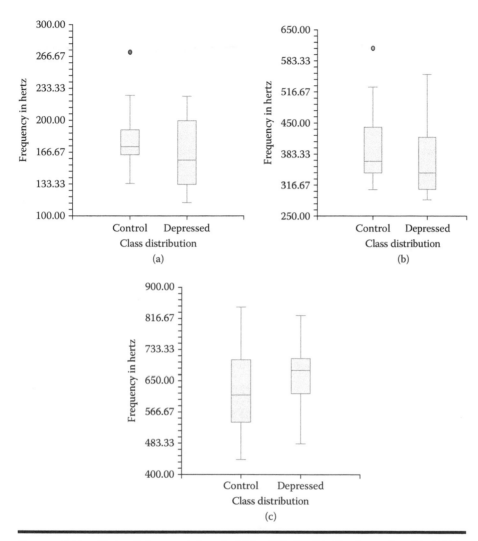

Figure 3.12 Plot of formant bandwidth (FBW) distributions for (a) FBW$_1$, (b) FBW$_2$, and (c) FBW$_3$.

that both SE and SP are high. Both SP and SE are found to be high for jitter factor, percentage power at PSD$_1$ and PSD$_2$.

We also tested the discriminating power of the combined feature set. Percentage powers at PSD$_1$ though PSD$_4$ are considered parts of a single feature vector (denoted as PSD$_{1-4}$). Similarly, the three formant frequencies form a feature vector (denoted as F$_{1-3}$) and so do the three formant bandwidths (denoted as FBW$_{1-3}$). The classification scores for these feature vectors are shown in Table 3.4.

Table 3.2 Results of *t*-Test

Parameter	Group	Mean	Standard Deviation	95% Confidence Intervals
Jitter	Control	1.11	0.21	(1.01, 1.22)
	Depressed	1.58	0.20	(1.47, 1.68)
% Power in PSD_1	Control	70.49	6.52	(67.14, 73.85)
	Depressed	61.82	4.57	(59.47, 64.17)
% Power in PSD_2	Control	21.29	5.12	(18.66, 23.92)
	Depressed	30.95	4.68	(28.54, 33.36)
% Power in PSD_3	Control	6.32	4.27	(4.12, 8.51)
	Depressed	4.04	1.83	(3.10, 4.98)
% Power in PSD_4	Control	1.88	0.81	(1.46, 2.30)
	Depressed	3.15	1.73	(2.25, 4.04)
F_1 (in Hz)	Control	291.53	36.72	(272.65, 310.41)
	Depressed	316.41	36.21	(297.79, 335.02)
F_2 (in Hz)	Control	722.94	75.16	(684.29, 761.58)
	Depressed	701.82	36.43	(683.09, 720.55)
F_3 (in Hz)	Control	1693.1	57.5	(1663.5, 1722.7)
	Depressed	1662.2	44.7	(1639.2, 1685.2)
FBW_1 (in Hz)	Control	181.17	31.88	(164.8, 197.6)
	Depressed	163.7	35.55	(145.4, 181.9)
FBW_2 (in Hz)	Control	398.5	82.7	(355.9, 441)
	Depressed	370.1	75.8	(331.1, 409.1)
FBW_3 (in Hz)	Control	622.1	112.1	(564.5, 679.8)
	Depressed	661.5	85.9	(617.3, 705.7)

Table 3.3 Classification Scores for Individual Parameters

Parameter	SE (%)	SP (%)	PP (%)	NP (%)	Accuracy (%)
Jitter	82.35	88.23	87.50	83.33	85.29
% Power in PSD_1	82.35	76.47	77.78	81.25	79.41
% Power in PSD_2	82.35	82.35	82.35	82.35	82.35
% Power in PSD_3	76.47	58.82	65.00	55.55	67.65
% Power in PSD_4	64.71	76.47	73.33	68.42	70.59
F_1	52.94	76.47	69.23	61.90	64.71
F_2	58.82	58.82	58.82	58.82	58.82
F_3	58.82	58.82	58.82	58.82	58.82
FBW_1	64.71	52.94	57.89	60.00	58.82
FBW_2	58.82	41.18	50.00	50.00	50.00
FBW_3	58.82	64.71	62.50	61.11	61.96

Table 3.4 Classification Scores for the Combined Feature Sets

Parameter	SE (%)	SP (%)	PP (%)	NP (%)	Accuracy (%)
Jitter	82.35	88.23	87.50	83.33	85.29
% Power in PSD_{1-4}	88.23	82.35	83.33	87.50	85.29
F_{1-3}	64.71	58.82	61.11	62.50	61.96
FBW_{1-3}	58.82	88.23	83.33	68.18	73.53

3.5 Discussions

A study on the speech of patients suffering from clinical depression compared with that of local population is undertaken. The role of acoustic properties of speech, that is, vocal jitter, power distribution, formant frequencies, and bandwidths, are investigated by differentiating the depressed subjects from the healthy controls. Vocal jitter, which represents the short-time fluctuations in fundamental frequency, averaged higher values for speech from patients with depression in comparison with speech from controls. The fundamental frequencies in depressed speech are more erratic than those in normal speech.

Our findings also demonstrate a relatively higher percentage of power at higher frequencies in depressed speech. Various researchers (France et al. 2000; Tolkmitt et al. 1992) have reported similar results. Vocal jitter and PSD features emerge as superior indicators of depression compared to formant characteristics, as evidenced from the results shown in Tables 3.2 and 3.3. Although the features related to formant structure (formant frequencies and bandwidths) do not play a significant role in separating control patients from depressed patients, the trend of increased first formant frequency and decreased second and third formant frequencies observed here agree with the findings reported in previous studies (France et al. 2000; Hargreaves and Starkwearher 1965; Whitman and Flicker 1996). The general trend of decreased first and second formant bandwidths and the increased third formant bandwidth identified in our study appear to be features of depressed speech. The speech corpus needs to be expanded to create statistically significant results for possible clinical applications.

3.6 Conclusions

In this study, we have investigated the role of acoustic properties of speech, that is, vocal jitter, formant characteristics, and power distribution, in differentiating depressed subjects from healthy control subjects through student t-test and discriminant analysis. The results obtained demonstrate that vocal jitter and features describing the power distribution of the speech signal are good indicators of depression. The general trend of decreased first and second formant bandwidths and increased third formant bandwidth identified in this study correlate with the features of depressed speech reported earlier in the literature. Speech samples from male subjects comprising normal individuals and individuals diagnosed with major depressive disorder belonging to the local population (Kannada-speaking population) were considered in this study. The statistical analyses indicated that all the parameters studied are significantly different for depressive speech and control (normal) groups, and these parameters were found to be reliable indicators of clinical depression.

References

Depression Overview, http://www.emedicinehealth.com (accessed August 24, 2009).

France, D. J., R. G. Shiavi, S. Silverman, M. Silverman, and D. M. Wilkes. 2000. Acoustical properties of speech as indicators of depression and suicidal risk. *IEEE Trans Biomed Eng* 47:829–37.

Fukunaga, K. 1990. *Introduction to Statistical Pattern Recognition*. San Diego, CA: Academic.

Gullick, E. L., and L. J. King. 1979. Appropriateness of drugs prescribed by primary care physicians for depressed outpatients. *J Affect Disord* 1:55–8.

Hargreaves, W., and J. Starkwearher. 1965. Voice quality changes in depression. *Lang Speech* 7:218–20.

Heiberger, V., and Y. Horii. 1982. Jitter and shimmer in sustained phonation. *Speech Lang* 7:299–332.

Hogberg, J. 1997. Prediction of formant frequencies from linear combination of filter-bank and cepstral coefficients. *TMH-QPSR* 4:41–9.

Johnson, D. 1974. Study of the use of antidepressant medication in general practice. *Br J Psychol* 125:186–212.

Lee, M., J. V. Santen, B. Mobius, and J. Olive. 1999. *Formant Tracking Using Segmental Phonemic Information.* Murray Hill, NJ: Bell Labs, Lucent Technologies.

Ozdas, A., R. G. Shiavi, S. Silverman, M. Silverman, and D. M. Wilkes. 2004. Investigation of vocal jitter and glottal flow spectrum as possible cues for depression and near-term suicidal risk. *IEEE Trans Biomed Eng* 51:1530–40.

Pokorny, A. 1983. A prediction of suicide in psychiatric patients. *Arch Genet Psychol* 40:249–57.

Popkin, M. K., and A. L. Callies. 1987. Psychiatric consultation to inpatients with early onset type I diabetes mellitus in a university hospital. *Arch Genet Psychol* 44:169–71.

Quatieri, T. F. 2002. *Discrete-Time Speech Signal Processing.* Englewood Cliffs, NJ: Prentice Hall PTR.

Tolkmitt, E., H. Helfrich, R. Standke, and K. R. Scherer. 1992. Vocal indicators of psychiatric treatment effects in depressives and schizophrenics. *J Commun Disord* 15:209–22.

Welch, P. D. 1967. The use of fast Fourier transform for the estimation of power spectra: A method based on time averaging over short, modified periodograms. *IEEE Trans Audio Electro Acoust* 15:70–3.

Whitman, E., and D. Flicker. 1996. A potential new measurement of emotional state: A preliminary report. *Newark Beth Isr Hosp* 17:167–72.

Zolfaghari, P., and T. Robinson. 1996. *Formant Analysis Using Mixture of Gaussians.* Trumpington Street Cambridge CB2 1PZ: Cambridge University Engineering Department.

Chapter 4

Typicality Measure and the Creation of Predictive Models in Biomedicine

Mila Kwiatkowska, Krzysztof Kielan,
Najib T. Ayas, and C. Frank Ryan

Contents

4.1 Introduction and Motivation

Predictive models play an important role in all aspects of diagnosis, prognosis, and treatment. Although these models are widely utilized in medical decision making, they are often created using incomplete knowledge of the functioning of biological systems. Because biological systems are highly complex, and because our knowledge about them is often fragmented or even contradictory, the models by nature are context dependent and goal oriented. With the recent availability of electronic patient records and access to medical databases, the process of model creation is supported by machine-learning methods, which can automatically or semiautomatically induce predictive models, such as, for example, clinical prediction rules from data (Kononenko 2001). This data-driven approach takes advantage of the prospective availability of vast amounts of data stored in electronic patient records and numerous data sets gathered for biomedical research. Since the data have been already collected, processed, and stored for their primary use, the cost of reusing them is minimal. Thus, on the one hand, the secondary use of medical data reduces the cost of data acquisition and allows one to access data from diverse and large populations as well as the data of rare and exceptional medical cases. On the other hand, the secondary use of data is encumbered by several challenges (Druzdzel and Diez 2003; Owens and Sox 2001). These problems stem from the inherent complexity of medical data and from their specific roles in the medical decision-making process. In general, biomedical data are characterized by imprecision, heterogeneity, mixed granularity, uncertainty, incompleteness, time dependency, and problem orientation. Moreover, biomedical data may have high acquisition costs, may be stored using incompatible formats, and may have limited utilization because of data confidentiality. The sets of data collected in a course of medical practice and the sets of data used in clinical trials are characterized by several biases:

- Method bias: Clinical studies and patient records use diverse types and numbers of measurements and definitions of outcome. Therefore, predictors have varied specifications, granularities, and precisions.
- Sampling bias: Clinical studies use diverse acquisition methods, inclusion criteria, and sampling methods.

- Selection bias: Clinical data sets include patients with different demographics in terms of gender, age, and ethnicity.
- Referral bias: Most specialized clinical studies are based on patients referred to specialists by primary care practitioners; therefore, the data represent a preselected group with high prevalence of the disease.
- Clinical spectrum bias: Patient records include data on patients with varied severity of a disease and cooccurring medical problems.

Therefore, the secondary use of biomedical data requires knowledge-based data analysis as a part of the preprocessing step in the knowledge discovery (KD) process (Cios and Moore 2001). One of the integral parts of the KD process and one of the key aspects of model induction from data is the notion of *typicality*. Since secondary analysis of medical data uses data sets created mostly via nonrandom techniques, such as retrospective analysis of consecutive patient records, the use of medical data must address problems such as typicality or *atypicality* of individual records in the context of the represented population and representativeness of the entire sample. Typicality of a feature or a combination of features defines the extent of dissimilarity with "typical" or "normal" values and combination of values. The concept of typicality is crucial for the identification of errors and outliers. Representativeness describes the extent of similarities between the distribution of characteristics of the sample and that of the population. The concept of representativeness is crucial for the inference process from the sample to the population. In this chapter, we address these problems by introducing a *prescriptive* model with an explicit typicality measure. This prescriptive model is based on a semiotic approach for the contextual interpretation of predictors and a fuzzy logic for the representation of imprecision in measurements.

This chapter is organized as follows: Section 4.2 contains a discussion on the notion of typicality and three general approaches to categorization. Section 4.3 describes the applications of a typicality measure in KD. Section 4.4 contains a description of a representational framework based on fuzzy logic and semiotics. Section 4.5 contains an example of a prescriptive model for the anatomical typicality measure (ATM) used in knowledge-based preprocessing of clinical data, and Section 4.6 details our conclusions.

4.2 Typicality and Categorization

The notion of typicality has been studied in the context of several disciplines. The roots of its definition can be traced to cognitive psychology (Bruner et al. 1956; Rosch and Mervis 1975). In cognitive psychology, the term typicality is used to describe members within the same category. Thus, typicality is used as "a measure of how well a category member represents that category" (Reed 1996, p. 231). But before typicality can be defined more precisely, several fundamental questions about the nature of human classification and categorization must be answered. One of the key questions in cognitive psychology is as follows: How do people classify objects into categories? In turn, since the methods of classification are closely related to the mental representation of categories, the following question arises: How do people create categories? Answering these questions is not only essential for understanding human cognitive processes but also fundamental for the practical applications of machine learning, KD from data, and the creation of predictive models. The findings from cognitive psychology about human categorization and classification have significant bearing on the representational methods and algorithmic techniques used for classification, class discovery, and class prediction in machine learning. Naturally, machine learning and cognitive psychology strongly influence each other. Machine learning uses several approaches based on the cognitive theories of human learning and, vice versa, cognitive psychology uses many concepts originating from computer-based information processing. However, we should emphasize that the categorization and classification processes performed by humans and by machines are not parallel and, furthermore, they may employ different strategies. For example, humans can evaluate simultaneously only a limited number of hypotheses (Reed 1996), whereas machines can perform a large number of simultaneous evaluations. Humans, on the other hand, can operate at mixed abstraction levels.

4.2.1 Natural Categories and Classification

Many cognitive psychologists have noted that category learning and classification in the real world is different from the classification of artificial categories such as, for example, mathematical categories (Reed 1996). Thus, methods used for the classification of artificial categories may be different from the methods required for the classification of natural categories. In general, cognitive psychology uses three approaches to the mental representation of real-world, or natural, categories: (1) *classical* approach, (2) *prototype* approach, and (3) *exemplar* approach (Minda and Smith 2001; Nosofsky 1992).

In the classical approach, which hearkens back to the philosophies of Plato and Aristotle, objects are grouped based on their properties. Objects are either a member of a category or not, and all objects have equal membership in a category. Bruner et al. (1956) used this classical approach to describe the cognitive processes involved in the learning of new concepts. In the classical approach, a category can be represented by a set of rules, which can be evaluated as true (object belongs to the category) or false (object is not in the category).

In the prototype approach (Rosch and Mervis, 1975), the notion of a prototype is used to describe how categories are mentally represented and organized. The prototype approach maintains that a prototype exists as an ideal member of a category and the other members of the same category may share some of the features with the "ideal member." Therefore, the members are more or less typical of the category; in other words, they belong to the category to a certain degree. The prototype of the category usually represents the central tendency of the category and can be defined as the "average" of all the members of the category. The typicality of an object is measured by its similarity (e.g., geometrical distance measure) to the prototype.

In the exemplar approach (Medin and Schaffer 1978), all exemplars of a category are stored in a memory, and a new instance is classified based on its similarity to all the prior exemplars. This representation requires the specification of a similarity measure, as well as storage and retrieval of multiple exemplars. In the exemplar-based representation, a category is defined by all the exemplars belonging to that category, and typicality is measured by the similarity measure to all exemplars.

4.2.2 Typicality and Natural Categories

Natural categories present several challenges related to typicality:

- Typicality gradient: As described by Rosch and Mervis (1975), the members of a natural category may not be equally representative of a category; thus, the members may vary in their typicality—each member of a natural class has a typicality gradient.
- Typicality and family resemblance: As demonstrated by Wittgenstein (1953), the natural categories do not have to share all the attributes; a natural category may have some attributes that are common to many members and some attributes that only a few members may share. This observation is the basis for the distinction between *monothetic* classification and *polythetic* classification. In monothetic classification, objects possessing necessary and sufficient attributes define categories. In polythetic classification, each member may possess some, but not all, of the attributes. Wittgenstein (1953) also introduced the notion of *family resemblance* of an item as a measure of how many other members of the category share the item's attributes. Rosch and Mervis (1975) studied the commonality of the attributes among members of various categories. They calculated a measure of family resemblance for each member—a total number of members sharing the same attributes. They showed that typical examples of polythetic categories have high family resemblance scores.
- Typicality and hierarchical organization of categories: Natural categories have a hierarchical organization. The hierarchical organization can be described in terms of three levels: (1) *superordinate*, (2) *basic-level*, and (3) *subordinate* categories (Rosch et al. 1976). Superordinate categories share only a few attributes, whereas subordinate categories share many attributes. Rosch and Mervis (1975) showed that typical members of categories tend to share attributes with other members for all three levels of categories, and family resemblance scores can be used for predicting the typicality of members.
- Typicality and goal-derived categories: Barsalou (1982) argued that some natural categories are created not based on the similarity of their features or family resemblance. Instead, the objects are classified together to satisfy specific goals. These goal-oriented categories are formed based on different principles.

For example, a category of "foods for weight reduction" may include foods with fewer calories, foods with more fiber, and foods with a higher nutritional value. As shown by Barsalou (1982), the typicality of the foods will be based on the degree to which they satisfy their goal. Thus, the diet-oriented typicality of foods such as cucumbers or carrots is higher than that of foods such as mashed potatoes with gravy or eggs and bacon.

■ Thus, in real-world classification and categorization the notion of typicality has several measures from the central tendency through family resemblance to the degree of goal satisfaction. Furthermore, the typicality rating may be influenced by several factors, such as frequency of instances, context, or prior knowledge (Barsalou 1982; Reed 1996). To address some of these problems, Yeung and Leung (2006, p. 947) describe a formal ontology for the notion of typicality and context. They define typicality as "a measure of an individual's degree of membership in a concept depending on properties (both necessary and nonnecessary) that are shared by most of the instances of the concept." Furthermore, the authors state that the interpretation of concepts is context sensitive and, therefore, the typicality of an individual is defined with respect to a concept and its interpretation within a specific context.

4.3 Typicality Measure in Knowledge Discovery

The concept of typicality and its measure have been used in KD for data preprocessing (e.g., instance reduction, detection of outliers) and for the classification of objects, especially graphical and image objects. Recent studies describe a typicality measure for shapes, a typicality measure for colors, a geometric typicality measure for image segmentation and retrieval (Joshi et al. 2002; Pizer et al. 2003), and a semantic typicality measure for natural scene categorization (Vogel and Schiele 2004).

4.3.1 Typicality Measure in Data Preprocessing

A typicality measure is used in data preprocessing mostly to reduce noisy instances in instance-based learning (Zhang 1992) and to determine potential outliers using neural networks (Sane and Ghatol 2006).

Zhang (1992) used the concept of instance typicality for the creation of a typical instance-based learning (TIBL) algorithm. This algorithm saves the "central" instances and reduces the noisy instances. In TIBL, instance typicality is based on the entire training set and is calculated using the formula shown in Equation 4.1. In TIBL, the typicality of instance Y, $Typicality(Y)$, is defined as the ratio of *interconcept* distance to *intraconcept* distance, and is calculated as the ratio of an average distance $d(X_i, Y)$ to the members of the other classes (n) and its average distance to the members of its own class (p; Zhang 1992; Ting 1996):

$$\text{Typicality}(Y) = \frac{\left(\sum_{i=1}^{n} d(X_i, Y)\right)/n}{\left(\sum_{j=1}^{p} d(X_j, Y)\right)/p} \tag{4.1}$$

The aforementioned definition of typicality underlines both the similarity of instance Y with other members in the same category (or class) and the dissimilarity of instance Y with members of other categories (classes). We should emphasize here that this definition of the typicality measure differs from the notion of intraclass typicality as a representative of a single category.

4.3.2 Fuzzy Sets and Typicality Measure

The notion of typicality has been modeled using the fuzzy-logic approach, as well as the exemplar and prototype theories. For example, Pedrycz and Gomide (2007) described an application of a fuzzy set as a descriptor of the notion of typicality. Lesot et al. (2008) describe typicality using the prototype approach and, further, the intracategory and intercategory features. Thus, the typicality of an object x with respect to category C depends on two components: (1) similarity of the object with the other members of the same category and (2) its dissimilarity with the members of the other categories. The first component is called *internal* resemblance and the second *external* dissimilarity. Thus, the intrasimilarity, $R(x,C)$, and interdissimilarity, $D(x,C)$ can be computed using (4.2) as follows:

$$R(x,C) = \text{avg}(r(x,y), y \in C) \quad D(x,C) = \text{avg}(d(x,y), y \notin C) \tag{4.2}$$

Thus, the typicality of an object x with respect to category C can be defined by the aggregation function ϕ over the internal resemblance and external dissimilarity:

$$T(x,C) = \phi(R(x,C), D(x,C)) \tag{4.3}$$

The resemblance can be measured using, for example, the *Jaccard similarity coefficient* for two sets A and B (generalized to fuzzy sets), defined as follows:

$$r(A,B) = \frac{|A \cap B|}{|A \cup B|} \tag{4.4}$$

The external dissimilarity can be measured using, for example, *Minkowski's distance* (generalized to fuzzy sets), defined as follows:

$$d(A,B) = \left(\frac{1}{Z}\left(\int |f_A(X) - f_A(X)|^n \, dx\right)\right)^{\frac{1}{n}} \tag{4.5}$$

where Z is a normalizing factor, n is an integer, and $f_A(X)$ and $f_B(X)$ denote the membership functions for the fuzzy set X.

4.3.3 Typicality Measure in the Classification of Natural Scenes

Vogel and Schiele (2004) created a typicality-ranking scheme for categorizing natural scenes. They describe each scene in terms of nine semantic concepts: (1) sky, (2) water, (3) grass, (4) trunks, (5) foliage, (6) field, (7) rocks, (8) flowers, and (9) sand. Each scene is graded for the nine concepts using an occurrence vector. The occurrence of a specific concept is measured as a percentage from 0% to 100%. Thus, a scene containing some sky and mostly water can have an occurrence vector of 30% for the sky, 70% for the water, and 0% for the remaining seven concepts. Furthermore, the authors defined six scene categories using prototypical representation. A category prototype is the most typical example (vector of values) within the category. In the prototypical representation, this example may not exist as an instance. The prototypical representation of a category is based on the mean values (for each semantic concept) over the occurrences of all category members. The authors graded the members of each category by a *semantic* typicality measure, which is calculated as the Mahalanobis distance between the member occurrence vector and the prototypical representation of the category. The typicality measure is also used to categorize new instances. The classification is based on the similarity measure to the prototypical representation of each category.

4.4 Fuzzy-Semiotic Framework for Typicality

In this section, we present a new knowledge-based framework for the representation of typicality in a secondary analysis of biomedical data. Our framework is based on a semiotic approach for the contextual interpretation of predictors and a fuzzy logic for the representation of imprecision of measurements. We argue that the complex nature of the concept of typicality and the equivalently complex characteristic of the medical predictive models require a modeling language that explicitly represents context-sensitivity and fuzziness.

Typicality has several characteristics: *contextuality*, *gradability*, and *prototypicality*. Its most important characteristic is that typicality is context dependent (Sassoon 2007). Thus, typicality predicates such as "x has a typical weight" must be placed within a reference to a class or a specific subclass, that is, "x has a typical weight for an adult patient in clinical studies of obstructive sleep apnea." This predicate will be evaluated against the calculated central tendency of the patient population attending that respiratory clinic.

Typicality predicates can be divided into gradable and prototypical. Gradable predicates, for example, tallness or neck thickness, can be represented by membership functions in fuzzy logic (Pedrycz and Gomide 2007). Multidimensional gradable predicates such as "x is clinically depressed" are associated with a set of dimensions; each dimension can be gradable, but the dimension may be associated with different weights (weights may be specified differently in different contexts). However, some typicality predicates, such as "x is a good patient," are not easily gradable. They are based on a concept of a prototype of a "good patient."

Medical predictive models are based on classifications of diseases. In most cases, medical classifications are polythetic. Thus, specific diagnoses can be reached based on a subset of attributes but not all of them. This is the result of biological variability of the manifestation of a disease. For example, Krueger and Bezdjian (2009) discuss the classification of mental disorders in current versions of the *Diagnostic and Statistical Manual of Mental Diseases* (DSM-IV-TR) and the *International Classification of Disease* (ICD-10). The authors show that both general diagnostic criteria, DSM-IV-TR and ICD-10, represent mental disorders as polythetic and categorical concepts. Polythetic means that "specific mental disorders are defined by multiple symptoms, and not all listed symptoms are necessary to consider a mental disorder present in a specific individual" (Krueger and Bezdjian 2009, p. 3). Categorical means "all mental disorders in the DSM/ICD are binary 'either/or' concepts" (Krueger and Bezdjian 2009, p. 3). For example, depression is defined using more than 10 symptoms, which have varied definitions and which are used in different ways by the diagnostic criteria. The symptoms are generally grouped into three classes: (1) affective (crying, sadness, apathy), (2) cognitive (thoughts of hopelessness, helplessness, suicide, worthlessness, guilt), and (3) somatic (sleep disturbance, changes in energy level, changes in appetite, and elimination). Not all symptoms are present at the same time, and the severity of symptoms differs. Moreover, the symptoms may vary in their "directions." For example, two subtypes of depression are distinguished: (1) depression with vegetative symptoms (e.g., weight loss, insomnia, appetite loss) and (2) depression with reverse vegetative symptoms (e.g., weight gain, hypersomnia, appetite increase). The second subtype, according to epidemiological studies, is characteristic of one-fourth to one-third of all people with major depression, and it is more common among women. Furthermore, some mental disorders may be defined by several symptoms, which may not even overlap in diagnosed cases. For example, obsessive-compulsive personality may be diagnosed using DSM-IV based on 4 out of 8 symptoms, and conduct disorder may be diagnosed based on a threshold of 3 symptoms out of 15 (Krueger and Bezdjian 2009). In general, mental disorders have complex etiologies and present themselves with a variety of symptoms, which differ in frequency and severity in different patients. Thus, many categories specified by current versions of DSM and ICD are characterized by a high level of within-category heterogeneity. As a result, some patients who are diagnosed using the same category may have no common symptoms.

Therefore, medical predictive models and computerized systems supporting these models must be based on a conceptual framework, which provides for dimensionality, vagueness, and contextuality. Furthermore, these predictive models must represent medical concepts, which are inherently imprecise. The fuzzy set theory, proposed by Lotfi Zadeh in the paper "Fuzzy sets" (1965), provides mathematical representation for classes of objects that do not have precise (crisp) membership functions. Whereas in traditional logic and set theory an element belongs to a set or not, in a fuzzy set an element may belong to a set "partially" with some degree of membership. Set A in a universe of discourse X is defined by a characteristic function, $\mu_A(x) = 1$ when $x \in A$, and $\mu_A(x) = 0$ when $x \notin A$, whereas in fuzzy set theory an element x can be in one of three states: (1) the element x belongs completely to set A, (2) the element x belongs partially to set A, or (3) the element x does not belong to set A. Therefore, membership in a fuzzy set is described by continuous values from an interval [0,1], where the value 1 means that an element belongs completely to a set A, the values between 0 and 1 describe the element's "degree" of participation, and the value 0 means the element does not belong to the set A.

Since the 1970s, many authors have suggested and used fuzzy set theory to represent various imprecise concepts (Lakoff 1973; Oden 1977). In the 1980s, the use of fuzzy logic in the psychology of concepts was a subject of some dispute (Osherson and Smith 1981; Zadeh 1982; Osherson and Smith 1982). In the 1990s, Osherson and Smith (1997) discussed the issues of typicality and membership functions in their paper, "On typicality and vagueness." Recently, Belohlavek et al. (2009) clarified various misunderstandings and misconceptions about the utilization of fuzzy set theory in the psychology of concepts. The authors emphasized that fuzzy set theory is not a theory of concepts, but the role of fuzzy set theory is to represent and deal with concepts. Consequently, in our fuzzy-semiotic framework we utilize fuzzy set theory to represent fuzzy boundaries between typical and atypical values.

4.4.1 Fuzzy-Semiotic Framework

The fuzzy-semiotic framework provides two advantages: (1) calculation of a degree of typicality based on prior medical knowledge, and (2) contextual interpretation of typicality. The typicality measure quantifies the typicality (and atypicality) of a specific feature using a fuzzy rule–based system. The typicality measure has values from 0 to 1.0. Low values, for example, 0.1, represent atypical values, whereas high values, for example, 0.9, represent typical values.

4.4.2 Fuzzy Inference System for the Typicality Measure

The fuzzy inference system (FIS) for the typicality measure requires n input values (crisp values for predictors), $P1$, $P2$, ..., Pn, to calculate a particular output value of the typicality measure, TM. The FIS uses the Mamdani fuzzy inference process (Cox 2005) composed of four steps: (1) fuzzification of input variables, (2) rule evaluation, (3) aggregation of results, and (4) defuzzification. We have applied the Mamdani fuzzy inference process. The input values are fuzzified according to predefined membership functions for $P1$ and $P2$. Next, the fuzzy rules are executed in parallel, and the results are aggregated into a single fuzzy set. After that, the set is defuzzified using the centroid technique to produce a crisp output, a typicality measure, TM.

4.4.3 Semiotic Approach to Typicality Measure

We use a semiotic model to define the interpretative and goal-oriented nature of typicality. The term "semiotics" was introduced in the second century by the famous physician and philosopher Galen (129–199), who classified semiotics (the contemporary symptomatology) as a branch of medicine. The use of the term semiotics to describe the study of signs was developed by the Swiss linguist Ferdinand de Saussure (1857–1913) and the American logician and philosopher Charles Sanders Peirce (1839–1914). Semiotics is a discipline that can be broadly defined as the study of signs. Since signs and their interpretations are present in every part of human life, the semiotic approach has been used in almost all disciplines from mathematics through literary studies to library and information sciences. A semiotic paradigm is, on the one hand, characterized by its universality and transdisciplinary nature, but, on the other hand, it is associated with different traditions and with a variety of empirical methodologies.

We are using a framework based on the Peircean model of sign and semiosis as a process. Peirce defined a sign as any entity that carries some information and is used in a communication process. Peirce, and later Charles Morris, divided semiotics into three categories (Chandler 2002): (1) syntax (the study of relations between signs), (2) semantics (the study of relations between signs and the referred objects), and (3) pragmatics (the study of relations between signs and the agents who use the signs to refer to objects in the world). This triadic distinction is represented by a Peirce's semiotic triangle having the *representamen* (the form that a sign takes), *object*, and *interpretant*. The notion of interpretant is represented in this chapter by a set of pragmatic modifiers: agents (e.g., patients, health professionals, medical sensors, computer systems), perspectives (e.g., health-care costs, accessibility, ethics), biases (e.g., specific subgroups of agents), and views (e.g., variations in the diagnostic criteria used by individual experts or clinics).

4.5 Anatomical Typicality in Knowledge Discovery

We applied the fuzzy-semiotic framework, described in Section 4.4, to data preparation in the creation of a predictive model for the diagnosis of obstructive sleep apnea (OSA).

4.5.1 Obstructive Sleep Apnea

OSA is a common, serious respiratory disorder afflicting, according to conservative studies, 2%–4% of the adult population. The word apnea means "without breath" and occurs only during sleep; it is, therefore, a condition that might go unnoticed for years. This condition is caused by the collapse of soft tissues in the throat as the result of natural relaxation of muscles during sleep. The soft tissue blocks the passage of air, and the sleeping person literally stops breathing (apnea event) or experiences a partial obstruction in breathing (hypopnea event). It is associated with significant risk for hypertension, congestive heart failure, coronary artery disease, myocardial infarction, stroke, and arrhythmia. Furthermore, patients with OSA have higher risk during and after anesthesia, since their upper airway may be obstructed because of sedation. Since a typical symptom of OSA is daytime sleepiness, untreated OSA patients have higher rates of vehicle accidents.

The gold standard for the diagnosis of OSA is an overnight in-laboratory polysomnography (PSG). The most important score for OSA diagnosis is the apnea–hypopnea index (AHI), which is calculated as the number of apnea and hypopnea events per hour of sleep (Douglas 2002). In a diagnosis based solely on the AHI index, apnea is classified as mild for AHI between 5 and 14.9, moderate for AHI between 15 and 29.9, and severe for AHI ≥ 30.

Medical literature describes several risk factors (predictors) for OSA. For example, Douglas (2002) lists seven groups of risks:

1. Daytime symptoms (e.g., excessive daytime sleepiness, morning fatigue, headaches)
2. Nocturnal symptoms (e.g., snoring, witnessed apneas)
3. Anatomical signs (e.g., general obesity, upper-body obesity, large neck circumference [NC])
4. Lifestyle factors (e.g., sleep deprivation, shift work, smoking)
5. Demographic factors (age, gender, and ethnicity)
6. Coexisting medical conditions (e.g., hypertension, depressive disorders, diabetes mellitus)
7. Physiological factors (e.g., blood oxygen saturation, blood pressure, and heart rate)

In this study, we analyzed one of the anatomical predictors: large NC.

4.5.2 Anatomical Typicality Study

The main objective of this study was the integration of medical knowledge into the analysis of medical outliers. This experimental study involved an important OSA predictor, neck thickness, which is measured clinically as the NC in centimeters. Several studies indicate a short and fat neck as a characteristic sign of OSA (Dancey et al. 2003; Lam et al. 2005).

The study used two data sets A ($N = 239$) and B ($N = 185$) for the analysis of atypical values, statistical outliers, and monotonic dependencies. Data set A was collected and used in a clinical study of correlation between OSA and craniofacial features (Lam et al. 2005). Data set A has 164 Asian patients and 75 white patients; there are 199 males (83.3%) and 40 females (16.7%) in the set. The records have the following characteristics: a mean age of 48.47 (±12.1), (middle-aged patients) and a mean body mass index (BMI) of 29.23 (±5.7), (moderately obese). Data set

B (169 records) contains data from the archived records of an educational sleep disorders clinic. Set B has 53 (28.6%) females and 132 (71.4%) males. The records have the following characteristics: mean age 51.72 (±12.38; middle-aged patients) and a mean BMI of 33.88 (±7.04; moderately obese).

We used three variables from both data sets: (1) NC, (2) gender, and (3) BMI calculated as weight (in kilograms) divided by the squared height (in meters). Additionally, we used the variable ethnicity from data set A; however, data set B does not include ethnicity.

For the analysis of NC, we used 239 records from data set A and 147 valid records from data set B (38 records have missing NC values). For the analysis of predictor strength, we used an outcome variable (AHI in data set A) indicating the severity of OSA.

The study is organized into the following five phases:

1. Preparatory phase: In the preparatory phase, we first constructed a knowledge base (KB) comprising facts and rules obtained from the medical literature. Next, based on this KB, we generated a prescriptive model for ATM. Finally, we calculated anatomical typicality for all records from two data sets, and we selected anatomically atypical records based on a threshold value ATM ≤ 0.50.

2. Analysis of univariate outliers: In the second phase, we performed a simple statistical analysis of univariate outliers in the two data sets. First, we applied this method to the entire data sets (data-driven approach), and then we applied the facts stored in the KB to group the data by specific subpopulations (knowledge-driven approach). We examined if additional outliers could be identified based on the facts stored in the KB. Furthermore, we applied the prescriptive model to the identified outliers to measure their anatomical typicality with reference to general population and specific subpopulations (knowledge-driven approach) to see if the statistical outliers also had atypical values.

3. Analysis of bivariate outliers: In the third phase, we used data set A and performed the analysis of bivariate outliers using the residual analysis method and the Mahalanobis distance measure (data-driven approach); then, we applied the facts stored in the KB to group the data by specific subpopulations (knowledge-driven approach) to determine if additional outliers could be identified based on KB facts.

4. Analysis of monotonic patterns: In the fourth phase, we used a knowledge-driven approach to test the monotonic patterns defined in the KB on both data sets. First, we performed correlation analysis for the entire data sets. Second, we analyzed correlations for specific subgroups. Third, we analyzed correlations based on complete data sets and on data sets after the removal of outliers.

5. Analysis of the strength of the predictor: In this phase, we applied the knowledge facts and studied the correlation between the predictor and the outcome variable, AHI. We performed correlation analysis for the entire data set, subgroups, and data sets with single outliers removed.

4.5.2.1 Phase 1: Building the Knowledge Base for Neck Thickness

The KB for neck thickness has two components: (1) fact repository (FR) and (2) typicality measure system (TMS). The FR contains four types of facts:

1. Purpose (e.g., diagnostic, prognostic, treatment evaluation)
2. Context (e.g., specific subgroups)
3. Bias (e.g., typical dependencies between predictors)
4. View (e.g., diagnostic criteria used by the clinics)

We constructed the FR based on 10 published medical studies: #1: Dancey et al. 2003; #2: Ferguson et al. 1995; #3: Lam et al. 2005; #4: Millman et al. 1995; #5: Ryan and Love 1996; #6: Rowley et al. 2000; #7: Rowley et al. 2002; #8: Jordan and McEvoy 2003; #9: Whittle et al. 1999; and #10: Young et al. 2002. The FR for neck thickness contains the following knowledge facts:

KB1 (purpose): A large neck is a characteristic sign of OSA (#3–#10).

KB2 (context): In general, NC is significantly larger in men than in women (clinic-based population [#3, #1, #6], epidemiological studies [#10], normal volunteers [#9]).

KB3 (context): In general, Asians tend to have smaller necks than whites (#3).

KB4 (bias): Heavier people are expected to have larger necks. However, there is gender-specific regional distribution of fat. The BMI appraises generalized obesity, whereas a larger neck reflects upper-body obesity, which is typical for men (#1). The NC was shown to be smaller in women despite a larger BMI (#7).

KB4a (bias): A study (#2) of 169 male patients reports that NC is correlated with BMI ($r^2 = 0.66$, $p < .005$).

KB5 (bias): Taller people are expected to have larger necks (#1).

KB6 (view): Clinically, NC is measured at the level of the cricothyroid membrane using a measuring tape (#3).

KB7 (view): The NC ranges from 25 to 65 cm for adults. Typical male NCs range from 42 to 45 cm (#2). Mean NC for women is 36.5 ± 4.2 (based on $N = 1189$) and 41.9 ± 3.8 for men (based on $N = 2753$; #1). Male NCs can be divided into three groups (ranges were selected for the diagnosis of OSA): (1) small to normal (<42), (2) intermediate (42–45), and (3) large (>45; #2).

4.5.2.2 Typicality Measure System

We constructed a TMS using medical knowledge represented in the FR for neck thickness. The TMS calculates ATM based on a relationship between NC and BMI, defined by a knowledge fact KB4 for monotonic dependencies. Furthermore, TMS uses gender and ethnicity information based on subpopulation differences (knowledge facts KB3 and KB4). The TMS system comprises a FIS, a set of membership functions, and a set of fuzzy rules. We implemented TMS using functions from a Fuzzy Logic Toolbox for MATLAB® 7.2.

The FIS system has four inputs (NC, BMI, gender, and ethnicity) and three outputs (ATM, ATM for males [ATM_m], and ATM for females [ATM_f]). Using the Mamdani inference process, the input values are fuzzified according to predefined membership functions. Subsequently, the fuzzy rules are executed and the results are aggregated into a fuzzy set. The resulting fuzzy set is defuzzified (using the centroid technique) to produce ATM, ATM_M, and ATM_F values.

We defined predictor neck thickness as the linguistic variable PNC = <NC, {small, typical, large, atypical}, [25, 65], and M>, where NC is the name of the variable, the set {small, typical, large, atypical} is a set of possible terms for the variable PNC, the interval [25, 65] is the domain of possible values (in centimeters) for NC, and M is a set of membership functions defining small, typical, large, and atypical NCs, $M = \{\mu_{small}, \mu_{typical}, \mu_{large}, \mu_{atypical}\}$. We defined the predictor BMI as the linguistic variable PBMI = <BMI {low, normal, obese}, [15, 70], M>, where BMI is the name of the variable, the set {low, normal, obese} is a set of possible terms for the variable BMI, the interval [15, 70] is the domain of possible values (in kg/m^2) for BMI, and M is a set of membership functions defining the low, normal, and obese terms for BMI, $M = \{\mu_{low}, \mu_{normal}, \mu_{obese}\}$.

We created gender-specific membership functions for NC. Based on the knowledge fact KB2, which states that males have significantly larger necks than females, we created two additional sets of membership functions for males and females. Using these gender-specific membership functions, FIS calculates ATM_M and ATM_F.

4.5.2.3 Typicality Measure Results

The NC values for data sets A and B were analyzed for their anatomical typicality, in terms of their relationship to BMI. The FIS for the neck thickness calculated the ATMs: ATM, ATM_M, and ATM_F. We identified six anatomically atypical NC values using a threshold of ATM ≤ 0.50. Table 4.1 shows the atypical records. In the table, IDs are unique numbers within each data set.

4.5.2.4 Phase 2: Analysis of Univariate Outliers

We used statistical analysis of outliers based on a median value and an interquartile range (IQR). We used a box plot, which is a practical and simple method for identifying univariate outliers in numerical continuous variables. Minor outliers, located 1.5 IQRs from the lower and the upper edges of the box, are shown as small circles above or below the boxes and are identified by numbers (case IDs), which are sequential numbers within each data set. Figures 4.1 and 4.2 represent the two steps in the analysis of univariate outliers in data set A. Figure 4.1 shows the median and quartile values for NC for all patients, whereas Figure 4.2 shows the box plots for NC by gender and ethnicity.

Table 4.1   Atypical NC Records (ATM ≤ 0.50) in Data Sets A and B

Data Set	ID	Gender	BMI (kg/m²)	NC (cm)	ATM	ATM$_M$	ATM$_F$
A	16	M	22.6	30.8	0.47	0.47	0.47
	239	M	34.5	50.0	0.50	0.12	0.50
	217	F	33.2	43.0	0.50	0.12	0.50
	32	F	40.8	39.0	0.50	0.50	0.50
B	101	F	23.8	53.0	0.12	0.12	0.12
	106	F	50.1	37.0	0.44	0.43	0.47

Source: Kwiatkowska, M., M. S. Atkins, N. T. Ayas, and C. F. Ryan, Knowledge-based data analysis: First step toward creation of clinical prediction rules using a new typicality measure, *IEEE Transactions on Information Technology in Biomedician*, Vol. 11, No. 6. 2007. © 2007 IEEE. With permission.

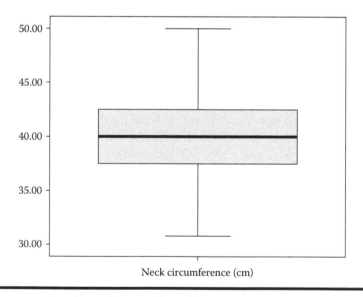

Neck circumference (cm)

Figure 4.1   Neck circumference for all patients from data set A. (From Kwiatkowska, M., M. S. Atkins, N. T. Ayas, and C. F. Ryan, Knowledge-based data analysis: First step toward creation of clinical prediction rules using a new typicality measure, *IEEE Transactions on Information Technology in Biomedician*, Vol. 11, No. 6. 2007. © 2007 IEEE. With permission.)

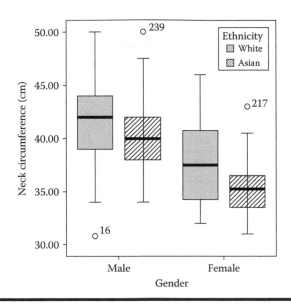

Figure 4.2 **Neck circumference by gender and ethnicity for patients from data set A. (From Kwiatkowska, M., M. S. Atkins, N. T. Ayas, and C. F. Ryan, Knowledge-based data analysis: First step toward creation of clinical prediction rules using a new typicality measure,** *IEEE Transactions on Information Technology in Bio-medician,* **Vol. 11, No. 6. 2007. © 2007 IEEE. With permission.)**

4.5.2.5 Typicality Analysis for Statistical Outliers

We calculated the ATM for the statistical outliers identified in data sets A (six outliers) and B (four outliers). Table 4.2 shows the ATM for the outliers identified in data set A: IDs 16, 75, and 239 for males; and IDs 70, 73, and 217 for females. The lowest typicality (ATM = 0.47) is assigned to a case of a male patient with a very small neck, but a low BMI of 22.6. On the other hand, the IDs 75, 70, and 73 have high BMI and large neck values with correspondingly high typicality. Case ID 217 of an Asian female has a medium typicality of 0.5, a result of a relatively large neck and above normal BMI. The typicality measure (ATM ≤ 0.50) indicates three outliers identified by grouping both gender and ethnicity factors.

Furthermore, Table 4.2 shows the ATM for all the outliers identified in the statistical analysis for data set B. The records for IDs 21, 129, and 137 have relatively high typicality. The NC above 50 cm is high; however, it is possible for tall (height above 180 cm), heavy (BMI = 39), and severely obese (BMI > 50) males. However, the record for ID 101 of a slim (BMI < 25) female has a very low ATM (0.12), indicating that NC = 53 cm is an atypical value for a slim female. We observed that ATM values are not the same for all outliers identified by the statistical approach. The lower typicality values indicate the outliers identified by additional grouping of data. In data set A, the outliers with lower typicality, IDs 16, 217, and 239, were

Table 4.2 Statistical Outliers in Data Sets A and B and Their ATM

Data Set	ID	Gender	Weight (kg)	Height (cm)	BMI (kg/m²)	NC (cm)	Anatomical Typicality
A	16	M	70.0	176.0	22.6	30.8	0.47
	75	M	126.0	169.0	44.1	50.0	0.89
	239	M	95.0	166.0	34.5	50.0	0.50
	70	F	109.0	148.0	49.8	45.0	0.83
	73	F	140.0	155.0	58.3	46.0	0.89
	217	F	86.0	161.0	33.2	43.0	0.50
B	21	M	181.0	184.0	52.9	55.0	0.89
	129	M	151.0	186.5	39.0	55.0	0.83
	137	M	181.8	188.0	51.4	58.4	0.88
	101	F	70.5	172.0	23.8	53.0	0.12

identified using subgroups of gender and ethnicity. In data set B, the outlier ID 101 was identified in subgroups based on gender.

4.5.2.6 Phase 3: Analysis of Bivariate Outliers

We used two statistical methods for bivariate outlier analysis: (1) residual analysis and (2) Mahalanobis distance. We performed the analysis for the larger data set, that is, data set A.

4.5.2.6.1 Residual Analysis

We constructed a simple linear regression model for NC, $\hat{y} = 28.5 + 0.392(\text{BMI})$, based on BMI as a dependent (predictor) variable. Next, we performed residual analysis using the least-squares method. Residual is defined as the difference between the observed value y and the fitted value $\hat{y}$: $e_i = y_i - \hat{y}_i$. The error mean square or the *residual mean square* (MS$_E$) was calculated as $\text{MS}_E = \dfrac{\sum (e_i - \bar{e})^2}{n - 2}$ and standardized residual d_i as $d_i = \dfrac{e_i}{\sqrt{\text{MS}_E}}$.

Potential outlier values were identified by the residuals that are larger in their absolute values than the majority of the other residuals. Figure 4.3 shows a box plot for the standardized residuals of NC against BMI with three marked outlier IDs: 239, 99, and 38.

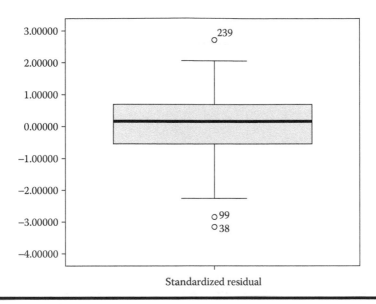

Figure 4.3 Box plot for standardized residuals of neck circumference against body mass index (data set A).

Figure 4.4 shows the distribution of residuals. The columns to the left of 0.0, especially for the residual values less than −5.0, reflect the overestimation of NC, and the columns to the right indicate the underestimation of NC based on BMI. This distribution is not normal; so we performed further analysis using facts from the KB.

Based on KB2, we grouped the residuals by gender. Figure 4.5 shows box plots of residuals of NC against BMI grouped by gender. The median for male records is close to 0.0; however, the median for female records is −5.0, indicating the model is overestimating NC based on BMI for females.

From the box plots in Figure 4.5, the grouping by gender identified six outliers: IDs 16, 126, 141, and 239 (males); and IDs 38 and 217 (females). The upper outlier ID 239 and the lower outlier ID 38 were already identified by the box plot without grouping. However, the analysis of outliers within the gender subpopulations discovered four additional outliers and changed the status of ID 99 (female) from outlier within the entire population to a value falling within the lower limits for females.

4.5.2.6.2 Mahalanobis Distance Measure

Since the variables NC and BMI are correlated and the residual analysis shows nonnormal distribution (for data set A, $r = 0.60$ and $p < .001$), we used Mahalanobis distance for multivariate outlier analysis. The Mahalanobis distance standardizes data using the Σ covariance matrix between variables. We used Mahalanobis distance to measure the dissimilarity between two vectors x_i and x_j as

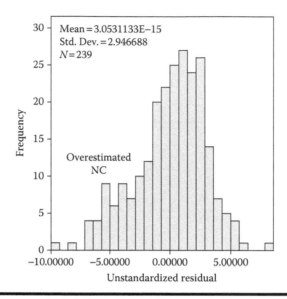

Figure 4.4 **Histogram of unstandardized residuals for neck circumference against body mass index (data set A).**

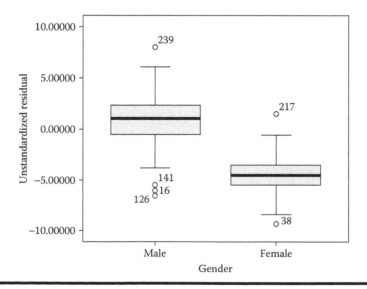

Figure 4.5 **Box plot for unstandardized residuals of neck circumference against body mass index by gender (data set A).**

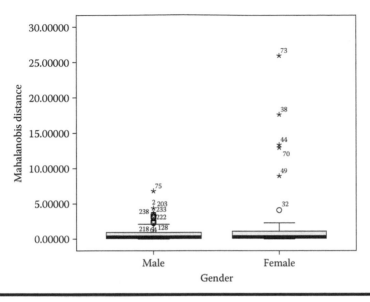

Figure 4.6 Box plot for Mahalanobis distance for neck circumference and body mass index grouped by gender.

$d_{\mathrm{MH}}(x_i - x_j) = \sqrt{(x_i - x_j)^T \Sigma^{-1}(x_i - x_j)}$, where T is the transposed vector and Σ is the covariance matrix.

Figure 4.6 shows the box plot for Mahalanobis distance for NC and BMI grouped by gender. There is a visible difference between genders. The female extreme outliers (IDs 73, 38, 44, 70, and 49) are further away than the male outliers (IDs 75 and 239). This tendency indicates that the regression model obtained from all the records fits the male records better than the female records.

Table 4.3 shows the outliers for female patients identified (indicated by "Yes") by univariate analysis, residual analysis, and Mahalanobis distance. The table also lists the general anatomical typicality, ATM, and the female anatomical typicality, $\mathrm{ATM_F}$. Two of the extreme outliers (IDs 70 and 73) were identified by univariate analysis. They represent two instances of records with severe obesity and large NC. Both have relatively high typicality (ATMs are 0.83 and 0.89, respectively), since larger necks are typical of severe obesity. In contrast, ID 217 was identified by univariate analysis and residual analysis but not discovered using Mahalanobis distance. The outlier ID 38 was identified by residual analysis; however, the outliers ID 44 and ID 49 were identified by neither univariate analysis nor residual analysis.

4.5.2.7 Phase 4: Analysis of Monotonic Patterns

We analyzed the monotonic pattern based on the KB4 fact: "Heavier people are expected to have larger necks." We used BMI as an indicator of "heaviness" and NC

Table 4.3   Outliers for Female Patients Identified by the Three Methods (from Data Set A)

ID	Univariate Analysis	Residual Analysis	Mahalanobis Distance	BMI (kg/m²)	NC (cm)	ATM	ATM$_F$
32	No	No	Yes (4.1)	40.77	39	0.50	0.50
38	No	yes	Yes (17.6)	53.17	40	0.50	0.59
44	No	No	Yes (13.4)	50.07	42	0.82	0.82
49	No	No	Yes (8.9)	46.29	42	0.52	0.52
70	Yes	No	Yes (12.9)	49.76	45	0.83	0.87
73	Yes	No	Yes (25.9)	58.27	46	0.89	0.88
99	No	Yes	No (0.1)	27.85	31	0.50	0.77
217	Yes	Yes	No (0.5)	33.18	43	0.50	0.50

as an indicator of neck thickness. Our analysis had two goals: (1) to examine the difference in correlations between NC and BMI in the entire sampled population and in the specific subgroups created using the facts from the FR, and (2) to examine the difference in correlations between NC and BMI for all the data and data after the removal of outliers detected by knowledge-driven methods. The relationship between BMI and NC was studied using the Pearson's correlation coefficient r and the coefficient of determination r^2. We used r^2 as a measure of the meaningfulness of r to describe the proportion of the variation in the observed values of NC, which can be explained by BMI. Furthermore, based on the facts KB2 and KB3, we grouped the data by gender and ethnicity and analyzed the same correlations in subgroups.

Figures 4.7 through 4.9 illustrate the relationship of BMI with NC for three groups, respectively: (1) all patients, (2) males only, and (3) females only.

The values of Pearson's correlation r and coefficient of determination r^2 are noticeably different for each of the studied groups. The BMI explains only 37% ($r^2 = 0.37$) of the variability of NC for all patients; however, the ability of BMI to explain NC variability increases when the data are grouped by gender. For the male group, the BMI values explain 57% of NC variability. For the female group, the BMI values explain 71% of NC variability. Furthermore, we analyzed r^2 for two ethnic groups: (1) Asian ($r^2 = 0.39$) and (2) white ($r^2 = 0.36$). In this case, the explanatory value of BMI was found to be similar for all patients and both ethnic groups.

We performed a t-test for gauging the significance of the correlations between BMI and NC for all patients, females, and males. We used a two-tailed test with $df = n - 2$ and a significance level of 0.001, and we calculated t using the following formula: $t = r\sqrt{(n-2)/(1-r^2)}$.

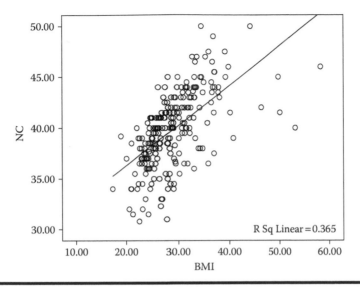

Figure 4.7 **Scatter plot for neck circumference and body mass index (all patients, data set A),** $r = 0.604$.

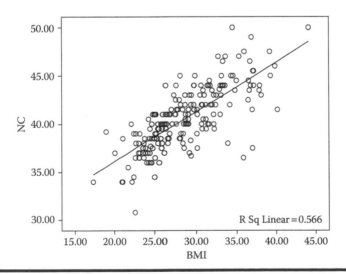

Figure 4.8 **Scatter plot for neck circumference and body mass index (male patients, data set A),** $r = 0.752$.

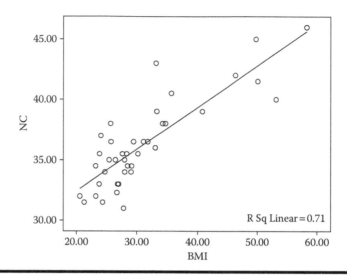

Figure 4.9 **Scatter plot for neck circumference and body mass index (female patients, data set A),** $r = 0.843$.

Table 4.4 **Correlations between BMI and NC (Data Sets A and B)**

Data Set	Group	R	N	T	t*
	All patients	0.60	239	11.7	3.29
A	Women	0.84	40	9.7	3.65
	Men	0.75	199	16.0	3.29
	All patients	0.54	147	7.7	3.3
	One outlier removed (ID 101)	0.57	146	8.2	3.3
B	Women	0.53	43	4.0	3.6
	One outlier removed (ID 101)	0.71	42	6.3	3.6
	Men	0.76	104	11.63	3.4

Table 4.4 shows that the correlations between BMI and NC for all groups are statistically significant at the level 0.001 (t^* represents critical values in the table).

We compared the values of the coefficient of determination, r^2, for data sets A and B for all patients, females and males, as summarized in Table 4.5. Additionally, we included the r^2 values for females in set B after the removal of one outlier case, ID 101. There are several interesting observations: The r^2 value for males is the

Table 4.5 Coefficient of Determination for BMI and NC in Data Sets A and B

	Data Set A (N = 239)	*Data Set B (N(Valid) = 147)*
All patients	$r^2 = 0.37$	
Asian (164)	$r^2 = 0.39$	$r^2 = 0.29 \ (0.32)$[a]
White (75)	$r^2 = 0.36$	
Women	$r^2 = 0.71$	
Asian (24)	$r^2 = 0.48$	$r^2 = 0.28 \ (0.50)$[a]
White (16)	$r^2 = 0.85$	
Men	$r^2 = 0.57$	
Asian (140)	$r^2 = 0.48$	$r^2 = 0.57$
White (59)	$r^2 = 0.73$	

[a] The values in parentheses indicate r^2 after the removal of the outlier, ID 101.

same in both the sets, whereas the r^2 values for females are very different. This difference can be partially explained by the fact that r^2 values are sensitive to outliers, especially in small data sets such as set B that has just 43 females. Therefore, these results must be interpreted with caution, since the magnitude of r^2 depends on the spread of values; hence r^2 may be small because the range of values is small.

The removal of outlier ID 101 changed considerably the strength of linear correlation between BMI and NC for the female group in data set B: The value of r^2 increased from $r^2 = 0.28$ to $r^2 = 0.50$ (as shown in Table 4.5). On the other hand, the removal of ID 101 changed only minimally the strength of linear correlation between BMI and NC for all patients from data set B: The value of r^2 increased from $r^2 = 0.29$ to $r^2 = 0.32$.

4.5.2.8 Phase 5: Analysis of the Strength of the Predictor

In this phase, we study the strength of a feature (NC) as a predictor of AHI. We used only data set A because this data set includes the clinically measured NC and the AHI value obtained from a standard diagnosis by PSG. We studied the correlation between neck thickness and AHI based on the knowledge fact, KB1: "A large neck is a characteristic sign of OSA." Our analysis had two goals: (1) to examine the difference in correlations for the sampled population and the specific subgroups created based on facts from the FR and (2) to examine the difference in correlations for the complete data and data after the removal of outliers detected by knowledge-based methods. We analyzed two outliers, which were difficult to identify using statistical methods alone, to demonstrate the importance of our knowledge-based methods for outlier detection.

We calculated Pearson's r and built simple linear regression models for AHI using NC as a predictor. Furthermore, we recalculated r for the subset of female and male patients after the removal of a single outlier. We selected outlier ID 217 (gender = female, ethnicity = Asian, weight = 86 kg, height = 161 cm, BMI = 33.18, NC = 43 cm) for two reasons: (1) low anatomical typicality value and (2) difficulty in detection. Note that the female outlier ID 217 was not detected by a simple univariate outlier analysis; however, it was identified after dividing the data sets into gender groups and further splitting them into ethnicity subgroups, as shown in Figure 4.2. Using the TMS system, we calculated the three ATMs for ID 217: ATM = 0.50, $ATM_F = 0.50$, and $ATM_M = 0.12$. Furthermore, ID 217 was identified by residual analysis, but, again, only after applying our knowledge-based method of subgroup analysis. However, ID 217 was not identified by the Mahalanobis distance measure (see Figure 4.6).

Similarly, the male outlier ID 16 (gender = male, ethnicity = white, weight = 70 kg, height = 176 cm, BMI = 22.6, NC = 30.8 cm) was not detected by a simple univariate outlier analysis; however, it was identified by our knowledge-based univariate analysis, as shown in Figure 4.1. It was detected after dividing the data sets into gender groups and after creating four subgroups based on gender and ethnicity as shown in Figure 4.2. Using the TMS system, we calculated the three ATMs for ID 16: ATM = 0.47, $ATM_F = 0.47$, and $ATM_M = 0.47$. Furthermore, ID 16 was identified by residual analysis as shown in Figure 4.5, but, again, only after applying our knowledge-based method of subgroup analysis. However, ID 16 was not identified by the Mahalanobis distance measure (Figure 4.6).

Table 4.6 shows values for Pearson's correlation r and coefficient of determination r^2 for data set A, the female group, the male group, and the male group divided by ethnicity.

The Pearson's correlation r is found different for all patients and subgroups in data set A. For example, for women $r = 0.37$, for men $r = 0.19$ and for white men $r = 0.41$. The correlation r is found increased in both cases of a single outlier removal. However, we were not able to show that the differences between subgroups in data set A are statistically significant, as tested by a two-tailed z-test for significance at 95% confidence. There are several possible reasons for not achieving statistical significance. The first reason could be a low ratio of females ($n = 40$) to males ($n = 193$) and furthermore unbalanced ratio of ethnic groups whites ($n = 75$) and Asian ($n = 164$). A second reason could be the smaller spread of values for NC among the Asian group. However, r^2 values are very small. Therefore, our discussion on the coefficient of determination r^2 is informal. In general, the NC values explain only a small percentage of the variability of AHI; however, Table 4.6 shows some interesting differences between the subgroups. The NC explains 8% ($r^2 = 0.08$) variability of AHI for all patients and only 4% ($r^2 = 0.04$) for male patients.

Table 4.6 Pearson's Correlations and Linear Regression between NC and AHI (Data Set A)

Data Set A (N(Valid) = 233)	Pearson's r	r^2
All patients (N = 233)	0.29^a	0.08
Women (n = 40)	0.37^b	0.14
ID 217 removed (n = 39)	0.40^b	0.16
Men (n = 193)	0.19^a	0.04
ID 16 removed (n = 192)	0.20^a	0.04
Asian (n = 160)	0.28^a	0.08
White (n = 72)	0.41^a	0.16

[a] Correlation is significant at the 0.01 level (two tailed).
[b] Correlation is significant at the 0.02 level (two tailed).

However, NC explains 14% (r^2 = 0.14) variability of AHI for female patients and 16% (r^2 = 0.16) for white male patients.

The removal of a single outlier increases r^2 minimally for the female group from r^2 = 0.14 to r^2 = 0.16 (after the removal of ID 217) and for the male group from r^2 = 0.19 to r^2 = 0.20 (after the removal of ID 16). Again, these results should be interpreted cautiously, since the sample size for females is small.

In this phase, we analyzed the correlation between NC and AHI, and we observed that the correlation is noticeably different between subgroups and minimally increased by the removal of the outliers. These results should be discussed in light of two study limitations: (1) imbalanced data with respect to gender and ethnicity, as well as distribution of the values for NC, and (2) the natural variability of patients and differences in symptoms and signs.

4.5.3 Study Discussion

In this study, we showed that the analysis of medical outliers requires a combination of a knowledge-driven approach and a statistical approach. The knowledge-driven approach was applied in three ways: (1) to find medically atypical records, (2) to perform outlier analysis in subgroups based on a stored KB, and (3) to evaluate medical typicality of the statistical outliers that were detected. The fuzzy-semiotic framework presented in this chapter detected several outliers that were not identified by the basic univariate analysis or, in some cases, by the basic bivariate analysis using residual analysis and the Mahalanobis distance method. Interestingly, we found six atypical records (ATM ≤ 0.50) using the FIS system and identified the same six records through univariate and bivariate outliers analysis with subgrouping.

We used the medical typicality measure to select atypical statistical outliers for testing and potential removal. We demonstrated that our medical typicality measure, calculated by a fuzzy logic–based TMS system, provides a consistent quantification of typicality based on prior medical knowledge. This typicality value is not sensitive to the particular distribution of the data in a single data set. For example, the univariate outlier ID 239 from data set A would not be detected by univariate analysis in data set B. This outlier would be detected by residual analysis, but only after grouping by gender. However, the outlier ID 239 would have the same ATM value in any set of data.

Our study confirmed the knowledge fact KB1 (A large neck is a characteristic sign of OSA) and showed that the linear correlation between NC and AHI is statistically significant at the 0.01 level. We demonstrated that the correlation between predictors should be analyzed in the context of knowledge. We have shown that the coefficient of determination r^2 significantly changes for the three groups created based on KB2 and KB3: (1) the entire population, (2) males, and (3) females. Furthermore, the values r and r^2 are sensitive to outliers. For example, we found that by removing one outlier (ID 101) from the data set B, the value r^2 increased from $r^2 = 0.28$ to $r^2 = 0.50$. Our study had two limitations: (1) small data set size, especially for females ($n = 43$), and (2) high natural variability in the manifestation of OSA symptoms and signs.

4.6 Conclusions

In this chapter, we examined the notion of typicality and its role in the creation of predictive models. We discussed typicality from a broad perspective of cognitive psychology, and described three approaches: (1) classical, (2) prototypical, and (3) exemplar. We demonstrated that an analysis of typicality must be contextualized and that an analysis of the typicality of features must be a part of the preprocessing step in KD. To address these issues, we presented a conceptual framework for explicit modeling of a typicality measure in the context of medical data. This framework has its theoretical foundations in fuzzy logic and semiotics. Fuzzy logic provides representational constructs for reasoning with imprecise values. Semiotics provides the modeling constructs for the description of the concept, its representation, and its interpretation. We applied the fuzzy-semiotic framework to the analysis of predictors used in the diagnosis of OSA. In the first step, we constructed a medical KB for NC, and we used the KB for the analysis of two data sets obtained from a clinical research study and from a database of patient medical records. In this study, we concentrated on the analysis of data using an ATM, detection of statistical outliers, and comparison of atypical values and statistical outliers. Our results show that traditional statistical data analysis is not sufficient for the identification of atypical values in heterogeneous data sets obtained through various data collection methods and nonrandom sampling techniques. The analysis of medical outliers should use knowledge-based methods to identify all potential outliers in conjunction with statistical univariate and multivariate analyses. Identified outliers should be analyzed on a case-by-case basis, for example, using a medical typicality measure. Furthermore, the data analysis should be performed with and without outliers in order to study the effects of the exclusion of specific outliers. We argue against any simple outlier removal procedures, such as trimming (removing a specific number or percentage of the lowest and highest values) or Winsorizing (replacing the highest and lowest values by values of the nearest observation). Trimming may eliminate not only obvious outliers but also many medically significant cases. Furthermore, trimming by 5%, for example, reduces the range of values and the size of available data sets (especially in the case of small sample size as in that of female OSA patients). Similarly, Winsorizing may reduce the range of values and may exclude unusual cases from the analysis.

We are planning to integrate and formalize the proposed fuzzy-semiotic framework and build a comprehensive model for the concept of typicality. We will apply the model to other well-known predictors of OSA such as hypertension, diabetes, and excessive daytime sleepiness. We also plan to use the combined fuzzy logic and semiotic approach to analyze the diagnostic criteria for clinical depression.

References

Barsalou, L. W. 1982. Context-independent and context-dependent information in concepts. *Mem Cogn* 10(1):82–93.

Belohlavek, R., G. J. Klir, H. W. Lewis III, and E. C. Way. 2009. Concepts and fuzzy sets: Misunderstanding, misconceptions, and oversights. *Int J Approx Reason* 51:23–43.

Bruner, J. S., J. J. Goodnow, and G. A. Austin. 1956. *A Study of Thinking*. New York: Wiley.

Chandler, D. 2002. *Semiotics: The Basics*. New York: Routledge.

Cios, K. J., and G. W. Moore. 2001. Medical data mining and knowledge discovery: Overview of key issues. In *Medical Data Mining and Knowledge Discovery*, ed. K. J. Cios, 1–20. Heidelberg: Springer Verlag.

Cox, E. 2005. *Fuzzy Modeling and Genetic Algorithms for Data Mining and Exploration*. Amsterdam: Morgan Kaufmann Publishers.

Dancey, D. R., P. J. Hanly, C. Soong, B. Lee, J. Shepard, and V. Hoffstein. 2003. Gender differences in sleep apnea: The role of neck circumference. *Chest* 123:1544–50.

Douglas, N. J. 2002. *Clinicians' Guide to Sleep Medicine*. London: Arnold.

Druzdzel, M. J., and F. J. Diez. 2003. Combining knowledge from different sources in causal probabilistic models. *J Mach Learn Res* 4:295–316.

Ferguson, K. A., T. Ono, A. A. Lowe, C. F. Ryan, and J. A. Fleetham. 1995. The relationship between obesity and craniofacial structure in obstructive sleep apnea. *Chest* 108(2):375–81.

Jordan, A. S., and R. D. McEvoy. 2003. Gender differences in sleep apnea: Epidemiology, clinical presentation and pathogenic mechanisms. *Sleep Med Rev* 7(5):377–89.

Joshi, S., S. Pizer, P. T. Fletcher, P. Yushkevich, A. Thall, and J. S. Marron. 2002. Multiscale deformable model segmentation and statistical shape analysis using medical descriptions. *IEEE Trans Med Imaging* 21(5):538–50.

Kononenko, I. 2001. Machine learning for medical diagnosis: History, state of the art and perspective. *Artif Intell Med* 23(1):89–109.

Krueger, R. F., and S. Bezdjian. 2009. Enhancing research and treatment of mental disorders with dimensional concepts: Towards DSM-V and ICD-11. *World Psychiatry* 8:3–6.

Lakoff, G. 1973. Hedges: A study in meaning criteria and the logic of fuzzy concepts. *J Philos Logic* 7:458–508.

Lam, B., M. S. M. Ip, E. Tench, and C. F. Ryan. 2005. Craniofacial profile in Asian and white subjects with obstructive sleep apnea. *Thorax* 60(6):504–10.

Lesot, M. J., M. Rifqi, and B. Bouchon-Meunier. 2008. Fuzzy prototypes: From a cognitive view to a machine learning principle. In *Fuzzy Sets and Their Extensions: Representation, Aggregation and Models*, eds. H. Bustince, F. Herrera and J. Montero, 431–52, Heidelberg: Springer Verlag.

Medin, D. L., and M. M. Schaffer. 1978. Context theory of classification learning. *Psychol Rev* 85:207–38.

Millman, R. P., C. C. Carlisle, S. T. McGarvey, S. E. Eveloff, and P. D. Levinson. 1995. Body fat distribution and sleep apnea severity in women. *Chest* 107(2):362–6.

Minda, J. P., and J. D. Smith. 2001. The effects of category size, category structure and stimulus complexity. *J Exp Psychol Learn Mem Cogn* 27:755–99.

Nosofsky, R. M. 1992. Exemplars, prototypes, and similarity rules. In *From Learning Theory to Connectionist Theory: Essays in Honour of William K. Estes,* ed. A. Healy, S. Kosslyn and R. Shiffrin, vol. 1. Hillsdale, NJ: Erlbaum.

Oden, G. C. 1977. Fuzziness in semantic memory: Choosing exemplars of subjective categories. *Mem Cogn* 5(2):198–204.

Osherson, D. N., and E. E. Smith. 1981. On the adequacy of prototype theory as a theory of concepts. *Cognition* 9:35–58.

Osherson, D. N., and E. E. Smith. 1982. Gradedness and conceptual combination. *Cognition* 12:299–318.

Osherson, D. N., and E. E. Smith. 1997. On typicality and vagueness. *Cognition* 12:299–318.

Owens, D. K., and H. C. Sox. 2001. Medical decision-making: Probabilistic medical reasoning. In *Medical Informatics: Computer Applications in Health Care and Biomedicine*, 2nd ed., ed. E. H. Shortliffe and L. E. Perreault, 76–129. New York: Springer.

Pedrycz, W., and F. Gomide. 2007. *Fuzzy Systems Engineering: Toward Human-Centric Computing*. New Jersey: John Wiley & Sons.

Pizer, S., P. T. Fletcher, S. Joshi, A. Thall, J. Z. Chen, Y. Fridman, et al. 2003. Deformable M-reps for 3D medical image segmentation. *Int J Comput Vis* 55(2/3):85–106.

Reed, S. K. 1996. *Cognition: Theory and Applications.* 4th ed. Pacific Grove, CA: Brooks/Cole Publishing.

Rosch, E., and C. B. Mervis. 1975. Family resemblances: Studies in the internal structure of categories. *Cogn Psychol* 7:573–605.

Rosch, E., C. B. Mervis, W. D. Gray, D. M. Johnsen, and P. Boyes-Braem. 1976. Basic objects in natural categories. *Cogn Psychol* 8:382–440.

Rowley, J. A., L. S. Aboussouan, and M. S. Badr. 2000. The use of clinical prediction formulas in the evaluation of obstructive sleep apnea. *Sleep* 23(7):929–38.

Rowley, J. A., C. S. Sanders, B. R. Zahn, and M. S. Badr. 2002. Gender differences in upper airway compliance during NREM sleep: Role of neck circumference. *J Appl Physiol* 92:2535–41.

Ryan, C. F., and L. L. Love. 1996. Mechanical properties of the velopharynx in obese patients with obstructive sleep apnea. *Am J Respir Crit Care Med* 154(3):806–12.

Sane, S. S., and A. A. Ghatol. 2006. Use of instance typicality for efficient detection of outliers with neural network classifiers. In *Proceedings of the Ninth International Conference on Information Technology* (Washington, DC, December 18–21, 2006), 225–8.

Sassoon, W. G. 2007. *Vagueness, Gradability and Typicality, A Comprehensive Semantic Analysis*. Unpublished doctoral dissertation, Tel Aviv University.

Ting, K. M. 1996. Decision combination based on the characterisation of predictive accuracy. In *Proceedings of the Third International Conference on Multistrategy Learning*, (Harpers Ferry, West Virginia, May 23–25, 1996). MSL-96, 186–197.

Vogel, J., and B. Schiele. 2004. A semantic typicality measure for natural scene categorization. In *Pattern Recognition. LNCS 3175*, ed. C. E. Rasmussen, H. H. Bulthoff, M. E. Giese and B. Scholkopf, 195–203. Berlin: Springer-Verlag.

Whittle, A. T., I. Marshall, I. Mortimore, P. K. Wraith, R. J. Sellar, and N. J. Douglas. 1999. Neck soft tissue and fat distribution comparison between normal men and women by magnetic resonance imaging. *Thorax* 54:323–8.

Wittgenstein, L. 1953. *Philosophical Investigations*. Oxford, England: Blackwell.

Yeung, C. -M. A., and H. -F. Leung. 2006. Formalizing typicality of objects and context-sensitivity in ontologies. In *Proceedings of the Fifth International Joint Conference on Autonomous Agents and Multiagent Systems* (Hakodate, Japan, May 08–12, 2006). AAMAS '06, ed. H. Nakashima, 946–8. New York: ACM.

Young, T., P. E. Peppard, and D. J. Gottlieb. 2002. Epidemiology of obstructive sleep apnea: A population health perspective. *Am J Respir Crit Care Med* 165(9):1217–39.

Zadeh, L. A. 1965. Fuzzy sets. *Inf Control* 8(3):338–53.

Zadeh, L. A. 1982. A note on prototype theory and fuzzy sets. *Cognition* 12:291–7.

Zhang, J. 1992. Selecting typical instances in instance-based learning. In *Proceedings of the Ninth International Workshop on Machine Learning*, (Aberdeen, Scotland, United Kingdom), ed. D. Sleeman and P. Edwards, 470–9. San Francisco, CA: Morgan Kaufmann Publishers.

Chapter 5

Gaussian Mixture Model–Based Clustering Technique for Electrocardiogram Analysis

Roshan Joy Martis, Chandan Chakraborty, and Ajoy Kumar Ray

Contents

5.1 Introduction

In this chapter, we explain automated feature extractors and classifiers for electrocardiogram (ECG) data. The ECG data is generated by the rhythmic electrical activity of the heart. This activity manifests a potential difference on the body surface, which is measured by a transducer and digitized by appropriate signal conditioning and an analog-to-digital converter (ADC). This voltage is a representation of underlying cardiac functionality and structure, and has a normal pattern. Departure from this normal pattern, which can be due to underlying anatomical and physiological changes, indicates different abnormalities. The aim of automated ECG abnormality detectors is the detection of such abnormalities automatically by a computer or the machine.

Before classification of data, valid features that represent the data must be extracted and selected. A feature extraction method finds an appropriate linear or nonlinear subspace of lower dimension than that of the data. There are many feature extraction methods, like principal component (PC) analysis (PCA; Duda et al. 2000), linear discriminant analysis (LDA; Duda et al. 2000), independent component analysis (ICA; Pierre 1994), and projection pursuit (Friedman and Tukey 1974). A well-known linear feature extractor is PCA or the Karhunen–Loeve transform, which uses eigenvalues and eigenvectors of the covariance matrix to project in the directions of maximum variability of the data.

Also, there are many nonlinear feature extraction techniques, like kernel PCA (KPCA; Scholkopf et al. 1996) and multidimensional scaling (MDS; Borg and Groenen 2005), which project the data on a nonlinear surface or manifold and find a representation of the data in a nonlinear space. There are also integrated feature extraction and classification strategies like error backpropagation neural network (Bishop 1995) and self-organizing feature map (Ultsch 2007), which are reported effectively in the literature.

Once the features that represent the data are extracted, it is necessary to select the features so that highest classification accuracy can be obtained. Once a set of d features are extracted, a subset of m features are selected to achieve the smallest classification error. Generally, feature selection is done off-line due to the time-consuming nature of feature selection algorithms. There are many feature selection algorithms including exhaustive search and branch and bound search that provide optimal feature subsets. There are also some feature selection algorithms, for example, sequential forward selection and sequential backward selection, that provide suboptimal features because of their simplicity and because they do not search over the entire search space.

Once valid features are extracted and selected such that they yield better accuracy, a suitable classifier must be designed. There are many ways to design a classifier. Most of the clustering methods use a centroid measure called "mean vector" for every class of data. There are two kinds of central clustering algorithms: (1) partitioned and (2) hierarchical clustering algorithms. Partitioned clustering algorithms use a fixed number of classes in the data, whereas hierarchical algorithms optimize the number of classes in the data using an iterative algorithm. Orthogonally, a clustering algorithm can be a model-based or a non-model-based algorithm. The *k*-means clustering algorithm (MacQueen 1967) is a non-model-based algorithm belonging to the partitioned central clustering category. Many variants of *k*-means clustering algorithms (Jain et al. 2000) have been proposed in the literature.

5.2 Segmentation of Electrocardiogram

The Pan–Tompkins algorithm (Pan and Tompkins 1985) is used to detect R points in the ECG, and a window of 200 samples is selected for further feature extraction using the PCA technique.

5.3 Feature Extraction

Usually, the observed pattern vector data in a pattern space contains redundant information and has a large dimensionality. Hence, it places a large computational burden on the algorithm, and large amounts of memory are required in the implemented platform. Most of the directions in the pattern space are redundant. They do not account for the data generation phenomenon. The measurement space or the observation space is blind to the underlying phenomenon by which the data is generated. Thus, it is necessary to perform optimization in the space in order to find a new representation of the data that is equivalent to the original representation and gives almost the same amount of information as the original data, but with fewer dimensions.

The problem that arises from having a large number of features is often called the "curse of dimensionality" due to many reasons. As the dimension increases, the complexity or running time of the succeeding classification algorithm will increase with dimension. Most of the methods or algorithms have a computational cost proportional to the square of dimension. As the data dimension increases, the number of samples will be too small for accurate parameter estimation. For this step, sample size should be much larger than the square of dimension; otherwise, the model will be complicated and will not represent the new unseen data. In other words, the model is said to be overfitted, since it is able to model the data patterns in the training set, but not the testing data patterns. In addition, the model is of higher order than usual and if a new pattern is added then the same model may not represent the new data. Hence, such a model lacks generalization ability.

There are many such optimization strategies in the pattern space. They are PCA, Fisher's LDA (FLDA), ICA, kernel PCA (KPCA), and kernel FLDA. However, in this chapter only PCA is discussed. This is used to derive a new representation in a space that is more representative of the underlying data generation process (Duda et al. 2000). The technique is explained in Section 5.3.1.

5.3.1 Principal Component Analysis

PCA is mathematically defined as an orthogonal linear transformation that transforms data to a new coordinate system so that the greatest variance of the data is retained by the projection of the data on a direction called the "first principal component" (first PC). Then, the second largest variance is retained on the projection on a similar axis called the "second principal component" (second PC), and so on. The motive for retaining the directions of maximum variations is that, generally, the largest variation in data corresponds to structure and the smaller variations correspond to noise in the data. Hence, PCA is a variance-preserving transformation.

The covariance matrix for the observed data is positive definite and all its eigenvalues are positive or zeroes and not negative. These eigenvalues give the variability of data in the directions defined by the corresponding eigenvectors. The highest eigenvalue and its corresponding eigenvector gives the first PC, the next highest eigenvalue indicates the second PC in the direction of the corresponding eigenvector, and so on. Finally, the data needs to be projected on the eigenvector directions. The detailed computational approach for computing PCs is as follows:

Step 1: Compute the sample mean as

$$\hat{x} = \frac{1}{n}\sum_{i=1}^{n} x_i \tag{5.1}$$

Step 2: Subtract the sample mean from the data as

$$z_i = x_i - \hat{x} \tag{5.2}$$

Step 3: Compute the covariance or scatter matrix as

$$S = \sum_{i=1}^{n} z_i z_i^{T} \tag{5.3}$$

Step 4: Compute eigenvectors e_1, e_2, e_3, ..., e_k such that they represent the k largest eigenvalues of S.
Step 5: Make e_1, e_2, e_3, ..., e_k as the k columns of matrix $E = [e_1, e_2, e_3, ..., e_k]$.
Step 6: Compute the PCs as

$$y = E^{T} z \tag{5.4}$$

In consideration, PCA gives those directions responsible for representing the data better in the vector space but not those directions responsible for discriminating the data classes.

5.4 Other Unsupervised Classifiers

Before going into the details of GMM, some of the commonly used classifiers are discussed. They are *k*-means and fuzzy *c*-means classifiers. The GMM is discussed subsequently.

5.4.1 k-Means Clustering

The motive for using the *k*-means clustering algorithm is that it starts with an initial partition and then assigns different patterns to different clusters to reduce mean-squared error. The mean-squared error reduces as the number of clusters increases. However, the mean-squared error can be minimized only for a fixed number of classes in the data. The number of classes in the data must be known prior to implementation. The use of such a clustering algorithm is called "unsupervised" learning.

The *k*-means clustering algorithm belongs to the central clustering category. It uses the L-2 Minkowski metric, called the "Euclidian metric," between the centroid and a data point. It uses a codebook, which is a table that contains entries that are updated to model the data. Vectors can change flexibly during the training phase. The algorithm gets trained so that it minimizes (or optimizes) the total mean-squared error over all the observations and over all the classes as follows:

$$E_{k\text{-means}} = \sum_{n=1}^{N} \sum_{k=1}^{c} \|x_n - \mu_k\|^2 \tag{5.5}$$

Equation 5.5 defines an objective function to be minimized for the optimization problem called unsupervised *k*-means data clustering. A general algorithm for iterative, partitioned *k*-means clustering is given as follows:

Step 1: Initialization: Initialize the cluster centroid vectors to be $\{\mu_1, \mu_2, \dots, \mu_c\}$.
Step 2a: Data assignment: For a data vector x_n, set

$$y_n = \arg_k \min \|x_n - \mu_k\|^2 \tag{5.6}$$

Step 2b: Centroid estimation: For each cluster k, let $X_k = \{x_n \mid y_n = k\}$; the centroid is computed as

$$\mu_k = \frac{1}{|X_k|} \sum_{x \in X_k} x \tag{5.7}$$

Step 3: Test for convergence: Stop the algorithm if y_n does not change for all n. Otherwise, go to Step 2a.

The aforementioned steps are explained in detail as follows: A set of k seed points are specified as initial centroids (or mean vectors). The k seed points can be the first k patterns, or they can be chosen randomly. A set of k patterns that are well separated from each other can be obtained by taking the centroid of the data as the first seed point and selecting the successive seed points that are at least away by a certain distance from the initially chosen centroid. The choice of different initial centroids leads to different final clustering, since the algorithm is a local optimization strategy of the objective function defined in Equation 5.1. Therefore, for a given data set, different clustering accuracies are likely for different initializations. One way to overcome the convergence to local minima is by running the algorithm many times with different initializations. If some of these initializations lead to the same final clustering partition, we will have some confidence that the global minima have been reached.

Based on these initial centroids, the initial partition is defined by assigning every pattern to a class corresponding to the closest centroid using Euclidian distance. The term "pass" or "cycle" refers to the method of examining the cluster label of every pattern once. MacQueen (1967) defined a k-means pass as an assignment of all patterns to the closest cluster center. The center of the gaining cluster is recomputed after each new assignment in MacQueen's k-means method. Another method, called "Forgy's method" (Forgy 1965), recomputes cluster centers once all patterns are examined. Although Euclidean distance is used in the aforementioned algorithm as the distance metric, Mahalanobis distance can also be used as a distance metric.

The k-means clustering algorithm is terminated when the criterion function cannot be improved further. There is no guarantee that a partitional k-means clustering algorithm converges with a global minimum. Some algorithms stop when the cluster labels for all the patterns do not change for two successive iterations. A maximum number of iterations can be specified in order to prevent endless oscillations.

The computational complexity of the k-means clustering algorithm is of the order $O(NdkT)$, where N is the number of patterns, d is the number of features, k is the number of classes, and T is the number of iterations. The value of T depends on the initial cluster centers, distribution of patterns, and size of the clustering problem. Generally, the user has to specify an upper bound on T to stop the iterations. For a few hundred patterns, the computational complexity of the k-means algorithm is large.

5.4.2 Fuzzy c-Means Clustering

The k-means clustering algorithm assigns one cluster, and the patterns are partitioned into disjoint sets. Patterns in one cluster are supposed to be more similar to each other than to patterns in the different clusters. If the clusters are well separated then there is no ambiguity or uncertainty associated with assigning each pattern to a cluster. When the clusters touch or overlap, cluster boundaries are not sharp and the assignment of patterns to clusters is difficult. For this kind of problem, fuzzy clustering is useful. Here, a pattern belongs to a class with a grade of membership. The degree of membership takes a value in the interval [0, 1]. For ordinary clusters called "crisp" clusters, the membership grade for a particular cluster is 1 if the pattern belongs to the cluster and 0 if it does not. With fuzzy clusters, the pattern x_i has a grade of membership $u_{ji} \geq 0$ or degree of belonging to the jth cluster, where $\sum_{j=1}^{K} u_{ji} = 1$ and K is the number of clusters. The larger the u_{ji} value, the more confidence that x_i belongs to cluster j. If u_{ji} is 1, pattern x_i belongs to cluster j with absolute certainty. The interpretation of values like 0.25 is less clear. Membership grades are subjective and are based on definitions rather than measurements. The grade of membership is not the same as the probability that a pattern belongs to a cluster although the grades of membership and probability both take values in the range [0, 1]. Under a probabilistic framework, a pattern x_i can belong to one and only one cluster depending on the outcome of a random experiment. In fuzzy set theory, a pattern x_i can belong to two clusters simultaneously. The membership grades determine the degree to which the two cluster labels are applicable. Most of the algorithms based on fuzzy set theory are partitional.

The fuzzy c-means algorithm aims at minimizing the following optimization function:

$$E_{FCM} = \sum_{i=1}^{N} \sum_{j=1}^{c} u_{ji}{}^{m} \left\| p_i - V_j \right\|^2 \tag{5.8}$$

where u_{ji} is the fuzzy membership, having m as the weighting exponent and pattern p_i such that it can associate with the cluster j having centroid V_j. The algorithm is as follows:

Step 1: Choose the number of clusters c, $2 \leq c \leq N$, weighting exponent, $m > 1$. Initialize randomly a cluster membership matrix $U^{(0)} \in (R \cap [0, 1])^{c \times N}$. Each step of this algorithm is labeled r, where $r = 0, 1, 2, \ldots$.
Step 2: Calculate the c fuzzy centers $\{v_i^{(r)}\}$ as

$$V_j^{(r)} = \frac{\sum_{j=1}^{N} \left(u_{ji}^{(r-1)} \right)^m p_i}{\sum_{j=1}^{N} \left(u_{ji}^{(r-1)} \right)^m} \tag{5.9}$$

where p_i is the feature vector.

Step 3: Update $U^{(r)}$ as

$$u_{ji}^{(r)} = \frac{\left(\dfrac{1}{\left\| p_i - V_j^{(r)} \right\|}\right)^{\frac{2}{(m-1)}}}{\displaystyle\sum_{v=1}^{c}\left(\dfrac{1}{\left\| p_i - V_j^{(r)} \right\|}\right)^{\frac{2}{(m-1)}}} \tag{5.10}$$

Step 4: When $\left\| U^{(r)} - U^{(r-1)} \right\|_F < \varepsilon$, stop. Otherwise, set $r := r+1$ and go to step 2.
Step 5: Clustering decisions for the ith data point are made by maximization:

$$\hat{\omega}(p_i) = \arg_{\omega_j \in \Omega} \max \mu_{ji} \tag{5.11}$$

where $\hat{\omega}(p_i) \in \Omega = \{\omega_1, \omega_2, \ldots, \omega_c\}$ is a set of clusters.

5.4.3 Gaussian Mixture Model

The d-dimensional-featured ECG signal in metric space is modeled statistically as a Gaussian wide-sense stationary process described by the model $\lambda = \{\mu, \Sigma\}$, where μ is the mean vector and Σ is the covariance matrix. The probability density function (PDF) of this kind of model is given by multivariate Gaussian distribution density function as follows:

$$p(x \mid \lambda) = \frac{1}{(2\pi)^{d/2} |\Sigma|^{1/2}} \exp\left[-\frac{1}{2}(x-\mu)^T \Sigma^{-1}(x-\mu)\right] \tag{5.12}$$

The maximum likelihood estimates of μ and Σ are given by

$$\mu = \frac{1}{N}\sum_n x_n \tag{5.13}$$

and

$$\Sigma = \frac{1}{N}\sum_n (x_n - \mu)(x_n - \mu)^T \tag{5.14}$$

The model is called "wide-sense stationary" since the first two moments are considered. It is easy to see that the number of parameters in Σ grows quadratically with dimensionality of data, which makes model training or accurate parameter estimation for high-dimensional data a difficult task to perform, thus necessitating the employment of dimensionality reduction.

Examples of model-based clustering approaches are the Gaussian mixture model (GMM), hidden Markov model (HMM), and Markov random field (MRF).

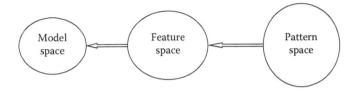

Figure 5.1 Bipartite graph view of clustering in different domains.

In conventional clustering schemes, the data is classified into feature centroids of different classes in the feature space based on the hill-climbing approach or any other optimization strategy, where, as in model-based approaches, data is classified based on an abstraction of a model space that is conceptually different from a feature space, as depicted in Figure 5.1.

The bipartite graph view shown in Figure 5.1 provides a good understanding and visualization of existing model-based clustering algorithms, and emphasizes that the generalized cluster centroids are in a model space that is conceptually separate from the feature space. The signal in the data space is mapped into the feature space by PCA, which is a linear transformation. Since the data in the data space is multivariate Gaussian distributed, after linear transformation it converts into another Gaussian with different mean and covariance matrix of lower dimension. This transformation is represented in the bipartite graph view shown in Figure 5.1. Several benefits from this bipartite graph can be immediately observed. This idea of representing clusters by models is radically more general than the standard *k*-means algorithm, in which both feature objects and cluster centroids are in the same feature space. The models also provide a probabilistic interpretation of clusters, which is a desirable feature in many applications.

5.4.4 Basic E-M Algorithm

The basic E-M algorithm is used to find the parameters of data having missing features. Let $D = \{x_1, x_2, \ldots, x_n\}$ be points taken from a single distribution with few features missing. Any sample can be written as $x_k = \{x_{kg}, x_{kb}\}$, that is, comprising "good" features and missing, or "bad," features. For notational convenience, these individual features are represented separately. We define the function given in Equation 5.15 as the expected value over the missing features:

$$Q(\theta; \theta^i) = E_{D_b}[\ln p(D_g, D_b; \theta) \,|\, D_g; \theta^i] \tag{5.15}$$

The use of semicolon denotes that $Q(\theta; \theta^i)$ is a function of θ, and θ^i is assumed to be fixed. On the right-hand side, the expected value is calculated over the space of missing features, assuming θ^i are the true parameters describing the full distribution.

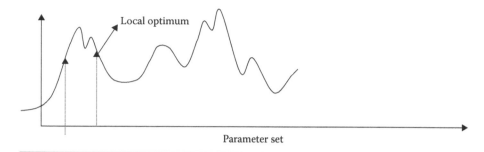

Figure 5.2 Training of model parameters by the expectation-maximization algorithm that converges to a local optimum of the defined objective function.

This equation is central in expectation maximization. The parameter vector θ^i is the current best estimate for the full distribution. The parameter θ is a candidate vector for an improved estimate. Given such a candidate θ, the right-hand side of Equation 5.15 calculates the likelihood of the data, including the unknown feature D_b marginalized with respect to the current best distribution, which is described by θ^i. Different candidate θs will lead to different such likelihoods. The E-M algorithm selects the best such candidate θ and calls it θ^{i+1}—the one corresponding to the greatest $Q(\theta; \theta^i)$.

The basic E-M algorithm is as follows:

1. Begin initialization of $\theta^0, T, i \leftarrow 0$
2. Do $i \leftarrow i + 1$
3. E-step: Compute $Q(\theta; \theta^i)$
4. M-step: $\theta^{i+1} \leftarrow \arg \max_{\theta} Q(\theta; \theta^i)$
5. Until $Q(\theta^{i+1}; \theta^i) - Q(\theta^i; \theta^{i-1}) \leq T$
6. Return $\hat{\theta} \leftarrow \theta^{i+1}$
7. End

For each iteration the expected value increases, the log likelihood increases monotonically, and the algorithm converges to a local optimum of the objective function as shown in Figure 5.2. The GMM algorithm is derived using the E-M algorithm.

5.4.5 Mathematical Form of GMM

The GMM is a parametric model, in which probabilities are defined by a small set of Gaussian parameters, that is, the mean and the covariance matrix. The GMM makes use of the following two assumptions:

1. There are K components. The ith component is w_i. Component w_i has an associated mean vector μ_i and a covariance matrix Σ.
2. Each component generates data from a Gaussian model with μ and Σ.

Each pattern is assumed to be drawn from one of K underlying populations or clusters. One clustering problem is to allocate each pattern to its correct population. Unlike density estimation or mode-seeking clustering algorithms, the form and number of underlying population densities are assumed to be known. The patterns are not labeled by population. If parameters of population densities can be estimated from the patterns, each pattern can be assigned to its appropriate cluster based on estimated probability densities. The clustering model is identical to the problem of unsupervised learning in statistical pattern recognition. The patterns are drawn from a population with a known number of clusters or classes. The underlying PDF for the class w_i is denoted by $p(x \mid w_i, \theta_i)$, where θ_i is a vector of unknown parameters w_i. If $P(w_i)$ is the priori probability of class w_i or the chance that a pattern comes from w_i, the mixture density can be written as

$$p(x \mid \theta) = \sum_{i=1}^{K} p(x \mid w_i, \theta_i) P(w_i) \tag{5.16}$$

where $\theta = (\theta_1, \theta_2, \ldots, \theta_K)$. The class conditional densities $p(x \mid w_j, \theta_j)$ are called "component densities," and the a priori probabilities $P(w_j)$ are called "mixing parameters." The mixing parameters follow the standard statistical properties as in Equation 5.17:

$$\sum_{i=1}^{K} P(w_i) = 1 \tag{5.17}$$

Using pattern vectors, the parameter θ has to be estimated such that the mixture can be decomposed into its component clusters, assuming that the mixture is identifiable. In the proposed algorithm, the expectation-maximization algorithm has been used to derive the parameter θ of the class.

Here, we have a binary classification problem of arrhythmia and ischemia classes using the probabilistic model GMM. The GMM aims at maximizing the following objective function:

$$P(X \mid \Lambda) = \sum_{n} \sum_{k} \alpha_k \, p(x_n \mid \lambda_k) \tag{5.18}$$

Step 1: Initialization: Initialize the model parameters Λ.

Step 2a: E-step: The posterior probability of model k, given a data vector x_n and current model parameters Λ, is estimated as

$$P(k \mid x_n) = \frac{\alpha_k p(x_n \mid \lambda_k)}{\sum_j \alpha_j p(x_n \mid \lambda_j)} \tag{5.19}$$

Step 2b: M-step: The maximum likelihood reestimation of model parameters Λ is given by

$$\mu_k^{(new)} = \frac{\sum_n P(k \mid x_n) x_n}{\sum_n P(k \mid x_n)} \tag{5.20}$$

$$\Sigma_k = \frac{\sum_n P(k \mid x_n)(x_n - \bar{x}_k)(x_n - \bar{x}_k)^T}{\sum_n P(k \mid x_n)} \tag{5.21}$$

$$\alpha_k = \frac{1}{N} \sum_n P(k \mid x_n) \tag{5.22}$$

Step 3: Stop if $P(k \mid x_n)$ converges; otherwise, go back to step 2a.

Step 4: For each data vector x_n, set

$$y_n = \arg_k \max(\alpha_k p(x_n \mid k)) \tag{5.23}$$

5.4.6 Estimation of Classification Error

The Chernoff bound provides a theoretical upper bound on the probability of misclassification. It gives prior information about the classifier performance. The maximum probability of the classification error is given by

$$P(\text{error}) \leq P^\beta(\omega_1) P^{1-\beta}(\omega_2) \int p^\beta(x \mid \omega_1) p^{1-\beta}(x \mid \omega_2) dx \quad \text{for} \quad 0 \leq \beta \leq 1 \tag{5.24}$$

The integral in Equation 5.24 is over the entire feature space. The conditional probabilities are evaluated analytically, yielding as the following equation:

$$\int p^{\beta}(x \mid \omega_1) p^{1-\beta}(x \mid \omega_2) dx = \exp[-k(\beta)] \tag{5.25}$$

where

$$k(\beta) = \frac{\beta(1-\beta)}{2}(\bar{x}_2 - \bar{x}_1)^t [\beta \Sigma_1 + (1-\beta)\Sigma_2]^{-1}(\bar{x}_2 - \bar{x}_1)$$
$$+ \frac{1}{2}\ln \frac{|\beta \Sigma_1 + (1-\beta)\Sigma_2|}{|\Sigma_1|^{\beta}|\Sigma_2|^{1-\beta}} \tag{5.26}$$

The Chernoff bound on $P(\text{error})$ is found by analytically or numerically finding the value of β that minimizes $\exp[-k(\beta)]$ and then substituting this number in Equation 5.25. This technique gives prior information about classifier performance. The Chernoff bound is loose for extreme values and tighter for the intermediate ones. Although the precise value of β depends on the parameters of distributions and prior probabilities, a computationally simpler but slightly less tight bound can be derived by simply setting $\beta = 1/2$. This value of β gives the Bhattacharya bound on the error given by Equations 5.27 and 5.28:

$$P(\text{error}) \leq \sqrt{P(\omega_1)P(\omega_2)} \int p(x \mid \omega_1) p(x \mid \omega_2) dx$$
$$= \sqrt{P(\omega_1)P(\omega_2)} \exp[-k(1/2)] \tag{5.27}$$

and

$$k(1/2) = 1/8(\bar{x}_2 - \bar{x}_1)^t \left[\frac{\Sigma_1 + \Sigma_2}{2}\right]^{-1} (\bar{x}_2 - \bar{x}_1) + \frac{1}{2}\ln \frac{\left|\frac{\Sigma_1 + \Sigma_2}{2}\right|}{\sqrt{|\Sigma_1||\Sigma_2|}} \tag{5.28}$$

The Chernoff bound is plotted for each dimension. Since the data dimension is 12, we can see a number of plots. The Chernoff bound is loose for extreme values, that is, near 1 and 0, and it is tight for intermediate values.

To assess the performance of the classifier on our test data, we use a diagnostic measure called "accuracy of the classifier," which is defined by Equation 5.29:

$$\text{Accuracy}(\%) = \left(1 - \frac{number\_of\_missclassifications}{total\_samples}\right) \times 100 \tag{5.29}$$

5.5 Result

Figure 5.3 shows the energy profile of PCs of ECG features. The logarithm of likelihood is plotted in Figure 5.4. The Chernoff and Bhattacharya bounds are plotted with respect to different dimensions in Figure 5.5. Table 5.1 shows the different Chernoff and Bhattacharya bounds for different initializations of the algorithm.

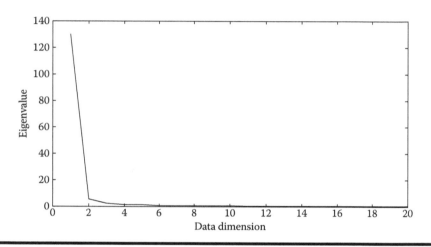

Figure 5.3 Energy profile of the principal components of electrocardiogram. (From Martis, R. J., C. Chakraborty, and A. K. Ray, *Pattern Recognition*, 42, 2979–88, 2009. With permission.)

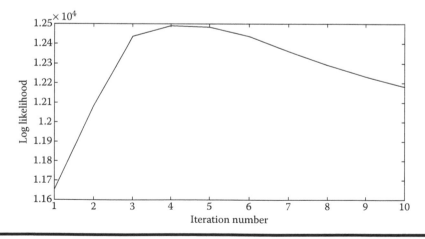

Figure 5.4 Stabilizing of log likelihood of Gaussian mixture model classifier with training. (From Martis, R. J., C. Chakraborty, and A. K. Ray, *Pattern Recognition*, 42, 2979–88, 2009. With permission.)

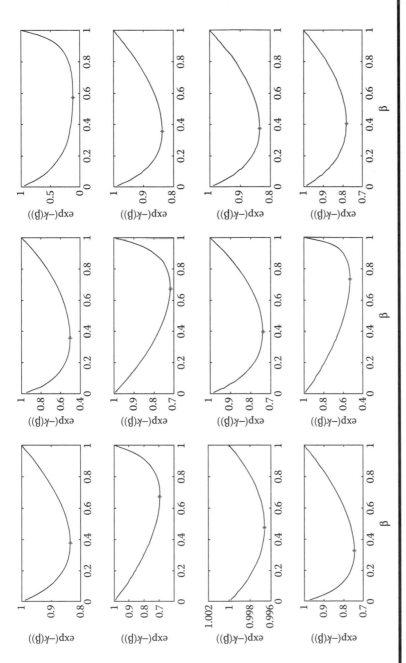

Figure 5.5 Chernoff and Bhattacharya bounds for the Gaussian mixture model classifier: The minimum of each curve shows the Chernoff bound, whereas the Bhattacharya bound corresponds to the star. (From Martis, R. J., C. Chakraborty, and A. K. Ray, *Pattern Recognition*, 42, 2979–88, 2009. With permission.)

Table 5.1 Chernoff Bounds and Accuracy of Classification for GMM Classifier

Deviation from Centroid (%)	Chernoff Bound	Battacharya Bound	Accuracy (%)
5	2.8406×10^{-09}	8.8826×10^{-10}	85.71
10	2.2653×10^{-12}	1.9924×10^{-11}	94.29
15	7.6864×10^{-12}	6.8548×10^{-11}	92.18
20	2.7287×10^{-09}	1.4955×10^{-09}	88.57
25	2.0029×10^{-08}	6.3308×10^{-09}	85.71
30	4.2960×10^{-08}	8.6274×10^{-09}	68.57
35	1.3134×10^{-08}	5.3380×10^{-09}	65.34
40	1.3065×10^{-08}	1.6748×10^{-09}	57.14
45	1.3686×10^{-07}	2.3315×10^{-08}	55.71

Source: Martis, R. J., C. Chakraborty, and A. K. Ray, *Pattern Recognition*, 42, 2979–88, 2009. With permission.

5.6 Discussion

In this chapter, GMM and its implementation details are explained, and it is shown that GMM performs better than *k*-means and fuzzy *c*-means clustering algorithms. The other algorithms, like *k*-means and fuzzy *c*-means, are conventional model-free approaches, whereas GMM is a model-based approach in which clustering takes place in the model space, which is representative of the observed patterns by the generative model for each class. The GMM models the data by its generative phenomenon. The metric used in GMM is not Euclidean distance but Mahalanobis distance, which scales the observations to unit variance and zero mean. This distance is the probabilistic distance between a sample observation and a given data point.

The dimensionality reduction technique must be applied before clustering. The PCA is used for this purpose; the energy profile of its PCs indicates that much energy is concentrated in only a few dimensions. These dimensions are sufficient for representation of the data. The PCA is used to extract these directions, and in these directions the data is projected. Then, the clustering algorithm is applied on these features. The GMM is found to be superior to *k*-means and fuzzy *c*-means clustering in terms of higher clustering accuracy and the low probability of misclassification. Chernoff and Bhattacharya bounds show that the GMM classifier gives less misclassification error and, hence, the classifier will classify the data almost without wrong classifications.

5.7 Conclusion

An ECG clustering framework is proposed in this chapter for classifying ECG into two groups, the arrhythmia and ischemia classes. Our experiment proves that the classification accuracy is good enough and is cross-validated so that it leads to a minimum error-rate classifier by Chernoff and Bhattacharya bounds. Our experiment proves that for an initialization of 10% deviation from the centroid, the overall classification accuracy is 94.29%. In the future, initialization-invariant optimization strategies for clustering, like genetic algorithms and genetic programming, must be developed.

References

Bishop, C. 1995. *Neural Networks for Pattern Recognition*. New York: Wiley Publishing.

Borg, I., and P. Groenen. 2005. *Modern Multidimensional Scaling: Theory and Applications*. 2nd ed. New York: Springer Verlag.

Duda, R., P. E. Hart, and D. G. Stork. 2000. *Pattern Classification*. 2nd ed. New York: Wiley Publishing.

Forgy, E. W. 1965. Cluster analysis of multivariate data: Efficiency vs. interpretability of classification. *Biometrics* 21:768.

Friedman, J. H., and J. W. Tukey. 1974. A projection pursuit algorithm for exploratory data analysis. *IEEE Trans Comput* C-23:881–90.

Jain, A. K., P. Robert, W. Duin, and J. Mao. 2000. *IEEE Trans Pattern Anal Mach Intell* 22:4–38.

MacQueen, J. B. 1967. Some methods for classification and analysis of multivariate observations. In *Proceedings of Mathematical Statistics and Probability*, vol. 1, 281–97.

Pan, J., and W. J. Tompkins. 1985. A real time QRS detection algorithm. *IEEE Trans Biomed Eng BME* 32:230–6.

Pierre, C. 1994. Independent component analysis: A new concept. *Signal Processing*. Elsevier 36:287–314.

Scholkopf, B., A. Smola, and K. R. Muller. 1996. Nonlinear component analysis as a kernel eigen value problem. Technical Report No. 44, Max Plank Institute, Germany.

Ultsch, A. 2007. Emerging self-organizing feature maps. In *Proceedings Workshop on Self-Organizing Maps, WSOM 07*. Bielefeld, Germany.

Chapter 6

Pattern Recognition Algorithms for Seizure Applications

Alan Chiu

Contents

6.1 Background

Epilepsy, one of the most common neurological disorders in the world, affects more than 60 million people (approximately 1% of the world's population). According to the Epilepsy Foundation, 200,000 new cases are diagnosed each year. Fifty percent of people with new cases of epilepsy have generalized onset seizures. The prevalence of active epilepsy is estimated at almost three million in the United States alone. Despite its relatively modest prevalence, one might argue that epilepsy has a higher social cost than stroke or other neurological disorders because of its unpredictable nature and the consequences of its hallmark symptom called seizure, which sometimes accompanies jerking; muscle rigidity; spasms; head turning; unusual sensations affecting vision, hearing, smell, taste, or touch; as well as memory or emotional disturbances.

Considerable attention has been given to the possibility of predicting seizure onset (Lehnertz and Elger 1998). Nonlinear analysis has been widely used to characterize the dynamics of state transitions leading to seizure onsets (Widman et al. 1999; Andrzejak et al. 2001). Most studies focus on finding the earliest possible time at which significant changes in the system dynamics may indicate an impending seizure. Reported prediction time can range from seconds (Osorio et al. 1998) to a few minutes (D'Alessandro, Esteller et al. 2003; D'Alessandro, Vachtsevanos et al. 2005) or even a few hours (Litt et al. 2001) depending on the methods and the recording locations. One aspect of research in novel preventative strategies for epilepsy involves the development of pattern recognition algorithms to detect or even predict the onset of seizures. This chapter is devoted to the development of feature extraction and supervised learning methods as well as their specific applicability to seizure detection along with the underpinning issues.

6.1.1 Current Treatment

Antiepileptic drugs are usually the first treatment option for epilepsy, and they have had success in preventing seizures in the majority of patients. Significant control of seizures can be achieved for substantial periods through this treatment. Another option called "ketogenic diet" can ameliorate intractable seizures in some pediatric cases. The third treatment option for epilepsy is surgery. It is successful in up to 90% of those carefully selected patient groups who are unresponsive to pharmacotherapy or ketogenic diet. This procedure is especially beneficial to patients who suffer from seizures associated with structural brain abnormalities. Unfortunately, the selection procedure excludes many patients with epileptic focus near regions of the brain that control motor, verbal, or cognitive functions.

6.1.2 Benefit of Seizure Anticipation

Electrical stimulations can be used to reduce the frequency of seizure episodes even though the mechanism behind their reported success is still relatively unknown (Theodore and Fisher 2004). A new therapeutic device for the treatment of epilepsy is envisioned as an implanted or externally mounted device that is able to collect electrical data from the brain, predict the onset of a seizure, and respond with the minimal number and intensity of preventative stimulation. Thus, anticipation of an approaching seizure onset is essential for the development of such a device. This possibility has both clinical and commercial significance, since about 15 million people (25% of the epileptic patients worldwide) have seizures that cannot be controlled by pharmacology or surgery. Successfully forecasting seizure onsets is a challenging task because of the relatively unknown and complex dynamics of the brain. In general, seizure time-surrogate analysis based on a Monte Carlo simulation or other methods are needed to test the performance of any seizure prediction strategy and to evaluate the statistical result against the null hypothesis of the existence of a "preictal" state (Andrzejak et al. 2003). Another option for evaluation is to compare the result of a sophisticated seizure prediction algorithm with the analytical results of naive or random approaches.

6.1.3 Prediction, Anticipation, and Early Detection

The amount of literature on seizure detection and anticipation is considerable, and it is important for the readers to understand the proper terms. As indicated by Mormann et al. (2007), some clarification is needed when studying the existing literature on seizure signal processing. Seizure prediction generally implies the ability of an algorithm to determine in advance a precise time window long before the electrographic seizure onset. The term seizure anticipation assumes that an ictal event will occur at a certain time in the future without pinpointing the time of onset in a precise manner. Finally, early seizure detection refers to the ability to find the electrographic seizure onset that is exhibited a few seconds before the first clinical symptoms are observed. The main benefit of early seizure detection is that patients can take appropriate actions to minimize the chance of injuries or initiate preventative treatments. At this moment, it is still unclear whether there exists an optimal time frame before an ictal onset (called the preictal period) such that any particular control action is more likely to abolish seizures (Iasemidis et al. 2003).

6.2 Feature Extraction of Electroencephalogram Data

Feature extraction is a procedure in which useful, sometimes hidden, information from the otherwise complex and noisy signals can be obtained using different pre-processing algorithms. In seizure detection applications, the typical data obtained is electrographic (measured electrical potential as a function of time) at various recording locations. Historically, qualified health-care professionals would identify features related to epileptic onsets through visual inspection. Over the last three decades, various analysis techniques have been proposed to extract seizure features from electroencephalogram (EEG) signals. Sometimes EEG data from different recording sites are considered independently, and sometimes the interactions between different brain regions are quantified. Furthermore, these techniques can be divided into univariate or multivariate approaches. As the need to determine the underlying properties that reflect preictal features through an automated process increases, appropriate signal processing strategies must be utilized.

6.3 Complexity Measures

Complexity analysis refers to the method of studying the complexity or information content of any nonlinear processes. It gives a quantitative interpretation of the organized, yet seemingly unpredictable behavior of the system.

6.3.1 Single-Variable Analysis

The search for chaos in neuronal systems began in the early 1980s. The most commonly used single-variable complexity analysis methods are correlation dimension (CD) and maximum Lyapunov exponent (Lmax) (Wolf et al. 1985). These techniques were first developed for the analysis of differential equations of nonlinear systems from their state space equations. A state space is generally a high-dimensional Euclidean space spanned by the degrees of freedom in a system. These techniques can also be modified to process time series, if a reconstructed state space of the system can be obtained. One possible method to construct the equivalent state space is Takens' theorem (Takens 1981), in which different time-delayed observations from a single-channel recording are embedded in a higher dimension. The reconstructed signal consists of M number of successive measurements from the original time series at a fixed time delay (T). The main question here is how to choose the appropriate values for M and T. One possible method is to set T equal to the time interval after which the autocorrelation of the recorded signal drops to a certain percentage of its initial value. This process is repeated for increasing values of *M* until the embedding process can no longer provide additional information. An improper state space reconstruction is a common source of error and can lead to gross mischaracterization of the system's dynamics. The analysis of EEG readings from patients with epileptic focal seizures has been reported. The analysis involved the attempt to detect unstable periodic orbits (UPOs) from the embedded time series between successive spike discharges (Le Van Quyen et al. 1997). It was found that a four-dimensional state space reconstruction is sufficiently complex to capture the dynamics of an EEG signal undergoing seizure transitions. An illustration of a similar concept using interpeak interval (IPI) as an estimate of state space is shown in Figure 6.1 (Khosravani et al. 2003). Here, a variable time delay is used for the

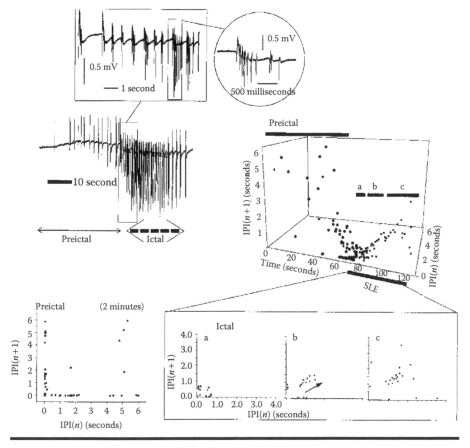

Figure 6.1 An illustration of reconstructed state space from the interpeak interval (IPI) undergoing transition from preictal to ictal state: By adding a third dimension (time) to the scatter plot, the evolution of successive IPIs can be seen. (Reprinted from *Biophys. J.*, 84, Khosravani, H., P. L. Carlen et al., The control of seizure-like activity in the rat hippocampal slice, 687–95, Copyright 2003, with permission from Elsevier.)

embedding, depending on the time delay between successive bursting phenomena. The transitions from preictal to ictal activities can be captured and visualized clearly in a simple two-dimensional embedding map.

Using these nonlinear analysis techniques, the dynamics of *petit mal* epileptic seizure was also investigated where the existence of an attractor in the state space reconstruction of the time series was found (Babloyantz and Destexhe 1986). This analysis suggests that epilepsy may correspond to a lower complexity system in comparison with healthy neuronal activity. It was also found that these lower-complexity seizure activities are often sandwiched between chaotic dynamics (Lopes Da Silva et al. 2003). Numerous reports also support the notion that the EEG signal was generated by nonlinear deterministic laws as "undoubted evidence of chaotic activity" in a biological neural network, which was demonstrated through mossy fiber stimulation (Hayashi and Ishizuka 1995). Unstable fixed points, positive Lmax, broadband power spectra, and correlation functions were found to support the claims of such reports (Iasemidis and Sackellares 1996).

The evolution of some seizure episodes involves preictal transitional states that are dynamically different from the interictal and ictal states. The implication of this finding could translate into the possibility of seizure detection, intervention, and control. Modification to the Lmax measurement using short data segments was used to analyze critical cortical regions (Iasemidis et al. 2005), illustrating convergence prior to seizure onsets (Figure 6.2). The use of this short-time Lmax (STLmax) measure is particularly useful when the interaction of the different brain regions changes over time. It showed great promise as a candidate for real-time implementation of the seizure prediction algorithm.

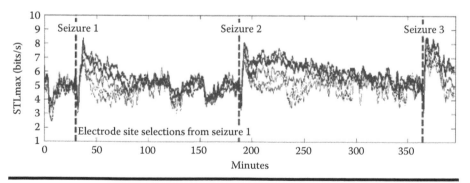

Figure 6.2 Short-time Lyapunov exponent analysis on a single channel of electroencephalogram from the critical cortical site illustrating a sudden change in complexity at seizure onset. (Reprinted from *Clin. Neurophysiol.*, 116, Iasemidis, L. D., D. S. Shiau et al., Long-term prospective on-line real-time seizure prediction, 532–44, Copyright 2005, with permission from Elsevier.)

6.3.2 Multivariable Analysis

Studies of state transitions into seizures using chaos theory principles are often limited to single-electrode recordings. This is in part due to the original definition of chaos as a point source phenomenon. A major assumption behind the use of chaos theory analysis is that an optimal location for seizure prediction must be found. A more general and robust approach to the prediction of seizure activities should involve recordings from multiple locations (Lehnertz et al. 2001). Unfortunately, complexity measures such as CD, Lmax, and STLmax, discussed in Section 6.3.1, cannot be readily applied to multivariable analysis. In addition to the embedding technique outlined in Section 6.3.1, spatial information (such as electrode locations) must be incorporated in the embedding. One possible implementation is to treat the measured value of each electrode (Martinerie et al. 1998) as an embedded dimension. Similar to the problems encountered in the single-variable case, questions such as the optimal distance between recording locations (d), optimal time delays (T), and minimum number of recording sites (n) must be answered. The transition into seizure can be traced using nonlinear indicators in the reconstructed state space. These indicators take into account the epileptic recruitment processes across different regions and over time. Using this method, seizure onsets were anticipated up to 6 minutes in advance when tested on 19 patients (Martinerie et al. 1998), suggesting the possibility of a common mode of seizure progression across different subjects.

6.3.3 Limitations on Current Complexity Measures

When analyzing the nonlinear dynamics of a time series using CD or Lmax, one must be aware that both these methods fall victim to an additional theoretical assumption that the signals must be stationary and infinite in length. Since none of these requirements can be fulfilled exactly in practice, these measures must be taken only as tentative indexes of EEG signal complexity. Numerous studies have identified numerous other factors (electrode properties, analog-to-digital conversion, amplifier and filter settings, and different recording montages) that can dramatically affect the final values of the complexity analysis as well. Furthermore, there has been much debate on the true nature of neurological or epileptic signals (Palus 1996). A sensible statement, perhaps, that can be used to describe the discrepancy in observation was offered by Slutzky et al. (2001, p. 615) that epileptiform bursting may contain "local islands of determinism within a globally stochastic sea." Here, the gap between the two schools of thought regarding the characterization of seizure dynamics may be filled.

6.4 Synchronization and Correlations

Neural synchronization is defined as the simultaneous oscillatory events that occur at different locations of the EEG recording sites. Although the term synchronization is used very often in seizure detection algorithms, a more correct terminology is "entrainment." There are two typical measures of synchronization: (1) phase and (2) lag synchronizations. It has been reported that EEG signals exhibit notable changes in spatial synchronization, sometimes even hours before the initiation of an ictal event. If such a time frame is common across subjects, this measure may be a powerful feature for long-term seizure predictability (Sabesan et al. 2008). This decrease in synchronization was also observed in the preictal state of intracranial EEG signals (Mormann et al. 2003). A significant decrease of synchrony in the focal area can be seen several minutes before the occurrence of seizures in the frequency band of 10–25 Hz (Chávez et al. 2003), and this decrease was used in early attempts of seizure prediction. In summary, the time windows of significant changes in synchronization prior to seizures differ drastically depending on the recording modality and seizure types.

6.4.1 Lag Synchronization

Two signals are said to be lag-synchronized if the dynamical variables of the two systems become synchronized with a time lag. If the cross-correlation function of two signals is maximized (close to a value of 1) at a time delay T, then the signals are said to be lag-synchronized by a phase of T. This is a more straightforward definition of synchronization, in which there is a one-to-one correlation in time shift. Experimentally, it has been observed that the changes in synchronization can be used as one of the features for the characterization of a preseizure state (Lehnertz et al. 2009).

6.4.2 Phase Synchronization

Phase coherence is one potential method for measuring phase synchronization. Inspired by the mathematical models of weakly coupled nonlinear equations that could generate phase-locking activities with chaotic variations in amplitude, the traditional approach of amplitude analysis of correlation may not be feasible. Data in medicine and biology such as EEG systems are inherently noisy where amplitude variations can mask the underlying entrainment characteristics. Shannon entropy and conditional probability have been proposed to quantify the strength of n-to-m phase locking (Tass 2007). The use of the multivariate linear discrimination method to discriminate seizures from interictal activities was also proposed (Jerger et al. 2005). The approach was based on multivariate synchrony measures to identify any possible preictal states. Cross-correlation and phase synchronization studies suggest that seizures are not simple reflections of increasing synchronization.

6.4.3 Nonlinear Interdependence

Nonlinear interdependence has been identified as one of the most reliable ways of assessing the extent of generalized synchronization in the measurement of electric field applications such as EEG (Krug et al. 2007). Nonlinear interdependence can be considered as the statistical characterization between two time series. It does not treat the phase separately from the amplitude information, as illustrated in Section 6.4.1. Furthermore, it provides additional information on the direction of interdependence or propagation. Reconstructed state space information is needed from two recorded EEG signals, and the points of one state space are mapped on to the other. It then provides a means to quantify the causal relationship across multiple recording locations. The interpretation of this analysis may be useful in determining the sequence of activated brain regions, leading to the full manifestation of seizure activities.

6.5 Time–Frequency Representation Analysis

Standard frequency analysis techniques such as Fourier transform (FT) entail the decomposition of a recorded signal into different harmonic components. It is applicable only to stationary signals. For nonstationary signals such as EEG recordings where frequency contents change over time, wavelet transform is often employed to localize the oscillatory activities in the time and frequency domains (Blanco, D'Attellis et al. 1996; Blanco, Figliola et al. 1998). A previous analysis on the use of wavelets demonstrated that slow-wave signals and sharp-wave signals can be detected with a sensitivity of 94%–100% and a specificity of approximately 89%–93% (Lopes 2007). Strong evidence for nonlinearities in epileptogenic regions of the brain was also found using wavelet analysis in a study on eight patients (Casdagli et al. 1996).

The Lmax measure described in Section 6.3.1 can also be applied to individual frequency components of an EEG signal after spectrum analysis (Chiu and Jahromi et al. 2006). Figure 6.3 illustrates how wavelet energies at different frequency components are varied over seizure numbers (the order of seizure episodes) and burst numbers (spike-to-spike variations in the preictal region).

By separating the EEG data into different frequency components, it was reported that CD and Lmax can effectively differentiate data obtained from healthy subjects and epileptic patents during the interictal versus ictal phase when the signals were sectioned into specific frequency subbands (Adeli et al. 2007). The use of a generic algorithm for the detection of seizures from the 8 channels of electrocorticogram from 14 subjects was proposed (Osorio et al. 2002). The analysis was performed offline using a band-pass wavelet filter in the 8–42-Hz range, and the relative median power of the signals demonstrated that a subject-dependent seizure detection approach results in less than 0.16 false alarms per hour with a delay of detection of less than 5 seconds.

6.5.1 Wavelet Entropy

Entropy is commonly used to evaluate the disorder of a system. Typically, probability measures are obtained solely from the raw amplitude of the EEG data. With the wavelet entropy quantifier (Rosso et al. 2001), probability terms are assigned as

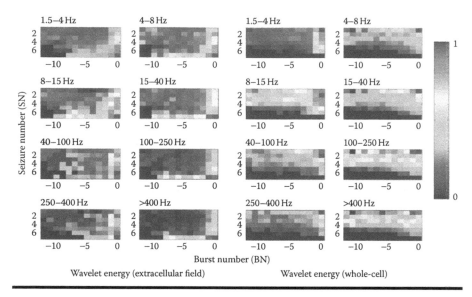

Figure 6.3 Plot of wavelet energies in different frequency bands: Each subplot denotes the normalized changes in a particular frequency band as functions of burst number (BN) and seizure number (SN). (From Chiu, A. W., S. S. Jahromi et al., *J Neural Eng*, 3:9–20, 2006. With permission.)

the relative energy of each frequency band of the signal. This relative energy gives information about the underlying dynamical process in the brain, more specifically about the synchrony and disorderly firing of different cell populations involved in the generation of epileptic field potentials. Wavelet entropy is also advantageous over Lmax or other complexity measures because it is parameter free and does not require a stationary time series. Wavelet entropy has been successfully applied to utilize coefficients of the discrete wavelet transformation of EEG signals. Significant differences were found between normal EEG data with seizure activity and epileptic EEG data, which appeared to have lower wavelet entropy values than a normal EEG (Ocak 2009). The lower entropy suggests more rhythmic firing pattern in the ictal state, which is a logical interpretation and a frequently observed phenomenon in seizure detection applications.

6.5.2 Wavelet Synchrony

Multisite analysis often suffers from an excessive amount of information; wavelet synchronization is one possible method to extract similar features from different EEG recording sites, by comparing the energy emerging from different signal wavelet coefficients. Combinations of different spatio–temporal EEG synchronization features such as cross-correlation, nonlinear interdependence, dynamical entrainment, and wavelet synchrony have been utilized for the analysis of seizures. These features were obtained from all channel pairs, and full frequency decomposition was performed. This information was combined to create a high-dimensional input space. Classifiers utilizing various machine learning techniques were created to discriminate interictal and preictal activities in a subject-dependent fashion (Mirowski et al. 2009). A reported 71% sensitivity was achieved with zero false alarms for all 15 subjects.

6.5.3 Matching Pursuit Algorithm

The matching pursuit (MP) algorithm is a time frequency analysis method that can provide continuous decompositions of recorded seizure activity so that changes in signal complexity can be monitored (Bergey and Franaszczuk 2001). The MP algorithm is an alternative approach to understanding the time and frequency resolution between different signal parts (Wilson et al. 2004). The EEG data is converted into localized "atoms" in time and frequency. Each atom can be considered as describing the time–frequency evolution of the EEG waveform. As the atom evolves in time, a seizure may be identified in one or more recording channels. If a feedforward artificial neural network approach is used to classify the EEG signal, it would imply that hidden units of the network be explicitly programmed using a rule-based approach to match the atoms. The MP method has been tested on 17 seizures from 12 patients, revealing that the most organized rhythmic seizure activity is more complex than limited cycle behavior in early seizures and that signal complexity increases as the seizure episodes progress.

6.6 Feature Selection and Channel Selection

Feature selection or dimension reduction in seizure detection algorithms can be implemented through sensitivity analysis. Feature selection is particularly important if a large number of EEG channels are recorded with a full spectrum of frequency features. Sensitivity analysis is a special form of pruning technique. It is implemented by first training the classifier using all the extracted features. At fixed time intervals of the training process, validation data is used to measure the discriminative powers of each input feature and irrelevant features are eliminated, resulting in a seizure classifier that requires only a limited number of EEG features. The advantage of this approach is that it selects the input features in a posteriori fashion with respect to initial classifier performance and not any a priori knowledge of the system. Automatic relevance determination (MacKay 1994) has been used to select input features in a recurrent neural network for seizure prediction (Chiu and Kang et al. 2006). Each input variable (wavelet coefficient feature) is associated with a hyperparameter that controls the range of variations of the synaptic weights of the artificial neural network (ANN) connected to that particular input unit. The classifier then decides which subsets of input features is the most appropriate. If the distribution (assumed to be a zero-mean Gaussian distribution) has a small standard deviation after training, that particular input unit is said to be less relevant. The performance of the network does not appear to change significantly when only relevant input features are used.

6.7 Evaluation of Pattern Recognition Tools

Sensitivity and specificity are statistical measures of binary classifiers and are typically the first reported measures for the performance of any seizure detector design. In seizure detection applications, sensitivity denotes the percentage of correctly identified ictal time windows, and specificity represents the percentage of correctly identified nonictal regions. In seizure prediction applications, sensitivity measures the percentage of correctly identified preictal states and specificity measures the percentage of correctly identified nonpreictal states. A test with a high specificity has a low type I error rate. A test with a high sensitivity has a low type II error rate. Like most supervised learning approaches, these measures are heavily dependent on the reliability of the target assignment, denoting the seizure onset times determined by the visual inspection of electrographic signals by physicians or neurosurgeons.

A similar tool for evaluating a seizure detector is the receiver operating characteristic (ROC) curve (Fawcett 2006). The ROC curves are obtained by plotting sensitivity against specificity at different threshold levels. They can be used to evaluate the effect of thresholds on types I and II errors. The area under the ROC curve is an important statistical tool to evaluate the probability that the classifier will rank a randomly chosen ictal data higher than a randomly chosen normal data.

False positive rate (FPR) is another common index to evaluate the clinical relevance of a seizure detection design. It is usually reported as the average number of false positives per hour. In a clinical environment for monitoring epileptic patients, FPR corresponds to the number of false alarms in an hour. A typical acceptable range for FPR should be less than 0.2/hour.

The computation time of a pattern recognition tool depends on whether seizure prediction or seizure detection is needed. Detection latency or delay is the time difference between the start of ictal activities and the announcement of ictal events. Ideally, a clinically relevant detector should respond to the onset of a seizure as quickly as possible. For the evaluation of seizure prediction algorithms, the signature of preictal events can occur from a few seconds to even a few hours.

6.8 Pattern Recognition Tools

The use of pattern recognition tools for seizure analysis began in the mid-1990s. The most direct approach for seizure detection is using a binary classifier based on a single variable. This approach is found to have limited success as it ignores other high-dimensional correlations between different recording locations and features. In this section, we will focus on the different machine learning algorithms using high-dimensional feature space or spatio–temporal EEG data. The first generation of binary seizure classifiers was able to discriminate preictal signals from interictal states (Mormann et al. 2006). The data from ictal and postictal states were usually discarded from the training process of a binary classifier. The eventual goal of this first-generation seizure detector was to warn the patient about future onsets by detecting potential preictal activities, so that the patient or the clinician can act accordingly.

6.8.1 Bayesian Learning

Large-amplitude rhythmic bursting activities were often confused with epileptiform activities (Khan and Gotman 2003). As a result, the Bayesian algorithm was proposed to capture the rhythmic nature of seizure discharges by comparing the fluctuations at different frequencies with the background signal. The method was evaluated on the test set obtained from 11 patients with FPR of approximately 0.3/hour. Spectral feature extraction, Bayes' theorem, and spatiotemporal analysis were also employed to classify seizure activities (Grewal and Gotman 2005). A priori information in this Bayesian approach was estimated using 407 hours of EEG signals from 19 patients having a total of 152 seizures. The test results yielded a sensitivity of 89.4%, an FPR of 0.22/hour, and a median delay time of 17.1 seconds. Another group attempted to utilize multiple features (such as peak-to-peak amplitudes, wavelet coefficient energy, standard deviation of the signal, relative signal power, and the average and relative cross-correlation functions) as independent variables for the Bayesian estimation (Kuhlmann et al. 2009). Preictal activities were identified by changes in synchronization over long-term intracranial EEG recordings using the k-nearest neighborhood (k-NN) algorithm (Le Van Quyen et al. 2005). The analysis was performed in 5 patients with over 300 hours of seizure data using phase synchronization input. A phase-locking measure was estimated using a sliding window analysis on frequency bands up to 30 Hz for different pairs of the existing 20 EEG channels. These observations were compared with other known preictal features. Synchronization can be observed several hours before the actual seizure onsets, and the changes were observed in the 4–15-Hz frequency band localized near the primary epileptogenic zone.

6.8.2 Multilayer Perceptron

Multilayer perceptron neural networks can be trained to recognize patterns and to estimate the nonlinear relationship of input features with respect to the classification of EEG data. Two types of multilayered networks (feedforward and recurrent networks) have been used to analyze EEG data pertaining to seizure phenomenon since the mid-1990s. One of the earliest attempts of seizure detection using ANNs was performed (Webber et al. 1996) using 31 simple input features, including amplitude, slope, curvature, and frequency spectrum, in 40 EEG channels to classify 8 distinct EEG patterns, including seizure, muscle movement, and other normal activities. The reported accuracy of seizure classification was in the range of 76% with an FPR of 1.0/hour. Around the same time, another group tested eight channels of EEG data from four human subjects (Pradhan et al. 1996). The specificity and sensitivity of their network were found to be 75% and 67%, respectively, which were acceptable but not exceptional for clinical tests. The filtered average time series was also proposed to construct spectrograms as input features to the ANN (Gabor 1998). The proposed algorithm achieved an accuracy of over 90% with an FPR of approximately 1.0/hour. One potential shortcoming of this approach is that it often misclassifies pseudosinusoidal patterns as epileptic data.

A proposed alternative approach was to include rule-based algorithms and cascade these rules in a neural network (Karayiannis et al. 2006). Seizure segments were taken from 12 channels of neonatal EEG and tested on 15 patients. Based on the power spectrum features such as dominant frequency, power, and amplitude stability, a rule-based algorithm separated short segments of pseudoharmonic EEG patterns from seizures. The trained neural network cascaded with the rule-based algorithm achieves 83% sensitivity and 79% specificity. Neural networks with adaptive activation functions were proposed to detect epileptic seizures. Three activation options were proposed: (1) sigmoid function, (2) sinusoidal function, or (3) wavelet function with free parameters. A 5-fold cross-validation technique was applied, and it demonstrated 100% average sensitivity, 100% average specificity, and approximately 100% average classification rate (Tezel and Özbay 2009). Wavelet neural networks (Subasi et al. 2005) have also been tested to classify epileptic and nonepileptic seizures in more than 600 episodes. The network utilized a combination of Morlet mother wavelet basis functions and linear basis functions. It achieved over 92% in specificity and over 93% in sensitivity. Autoregressive model and fast FT (FFT)-based feature extraction techniques were used to find the optimal input features of the ANN for seizure detection (Kiymik et al. 2004). Data were taken from the scalp electrodes of five patients. It was observed that ANN classification of EEG signals with autoregressive model preprocessing shows a higher sensitivity than simple frequency information (92% vs. 90%) and higher specificity (96% vs. 93%).

Another modification to existing multilayer perceptrons includes the ability to adjust the structure of a multilayer backpropagation network adaptively (Khorasani and Weng 1996). As opposed to standard pruning strategies in which network complexity must be overestimated initially to allow for the removal of irrelevant model units, this approach starts with a small network. The stabilized error is used as an index to determine whether a new neural unit is needed. The new units are placed at locations that contribute the most to the network error through the fluctuation of their corresponding weight vectors. It was reported that the number of training iterations is reduced by over 60% without jeopardizing the sensitivity and specificity of the seizure detector.

Finally, ANN can also be used to reduce FPR. Features associated with false positives of seizure detection can also be learned in a subject-dependent manner (Qu and Gotman 1993). An important problem in automatic seizure detection during long-term epilepsy monitoring is the frequency of false positive alarms. The program learns about the false detections occurring in the first day of a prolonged monitoring session and attempts to prevent similar patterns from being labeled as positive events during the remainder of the session. Features such as recording locations, amplitudes, average frequency, and coefficients of variation of wave duration are selected. The Euclidean distance between points in feature space and their probability distribution were studied using signals from 20 patients with 134 EEG sessions recorded from scalp and depth electrodes. False detections were reduced by 61% with a 5% decrease in sensitivity. The average FPR dropped to 1.26/hour from 3.25/hour (Qu and Gotman 1993).

6.8.3 Recurrent Neural Networks

Recurrent neural networks are capable of representing and encoding an arbitrary number of previous inputs. They can perform highly nonlinear dynamic mappings, whereas feedforward networks are confined to performing static signal classification. A recurrent network was been tested on 32-channel EEG data from 2 patients who underwent long-term electrophysiological monitoring for epileptic implants (Petrosian et al. 2000). The leave-one-out testing method was employed to evaluate six scalp recordings. The EEG time series and their wavelet transform coefficients, consisting of 2 input units, 10 or 15 recurrent hidden neurons, and 1 output unit, were fed into the recurrent network. The wavelet decomposition method appeared to achieve good classification results. Researchers were able to demonstrate the existence of preictal activities a few minutes before ictal onset. The shortcoming of this particular approach was that this network had been trained only to process each EEG channel independently. The network did not utilize any multivariate measurements and the structure of the network was dependent on the sampling frequency of the data.

Elman recurrent neural networks (illustrated in Figure 6.4) have also been proposed for seizure detection using Lmax values as one of the input features (Güler et al. 2005). An explicit memory of the hidden unit outputs was created using the so-called context layer. These outputs were then used as an extra input feature for subsequent computations. The EEG signals from a healthy control group and from epilepsy patients were discriminated using a single electrode at the epileptogenic region and 128 Lyapunov exponents. The complete data set consisted of

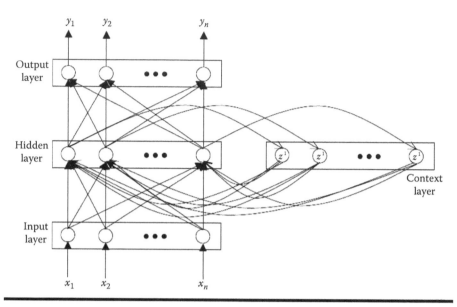

Figure 6.4 **Illustration of an Elman network where the context layer stores the hidden-layer outputs from the previous iteration and feeds these data back to the hidden-layer units as extra inputs. (Reprinted from *Expert Systems with Applications*, 29, Güler, N. F., E. D. Übeyli et al., Recurrent neural networks employing Lyapunov exponents for EEG signals classification, 506–14, Copyright 2005, with permission from Elsevier.)**

500 single-channel EEG signals. The network was able to classify healthy EEG segments, seizure-free epileptogenic zone segments, and epileptic seizure segments with accuracies 97.38%, 96.88%, and 96.13%, respectively.

6.8.4 Support Vector Machine

The advantage of using a support vector machine (SVM) for EEG signal classification is that it can minimize the expected risk of misclassifying unseen data (Burges 1998). The optimization process of an SVM involves solving constrained quadratic programming that estimates the hyperplane boundary between two data classes defined by the kernel functions. The general implementation for a binary SVM is the one-versus-one or the one-versus-all approach. A one-class SVM for normal EEG data has been reported (Gardner et al. 2006), in which a model is created for normal EEG data and other patient-specific parameters. The feature space was composed of three energy-based statistics: (1) mean curve length, (2) mean energy, and (3) mean Teager energy (as shown in Figure 6.5). It was reported that the SVM achieved 97.1% sensitivity on detecting normal EEG data. The average reported FPR of the algorithm was 1.56/hour and ictal onsets could be identified with a delay of 7.58 seconds.

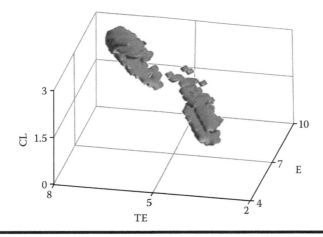

Figure 6.5 Volume unique to the one-class support vector machine for interictal data. (From Gardner, A. B., A. M. Krieger et al., *J Mach Learn Res*, 7:1025–44, 2006. With permission.)

Another proposed SVM for EEG analysis utilized the energies in its wavelet decompositions and the encoded spatial localization information as features shown in Figure 6.6 (Shoeb et al. 2004). Data from 36 subjects collected over 60 hours, including over 139 seizures episodes, were used in the analysis. Wavelet energy features from four levels of discrete wavelet decomposition was used. The reported average latency of seizure onset detection was approximately 8 seconds with more than 94% sensitivity. The reported FPR was approximately 0.25/hour. For any applications beyond a binary classifier, modification was made using a directed acyclic graph (Platt et al. 2000), which is a tree structure of one-versus-one approach that systemically eliminates incorrect classes. The directed acyclic graph implementation may be useful if the identification of multiple seizure states is required.

6.8.5 Genetic Program

The genetic program (GP) algorithm utilizes the genetic operators selection, crossover, mutation, and reproduction to come up with the highly nonlinear mathematical equation that best represents the decision boundary between different classes of data. The main difference between GP and previous approaches is that previous techniques are based on features calculated using conventional techniques. These techniques rely on the knowledge of some feature extraction formula or previous algorithms that were derived from intuition and the physical nature of the problem. Once the reconstructed state space had been created from EEG signals, an optimal nonlinear transformation was obtained using the GP algorithm. Operators including addition, subtraction, multiplication, division, square, square root, logarithm,

Radial basis nonlinear decision boundary

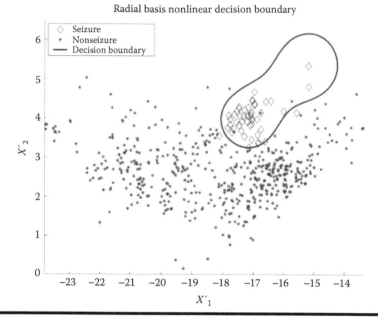

Figure 6.6 **An illustration of a radial basis nonlinear decision boundary used on two-dimensional projections of four-dimensional seizure and nonseizure feature vectors. (Reprinted from *Epilepsy & Behavior* 5, Shoeb, A., H. Edwards et al., Patient-specific seizure onset detection, 483–98, Copyright 2004, with permission from Elsevier.)**

absolute value, sine, and cosine were used to separate the baseline or the interictal events from the preictal events using the k-NN algorithm (as illustrated in Figure 6.7). It has been reported that good artificial features and the prediction horizons might vary from patient to patient. The GP algorithm automatically generates artificial features from raw data that fits well in a patient-specific scenario (Firpi et al. 2006). On an average of 6 patients, it was reported that the GP algorithm achieved 93% specificity and 79% sensitivity in classification. More work is needed to make the accuracy level of the GP algorithm clinically acceptable. Another problem with the genetic algorithm is that the features extracted are the so-called artificial features, which are computer-generated features that do not necessarily have any physiological significance.

6.8.6 *Markov Model*

The unsupervised approach for seizure detection and prediction offers three distinct advantages over the supervised techniques described in Sections 6.8.1 through 6.8.5: First, there is no need to perform supervised, patient-specific training task. Second, there is no need to assume homogeneity in the electrographic

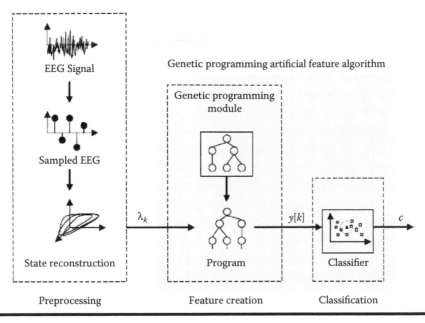

Preprocessing · Feature creation · Classification

Figure 6.7 **An illustration of the genetic algorithm utilizing reconstructed state space parameters as the input features for a seizure classifier for baseline versus ictal electroencephalogram seizures. (With kind permission from Springer Science & Business Media:** *Ann of Biomed Eng,* **On prediction of epileptic seizures by means of genetic programming artificial features, 34, 2006, 515–29, Firpi, H., E. Goodman et al.)**

signals of seizures. Last, there is no need to indicate the onsets and durations of seizure activities or other distinct preictal states for training. Similar to the recurrent neural network approach, the Markov model assumes that the current state in a seizure episode has some dependency on information from the previous time instance. The Markov model is particularly useful in modeling the progression of any time-varying phenomenon in which the observations are generated from underlying states that have unknown or hidden processes. Markovian dynamics have been successfully used in modeling the process of seizure generation in experimental animal models (Le et al. 1992; Sunderam et al. 2001). The hidden Markov model (HMM) approach was also proposed as a three-state model to detect seizures (Wong et al. 2007).

Another HMM implementation hypothesizes that the generation of an ictal process in a neuronal system can be categorized into sequential preictal states (Gadi et al. 2008). Wavelet coefficients were embedded into the feature space. The transition probability matrix was used as a validation tool to test that transitions during a seizure episode actually follow the progression from interictal (or baseline) and preictal to ictal, through a potential postictal state before returning to the interictal

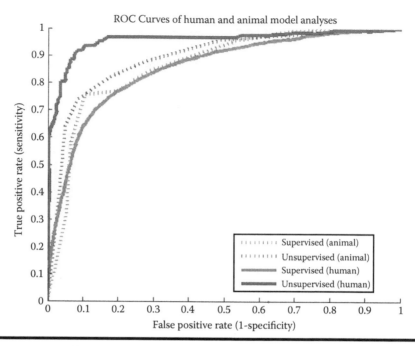

Figure 6.8 **Analysis of seizure detection using hidden Markov model compared with multilayer perceptron using identical features from the animal model and human intracranial electroencephalogram data.**

activity. Radial basis functions were used to perform clustering tasks on embedded information pertaining to seizure activities. The embedding strategy can be either temporal (time delays) or spatial or both. When the performance of the HMM model was compared with that of the supervised ANN classifier using identical input features, ROC curve analysis showed that the unsupervised HMM approach performs at least the same as, if not better than, the supervised approach, as shown in Figure 6.8.

6.8.7 Other Approaches

Rule-based fuzzy logic (Aarabi et al. 2009) has been applied to detect seizures in intracranial EEG recordings. The system was designed based on previously gathered temporal, spatial, and complexity information. Twenty-one subjects, consisting of 78 seizure episodes, were involved in the study. The proposed algorithm was able to achieve 98.7% sensitivity with an FPR of 0.27/hour. All but one of the 78 seizure onsets were detected with an average detection latency of 11 seconds. The algorithm had difficulty in distinguishing high-amplitude rhythmic (nonseizure) activities from ictal events.

Blind source identification (Corsini et al. 2006) is another approach to quantify the dynamical changes of the brain. It is said to perform better than principal component analysis (PCA) in artifact removal (De Clercq et al. 2005). Twenty sets of simultaneous intracranial and scalp EEG recordings from 20 patients were analyzed. When the epileptiform activity is common in the majority of the channels, the proposed method can be used to remove the artifacts. Chaotic behavior of the reconstructed source estimate can then be evaluated using Lmax to detect seizure onsets.

6.9 Limitations

Both the linear and nonlinear analysis techniques mentioned in Section 6.8 must be used with vigilance since many of them assume that the underlying dynamical systems are stationary and artifact free. The main difficulty of the previous EEG artifact removal strategy is that the visual selection of artifact components is a tedious process. A Bayesian classifier was proposed to distinguish epochs of seizure segments between EEG activity and artifacts (electromyography and 60-Hz noise) through independent component analysis (ICA) decomposition (LeVan et al. 2006). The classifier was trained using statistical, frequency and spatial features from 46 patients and 205 seizure episodes. The classifier was able to achieve a sensitivity of 82.4% and a specificity of 83.3%. The sensitivity to artifacts by the automated network was similar to the expert's performance. One should also realize the limitations to some of the methods discussed in Section 6.8, whether the detector is only subject specific or is applicable across different subjects, seizure types, or recording modality. It has been reported that the detections of seizure onsets with long EEG signals are patient specific (Qu and Gotman 1997). This hypothesis was evaluated in 12 patients having 47 seizure episodes. A modified nearest-neighbor classifier was designed using time and frequency features. The system was able to detect seizure onsets with 100% accuracy while maintaining an average delay of 9.35 seconds. The average FPR was minimal at 0.02/hour.

6.10 Beyond Detection

A limitation to all the methods discussed in Section 6.8 is their binary approach. A potential improvement would be the replacement of binary classification by regression analysis to estimate a function of the onset time of a seizure (Winterhalder, Maiwald et al. 2003; Winterhalder, Schelter et al. 2006). The "dynamical similarity index" was proposed and tested on 582 hours of intracranial EEG data from 88 seizure episodes. Changes in synchronization were used to predict seizures. The average sensitivity was close to 60% with an FPR of 0.15/hour. A prediction horizon of approximately 10 minutes was reported.

Another way to forecast generalized epileptic seizures from EEG signals was tested utilizing wavelet analysis and dynamic unsupervised fuzzy clustering technique (Geva and Kerem 1998). Features were extracted using wavelet transform. The features associated with the wavelet coefficients were entered to the unsupervised optimal fuzzy clustering algorithm. The unsupervised selection of the number of clusters overcame the challenge posed by a priori unknown and variable number of states. The classification succeeded in identifying behavior-related states such as sleep, rest, alert, and active wakefulness, as well as seizures. It was reportedly able to detect unspecified new states in a patient-specific manner. The average forecasting time was found to be 0.7–4 minutes.

It has been shown that wavelet ANNs are able to predict forthcoming ictal onsets. They are able to track the frequency variations in the preictal region, thereby predicting when a seizure onset is most likely to occur (Chiu et al. 2005). Preliminary results suggest that prediction can be made as early as 2 minutes prior to the onset with a mean accuracy of over 75%. The precision of prediction is found to be within a 30-second window (as shown in Figure 6.9).

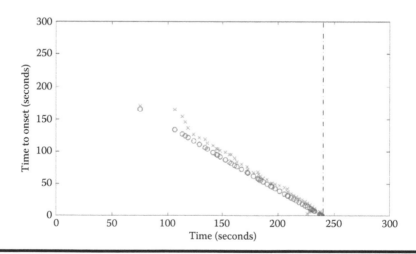

Figure 6.9 **Actual time to onset (O) and predicted seizure onset (X) using the wavelet artificial neural network predictor approach. (With kind permission from Springer Science & Business Media:** *Ann Biomed Eng,* **Prediction of seizure onset in an in-vitro hippocampal slice model of epilepsy using Gaussian-based and wavelet-based artificial neural networks, 33, 2005, 798–810, Chiu, A. W., S. Daniel et al.)**

References

Aarabi, A., R. Fazel-Rezai et al. 2009. A fuzzy rule-based system for epileptic seizure detection in intracranial EEG. *Clin Neurophysiol* 120:1648–57.

Adeli, H., S. Ghosh-Dastidar et al. 2007. A wavelet-chaos methodology for analysis of EEGs and EEG subbands to detect seizure and epilepsy. *IEEE Trans Biomed Eng* 54:205–11.

Andrzejak, R. G., K. Lehnertz et al. 2001. Indications of nonlinear deterministic and finite-dimensional structures in time series of brain electrical activity: Dependence on recording region and brain state. *Phys Rev E Stat Nonlin Soft Matter Phys* 64(6 I):061907.

Andrzejak, R. G., F. Mormann et al. 2003. Testing the null hypothesis of the nonexistence of a preseizure state. *Phys Rev E Stat Nonlin Soft Matter Phys* 67:109011–4.

Babloyantz, A., and A. Destexhe. 1986. Low-dimensional chaos in an instance of epilepsy. *Proc Natl Acad Sci U S A* 83:3513–7.

Bergey, G. K., and P. J. Franaszczuk. 2001. Epileptic seizures are characterized by changing signal complexity. *Clin Neurophysiol* 112:241–9.

Blanco, S., C. E. D'Attellis et al. 1996. Time-frequency analysis of electroencephalogram series. II. Gabor and wavelet transforms. *Phys Rev E Stat Phys Plasmas Fluids Relat Interdiscip Topics* 54:6661–72.

Blanco, S., A. Figliola et al. 1998. Time-frequency analysis of electroencephalogram series. III. Wavelet packets and information cost function. *Phys Rev E Stat Phys Plasmas Fluids Relat Interdiscip Topics* 57:932–40.

Burges, C. J. C. 1998. A tutorial on support vector machines for pattern recognition. *Data Min Knowl Discov* 2:121–67.

Casdagli, M. C., L. D. Iasemidis et al. 1996. Characterizing nonlinearity in invasive EEG recordings from temporal lobe epilepsy. *Physica D* 99:381–99.

Chávez, M., M. Le Van Quyen et al. 2003. Spatio-temporal dynamics prior to neocortical seizures: Amplitude versus phase couplings. *IEEE Trans Biomed Eng* 50:571–83.

Chiu, A. W., S. Daniel et al. 2005. Prediction of seizure onset in an in-vitro hippocampal slice model of epilepsy using Gaussian-based and wavelet-based artificial neural networks. *Ann Biomed Eng* 33:798–810.

Chiu, A. W., S. S. Jahromi et al. 2006. The effects of high-frequency oscillations in hippocampal electrical activities on the classification of epileptiform events using artificial neural networks. *J Neural Eng* 3:9–20.

Chiu, A. W., E. E. Kang et al. 2006. Online prediction of onsets of seizure-like events in hippocampal neural networks using wavelet artificial neural networks. *Ann Biomed Eng* 34:282–94.

Corsini, J., L. Shoker et al. 2006. Epileptic seizure predictability from scalp EEG incorporating constrained blind source separation. *IEEE Trans Biomed Eng* 53:790–9.

D'Alessandro, M., R. Esteller et al. 2003. Epileptic seizure prediction using hybrid feature selection over multiple intracranial EEG electrode contacts: A report of four patients. *IEEE Trans Biomed Eng* 50:603–15.

D'Alessandro, M., G. Vachtsevanos et al. 2005. A multi-feature and multi-channel univariate selection process for seizure prediction. *Clin Neurophysiol* 116:506–16.

De Clercq, W., B. Vanrumste et al. 2005. Modeling common dynamics in multichannel signals with applications to artifact and background removal in EEG recordings. *IEEE Trans Biomed Eng* 52:2006–15.

Fawcett, T. 2006. An introduction to ROC analysis. *Pattern Recognit Lett* 27:861–74.

Firpi, H., E. Goodman et al. 2006. On prediction of epileptic seizures by means of genetic programming artificial features. *Ann Biomed Eng* 34:515–29.

Gabor, A. J. 1998. Seizure detection using a self-organizing neural network: Validation and comparison with other detection strategies. *Electroencephalogr Clin Neurophysiol* 107:27–32.

Gadi, H., D. W. Moller et al. 2008. Spatial time-frequency analysis and non-parametric classification of human IEEG recordings using HMM. In *BMES Annual Fall Meeting*, St. Louis, MO.

Gardner, A. B., A. M. Krieger et al. 2006. One-class novelty detection for seizure analysis from intracranial EEG. *J Mach Learn Res* 7:1025–44.

Geva, A. B., and D. H. Kerem. 1998. Forecasting generalized epileptic seizures from the EEG signal by wavelet analysis and dynamic unsupervised fuzzy clustering. *IEEE Trans Biomed Eng* 45:1205–16.

Grewal, S., and J. Gotman. 2005. An automatic warning system for epileptic seizures recorded on intracerebral EEGs. *Clin Neurophysiol* 116:2460–72.

Güler, N. F., E. D. Übeyli et al. 2005. Recurrent neural networks employing Lyapunov exponents for EEG signals classification. *Expert Syst Appl* 29:506–14.

Hayashi, H., and S. Ishizuka. 1995. Chaotic responses of the hippocampal CA3 region to a messy fiber stimulation in vitro. *Brain Res* 686:194–206.

Iasemidis, L. D., and J. C. Sackellares. 1996. Review: Chaos theory and epilepsy. *Neuroscientist* 2:118–26.

Iasemidis, L. D., D. S. Shiau et al. 2003. Adaptive epileptic seizure prediction system. *IEEE Trans Biomed Eng* 50:616–27.

Iasemidis, L. D., D. S. Shiau et al. 2005. Long-term prospective on-line real-time seizure prediction. *Clin Neurophysiol* 116:532–44.

Jerger, K. K., S. L. Weinstein et al. 2005. Multivariate linear discrimination of seizures. *Clin Neurophysiol* 116:545–51.

Karayiannis, N. B., A. Mukherjee et al. 2006. Detection of pseudosinusoidal epileptic seizure segments in the neonatal EEG by cascading a rule-based algorithm with a neural network. *IEEE Trans Biomed Eng* 53:633–41.

Khan, Y. U., and J. Gotman. 2003. Wavelet based automatic seizure detection in intracerebral electroencephalogram. *Clin Neurophysiol* 114:898–908.

Khorasani, K., and W. Weng. 1996. An adaptive structure neural networks with application to EEG automatic seizure detection. *Neural Netw* 9:1223–40.

Khosravani, H., P. L. Carlen et al. 2003. The control of seizure-like activity in the rat hippocampal slice. *Biophys J* 84:687–95.

Kiymik, M. K., A. Subasi et al. 2004. Neural networks with periodogram and autoregressive spectral analysis methods in detection of epileptic seizure. *J Med Syst* 28:511–22.

Krug, D., H. Osterhage et al. 2007. Estimating nonlinear interdependences in dynamical systems using cellular nonlinear networks. *Phys Rev E* 76:041916.

Kuhlmann, L., A. N. Burkitt et al. 2009. Seizure detection using seizure probability estimation: Comparison of features used to detect seizures. *Ann Biomed Eng* 37:2129–45.

Le, N. D., B. G. Leroux et al. 1992. Exact likelihood evaluation in a Markov mixture model for time series of seizure counts. *Biometrics* 48:317–23.

Le Van Quyen, M., J. Martinerie et al. 1997. Unstable periodic orbits in human epileptic activity. *Phys Rev E Stat Phys Plasmas Fluids Relat Interdisc Topics* 56(3 B):3401–11.

Le Van Quyen, M., J. Soss et al. 2005. Preictal state identification by synchronization changes in long-term intracranial EEG recordings. *Clin Neurophysiol* 116:559–68.

Lehnertz, K., R. G. Andrzejak et al. 2001. Nonlinear EEG analysis in epilepsy: Its possible use for interictal focus localization, seizure anticipation, and prevention. *J Clin Neurophysiol* 18:209–22.

Lehnertz, K., S. Bialonski et al. 2009. Synchronization phenomena in human epileptic brain networks. *J Neurosci Methods* 183:42–8.

Lehnertz, K., and C. E. Elger. 1998. Can epileptic seizures be predicted? Evidence from nonlinear time series analysis of brain electrical activity. *Phys Rev Lett* 80:5019–22.

LeVan, P., E. Urrestarazu et al. 2006. A system for automatic artifact removal in ictal scalp EEG based on independent component analysis and Bayesian classification. *Clin Neurophysiol* 117:912–27.

Litt, B., R. Esteller et al. 2001. Epileptic seizures may begin hours in advance of clinical onset: A report of five patients. *Neuron* 30:51–64.

Lopes, H. S. 2007. Genetic programming for epileptic pattern recognition in electroencephalographic signals. *Appl Soft Comput* 7:343–52.

Lopes Da Silva, F. H., W. Blanes et al. 2003. Epilepsies as dynamical diseases of brain systems: Basic models of the transition between normal and epileptic activity. *Epilepsia* 44(12 Suppl.):72–83.

MacKay, D. 1994. *Bayesian Methods for Backpropagation Networks*. Berlin: Springer.

Martinerie, J., C. Adam et al. 1998. Epileptic seizures can be anticipated by non-linear analysis. *Nat Med* 4:1173–6.

Mirowski, P., D. Madhavan et al. 2009. Classification of patterns of EEG synchronization for seizure prediction. *Clin Neurophysiol* 120:1927–40.

Mormann, F., R. G. Andrzejak et al. 2003. Automated detection of a preseizure state based on a decrease in synchronization in intracranial electroencephalogram recordings from epilepsy patients. *Phys Rev E Stat Nonlin Soft Matter Phys* 67(2 Pt 1):021912.

Mormann, F., R. G. Andrzejak et al. 2007. Seizure prediction: The long and winding road. *Brain* 130(Pt 2):314–33.

Mormann, F., C. E. Elger et al. 2006. Seizure anticipation: From algorithms to clinical practice. *Curr Opin Neurol* 19:187–93.

Ocak, H. 2009. Automatic detection of epileptic seizures in EEG using discrete wavelet transform and approximate entropy. *Expert Syst Appl* 36(2 Pt 1):2027–36.

Osorio, I., M. G. Frei et al. 1998. Real-time automated detection and quantitative analysis of seizures and short-term prediction of clinical onset. *Epilepsia* 39:615–27.

Osorio, I., M. G. Frei et al. 2002. Performance reassessment of a real-time seizure-detection algorithm on long ECoG series. *Epilepsia* 43:1522–35.

Palus, M. 1996. Nonlinearity in normal human EEG: Cycles, temporal asymmetry, nonstationarity and randomness, not chaos. *Biol Cybern* 75:389–96.

Petrosian, A., D. Prokhorov et al. 2000. Recurrent neural network based prediction of epileptic seizures in intra- and extracranial EEG. *Neurocomputing* 30(1–4):201–18.

Platt, J., N. Cristianini et al. 2000. Large margin DAGs for multiclass classification. *Adv Neural Inf Process Syst* 12:547–53.

Pradhan, N., P. K. Sadasivan et al. 1996. Detection of seizure activity in EEG by an artificial neural network: A preliminary study. *Comput Biomed Res* 29:303–13.

Qu, H., and J. Gotman. 1993. Improvement in seizure detection performance by automatic adaptation to the EEG of each patient. *Electroencephalogr Clin Neurophysiol* 86:79–87.

Qu, H., and J. Gotman. 1997. A patient-specific algorithm for the detection of seizure onset in long-term EEG monitoring: Possible use as a warning device. *IEEE Trans Biomed Eng* 44:115–22.

Rosso, O. A., S. Blanco et al. 2001. Wavelet entropy: A new tool for analysis of short duration brain electrical signals. *J Neurosci Methods* 105:65–75.

Sabesan, S., L. Good et al. 2008. Global optimization and spatial synchronization changes prior to epileptic seizures. *Optim Med* 12:103–25.

Shoeb, A., H. Edwards et al. 2004. Patient-specific seizure onset detection. *Epilepsy Behav* 5:483–98.

Slutzky, M. W., P. Cvitanovic et al. 2001. Deterministic chaos and noise in three in vitro hippocampal models of epilepsy. *Ann Biomed Eng* 29:607–18.

Subasi, A., A. Alkan et al. 2005. Wavelet neural network classification of EEG signals by using AR model with MLE preprocessing. *Neural Netw* 18:985–97.

Sunderam, S., I. Osorio et al. 2001. Stochastic modeling and prediction of experimental seizures in Sprague-Dawley rats. *J Clin Neurophysiol* 18:275–82.

Takens, F. 1981. Detecting strange attractors in turbulence. In *Dynamical Systems and Turbulence, Warwick 1980*, ed. D. Rand and L. S. Young, 366–81. Berlin: Springer.

Tass, P. A. 2007. *Phase Resetting in Medicine and Biology: Stochastic Modelling and Data Analysis*. Berlin: Springer.

Tezel, G., and Y. Özbay. 2009. A new approach for epileptic seizure detection using adaptive neural network. *Expert Syst Appl* 36:172–80.

Theodore, W. H., and R. S. Fisher. 2004. Brain stimulation for epilepsy. *Lancet Neurol* 3:111–8.

Webber, W. R., R. P. Lesser et al. 1996. An approach to seizure detection using an artificial neural network (ANN). *Electroencephalogr Clin Neurophysiol* 98:250–72.

Widman, G., D. Bingmann et al. 1999. Reduced signal complexity of intracellular recordings: A precursor for epileptiform activity? *Brain Res* 836(1–2):156–63.

Wilson, S. B., M. L. Scheuer et al. 2004. Seizure detection: Evaluation of the Reveal algorithm. *Clin Neurophysiol* 115:2280–91.

Winterhalder, M., T. Maiwald et al. 2003. The seizure prediction characteristic: A general framework to assess and compare seizure prediction methods. *Epilepsy Behav* 4:318–25.

Winterhalder, M., B. Schelter et al. 2006. Spatio-temporal patient-individual assessment of synchronization changes for epileptic seizure prediction. *Clin Neurophysiol* 117:2399–413.

Wolf, A., J. B. Swift et al. 1985. Determining Lyapunov exponents from a time series. *Physica D* 16:285–317.

Wong, S., A. B. Gardner et al. 2007. A stochastic framework for evaluating seizure prediction algorithms using hidden Markov models. *J Neurophysiol* 97:2525–32.

Chapter 7

Application of Parametric and Nonparametric Methods in Arrhythmia Classification

---

Haseena H., K. Paul Joseph, and Abraham T. Mathew

Contents

7.1 Introduction

Recent census statistics compiled by the World Health Organization (WHO) reveal that the number of deaths due to sudden cardiac arrest (SCA) has overtaken mortality due to all cancers combined. Fifty-seven million Americans (Hudson and Cohen 1999) and six million Canadians (Thayer 2002) suffer from one or more forms of cardiovascular disease. Sudden death from cardiovascular disease accounts for more than 300,000 deaths per year in the United States and approximately 4,280 out of every 1 lakh from India every year. Heart disease is the leading cause of death in these two and many other countries. Monitoring heart rhythms to detect early problems and benign situations is an important task of cardiologists and related medical staff.

Automatic recognition and discrimination of electrocardiogram (ECG) rhythms play a vital role in the diagnosis and treatment of heart patients. The ECG signal provides key information about the electrical activity of the heart. Continuous monitoring of the signal helps to understand patients' conditions, and appropriate treatment enhances the quality of living for the patients. Since it is a nonstationary signal, the disorder in ECG may not be apparent at all times. This infrequent nature of ECG can create the need for observing the signal over several hours for the identification of disorders like arrhythmia. Naturally, there is some possibility for an analyst to miss (or misrepresent) vital information when visual examination is performed with the ECG chart. Reliable computational methods are required in conjunction with this voluminous data to produce precise results. Therefore, computer-based analysis and classification of rhythm can be useful for accurate diagnosis. Computerized interpretation is possible because of the availability of digitized ECG, to which the mathematical technique of signal processing can then be applied (Reddy 2005). Due to the large number of patients in intensive care units (ICUs) and the need for medical staff and doctors to continuously monitor them, several methods for automated arrhythmia detection have been developed in the past few decades to simplify the monitoring task (Willems and Lesaffre 1987; Coast et al. 1990; Lagerholm et al. 2000; Caswell et al. 1993; Thakor, Natarajan, and Tomaselli 1994; Afonoso and Tompkins 1995; Finelli 1996; Zhang et al. 1999; Chen 2000; Melo, Caloba, and Nadal 2000; Senhadji et al. 1995; Khadra, Al-Fahoum, and Al-Nashash 1997; Tian and Tompinks 1997).

In this chapter, the neural network approach is analyzed, and its performance in ECG classification is tested. Artificial neural networks (ANNs) play an important role in the detection and classification of abnormalities in ECG waveforms due to their inherent capability to perform pattern recognition (Owis et al. 2002; Silipo and Marchesi 1998; Hu et al. 1994; Niwas, Kumari, and Sadasivam 2005). Most research works are carried out with multilayered feedforward networks (MLFFNs) using a backpropagation algorithm (Haykin 1994). Although this method is simple from the implementation point of view, there is no guarantee of reaching the global minimum and the convergence speed is quite slow. The other problem with this method is that there is no specific rule to determine the exact structure of ANN. Features of ANN such as the number of layers, number of hidden neurons per layer, and selection of the learning rule that modifies the connection weights may vary. If the chosen structure does not correspond to what is necessary for the task, the network does not converge to a good result. To solve these problems, researchers are concentrating on other types of neural networks and hybrid structures. Hybrid structures are formed by combining ANN with other soft computing techniques, which improves the classification results (Al-Fahoum and Howitt 1999; Osowski and Linh 2001; Dokur and Olmez 2001; Engin and Demirag 2003; Ceylan and Ozbay 2007; Ozbay, Ceylan, and Karlik 2006). Radial basis function networks (RBFNs) have attracted a great deal of interest because their training and generalization characteristics are better than those of MLFFN. Among the different categories of RBFNs, probabilistic neural networks (PNNs) are particularly suitable for pattern classification tasks (Yu and Chen 2007).

The first step toward ECG pattern recognition is feature extraction. Features give the condensed representation of patterns. The extracted features are fed to neural network classifiers for training. Specific patterns are assigned to a particular class according to the characteristic features selected for the class. Then, features are used to represent signals that are not used for the training phase and are fed to the neural network classifiers for testing. Here, parametric modeling strategies, in combination with nonparametric methods, are optimal to classify different ECG beats. A comparative analysis of MLFFN and PNN for such classification has been carried out.

Modeling is necessary when the system is nonexistent and is in the design stage, or when it is not possible to experiment on the real system even if it exists as it may be too expensive or dangerous to shut down the system. Whenever a mathematical model of the system is not readily available, one can determine a model from the experimental data on the real system. This model should predict the behavior of the system within an acceptable tolerance. In most cases, the underlying process is assumed to be stochastic. The aim of model building is to reveal the underlying process from the data. Models estimated using statistical methods are capable of finding the regularities and dependencies that exist in the data. Statistical methods based on linear and nonlinear models have been effective in many applications (Weigend and Gershenfeld 1993). Among linear regression methods, the autoregressive (AR) and AR moving average (ARMA) models are the most popular (Box, Genkins, and Reinsel 1994). In practice, almost all measured processes are nonlinear to some extent and, hence, linear modeling methods are, in some cases, inadequate. Nonlinear methods became widely applicable with the growth of computer processing speed and data storage. Among the nonlinear methods, neural networks soon became very popular. Neural networks gained a lot of interest due to their ability to learn nonlinear dependencies from large volumes of data effectively (Bishop 1995).

In this study, simple AR model parameters along with spectral entropy (SE) of the signal have been used as features for discriminating various types of ECG beats. The AR modeling approach is used to classify physiological signals like ECG and electroencephalogram (EEG). The advantage of AR modeling is its simplicity, and it is suitable for real-time classification in the ICU or for ambulatory monitoring. The AR models are popular due to the linear form of the system, availability of simultaneous equations involving the unknown AR model parameters, and availability of an efficient algorithm for computing the solution. An AR model of order four is sufficient to model an ECG signal (Ge, Srinivasan, and Krishnan 2002). Fourth-order AR coefficients, together with SE, have been chosen as the parameters for classification. The concept of entropy stems from the irregularity of a stochastic signal, which can be represented by this measure.

7.2 Materials Used

In this study, 20 records were randomly selected from the Massachusetts Institute of Technology–Beth Israel Hospital (MIT-BIH) arrhythmia database (http://www.physionet.org/), which lists various types of arrhythmias. This database contains 48 half-hour excerpts of 2-channel ambulatory ECG recordings, as well as a set of their corresponding beat and rhythm annotations. The waveforms are sampled at 360 Hz with an 11-bit resolution over a 10-mV range. For consistency, all the signals recorded with Mason–Likar (ML) II leads system were employed for the analysis. For the present study, normal data as well as seven types of arrhythmia data were chosen. Out of the three basic waves, P, QRS, and T, in ECG, most of the information from a particular cardiac cycle belongs to QRS complex. Thus in this work, data around the R peak has been selected.

Arrhythmia (also called dysrhythmia) is a disorder in which regular rhythmic beating of the heart is affected. The heartbeat may be too fast or too slow, and may be regular or irregular. In this study, the cardiac classes that include both normal and other cardiac abnormalities are classified into eight categories:

1. Normal sinus rhythm (NSR)
2. Left bundle branch block (LBBB)
3. Right bundle branch block (RBBB)
4. Premature ventricular contraction (PVC)
5. Atrial premature beat (APB)
6. Paced beat (PB)
7. Ventricular flutter beat (VFB)
8. Ventricular escape beat (VEB)

This section gives a brief description of the different cardiac classes:

NSR: It occurs due to the continuous, periodic performance of the pacemaker and the integrity of neuronal conducting pathways. In the normal rhythm, the PR interval should not exceed 0.20 seconds and the QRS duration never exceeds 0.12 seconds. The P wave has a maximum duration of 0.08 seconds, and the T wave should have at least 0.20 seconds. For a healthy person, the heartbeat ranges from 60 to 100 bpm (beats per minute), so the R–R interval should be from 0.6 to 1 second (Clifford, Azuaje, and McSharry 2006).

LBB: In this condition, activation of the left ventricle is delayed, which results in the left ventricle contracting later than the right ventricle, and the condition results in a QRS interval greater than 0.12 second.

RBB: In this state, the two ventricles no longer receive the electrical impulse simultaneously. The duration of the QRS complex on the ECG is between 0.10 and 0.11 second (for incomplete RBBB) or 0.12 second or more (for complete RBBB), and has a prolonged ventricular activation time or QR interval of 0.3 second or more.

PVC: The heartbeat occurs earlier than expected and disrupts the regular rhythm of the heart. Irregularity of the rhythm, obscuring of P waves by the QRS complex or the T wave, widening of the QRS complex, and the opposing polarity of the T wave with respect to the R wave are important characteristic features of PVC.

APB: The APB is an electrical impulse that originates in the atria outside the sinus node. It is premature, occurring before the next expected sinus beat. Since it arises in the atrium (hence, it is close to the sinus node), it usually causes depolarization of the sinus node. Absence of the QRS complex followed by P wave is the main aspect of APB.

PB: An area in the excitable ventricular musculature tries to control the heartbeat conduction and slows the heart rate into the range 30–50 bpm. This pattern shows an abnormal, wide, and bizarre QRS complex.

VFB: The VFB is similar to ventricular tachycardia; but in VFB, the heart rate is faster. Ventricular flutter is more dangerous because there is virtually no cardiac output. A ventricular rate of 220–400 bpm, the absence of P waves, and occurrence of wide and regular sine wave–type QRS complex are the common aspects of VFB.

VEB: Ventricular escape beats occur when the rate of electrical discharge reaching the ventricles falls below the base rate determined by the ventricular pacemaker cells. This condition results in a heart rate of less than 100 bpm. In addition, it gives a wide and bizarre QRS complex and results in the absence of P waves.

The origin of the ECG beats with annotations and number of samples used for the training and testing of ANN is summarized in Table 7.1.

The portion of data where corresponding arrhythmia are dominant was identified from the annotations provided by the MIT–BIH arrhythmia database. The R peaks of

Table 7.1   Number of Samples Used for the Training and Testing of ANN

Sl. No.	Beat Type/ Annotation	MIT–BIH File No.	Training (Beats/File)	Testing (Beats/File)
1	N/N	100, 101, 103, 115, 219, 234	150	50
2	LBBB/L	109, 111, 214	150	50
3	RBBB/R	118, 124, 212, 231	150	50
4	PVC/V	208, 223	200	50
5	APB/A	209, 232	150	50
6	PB//	102, 107	150	50
7	VFB/!	207	150	50
8	VEB/E	207	50	30

the ECGs were detected using Tompkins' algorithm (Tompkins 1993). In this study, a sample size of 432 (1.2 seconds), consisting of 144 samples (0.4 seconds) before R peak and 288 samples (0.8 seconds) after R peak, was chosen because it is able to capture most of the information from a particular cardiac cycle.

7.3 Methods

The basic methodology of the proposed work depended on the analysis of digital ECG data. For analysis, mainly four steps namely (1) signal preprocessing (2) feature extraction (3) learning of neural networks, and (4) classification by neural networks have to be carried out.

7.3.1 Preprocessing

The ECG signal is normally contaminated with noises generated by biological and environmental resources. Different types of predominant noise that commonly contaminate the signal are baseline wander noise, waveform of an electrical activity of muscles (known as electromyographic interference), motion artifacts from electrodes, skin interference, and 50- or 60-Hz powerline interference. In order to remove these unwanted signals, some preprocessing activity has to be performed. The signals are filtered using a finite-impulse-response (FIR) band-pass filter and detrended to cancel low-frequency components. The main advantage of using an FIR filter is that it minimizes waveform distortion. To cancel the effects of both low- and high-frequency noise components, we considered a band-pass filter in the range of 1–100 Hz. The processing blocks used in the training and classification of ECG waveforms are shown in Figure 7.1.

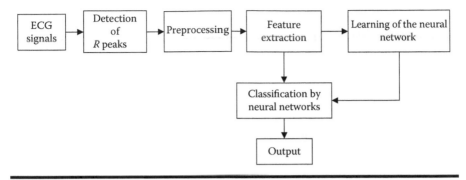

Figure 7.1 Processing blocks used in the classification.

7.3.2 Feature Extraction

Usually, features represent patterns with minimal loss of important information. The important features extracted in this chapter are the fourth order AR model coefficients and SE of ECG signals under analysis.

7.3.2.1 Autoregressive Modeling

The AR model was selected as the baseline global linear model for prediction, system identification, and control, since the 1960s. The properties of linear models are well-known, and efficient algorithms exist for parameter estimation (Box, Genkins, and Reinsel 1994).

An AR model of order p can be expressed as follows:

$$y(t) = a_1 y(t-1) + a_2 y(t-2) + \cdots + a_p y(t-p) + w(t) \tag{7.1}$$

where $y(t)$ is the current value of y, $w(t)$ is the zero-mean white-noise term, and $a_1, a_2, \ldots, a_p$ are the AR coefficients. The term "autoregression" signifies that the variable y is regressed on a previous value of itself. A number of techniques are available for the computation of AR coefficients. The two main categories are least-squares method and Burg's method. In the present study, the least-squares method has been used to calculate the AR coefficients. An important step in AR modeling is the selection of a parsimonious model—a model possessing maximum simplicity and a minimum number of parameter constants with representational adequacy. In the present study, the criteria used for determining the model order are Akaike's final prediction error (FPE) and normalized mean-square error (NMSE). The FPE is calculated as follows:

$$\text{FPE} = \frac{1 + \dfrac{d}{N}}{1 - \dfrac{d}{N}} \times V \tag{7.2}$$

where d is the total number of estimated parameters, N is the length of the data record, and V is the loss function (quadratic function) for the selected structure. According to Akaike's theory, we choose the model with the smallest FPE. This approach emphasizes the principle of parsimony in model building, since if two models are equally likely the one with fewer parameters is selected (Sinha and Kuszta 1983).

The NMSE is an estimate of the overall deviations between modeled and actual values. It is obtained by dividing the MSE between actual and modeled data with the mean square of actual data, as in

$$\text{NMSE} = \frac{\frac{1}{N}\sum_{i=1}^{N}(y(i) - \hat{y}(i))^2}{\frac{1}{N}\sum_{i=1}^{N}(y(i))^2} \tag{7.3}$$

In NMSE, the squares of the deviations are summed. For this reason, NMSE generally shows the most striking differences among models. If a model has a very low NMSE, then it performs well both in space and time.

7.3.2.2 Spectral Entropy

Entropy is a thermodynamic quantity that describes the amount of disorder in a system. It originates from a measure of information called "Shanon entropy." The application of Shannon's channel entropy gives an estimate of the SE of a process, where entropy is given by

$$H = \sum_f P_f \log\left(\frac{1}{P_f}\right) \tag{7.4}$$

where P_f is the probability density function (PDF) value at frequency f.

Entropy is interpreted as a measure of system complexity. Spectral entropy H describes the complexity of the heart rate variability signal (Anuradha and Reddy 2008).

7.3.3 Test for Validation

After a model is obtained, it is always necessary to test the goodness of fit in modeling. To check the suitability of the model, a validation test has to be performed. The adequacy of the model can be confirmed by the whiteness test. To conduct the test, the residual corresponding to the parsimonious model has to be calculated. The sequence can be considered as white within 95% confidence limits if $|\rho_i| \le \dfrac{1.98}{\sqrt{N}}$ for all $i \geq 1$, where ρ is the correlation coefficient of the sequence of residuals and N is number of data. In practice, it is sufficient to test this inequality for $i = 1$ to 20 (Sinha and Kuszta 1983).

7.3.4 Neural Network Classifiers

Neural networks are chemical and hormonal systems in our brain. An ANN model distinguishes itself from conventional computers in that it is a parallel, distributed computing network that mimics the information processing structure of a biological neural system (Silipo and Marchesi 1998). A neural network is characterized by nodes connected by links. These links have connection strengths, which can be excitory (positive weights) or inhibitory (negative weights). A well-suited structure for classification problems is the feedforward connectionist network: It does not contain feedback connections. In this structure, it is possible to distinguish an input layer, some hidden layers, and an output layer. The state s_i (or activation level) of the ith node is given by the inner product of the inputs x_j applied to it (x_j is the output of the jth node), and the strengths or weights of the links w_{ij} between node i and node j are as follows:

$$s_i = \sum_{ij} w_{ij} x_j \qquad (7.5)$$

Such a state is processed by an activation function F, which produces the output signal of the node:

$$y_i = F(s_i) \qquad (7.6)$$

The most used nonlinear function F is the so-called sigmoid function:

$$F(s) = \frac{1}{\left(1 + e^{-\frac{s}{k}}\right)} \qquad (7.7)$$

where k is the inverse of the sigmoidal slope. In the network, the weights are initially set in a random way, and the training phase consists of adjusting such weights according to a predetermined procedure by sequentially applying input vectors. Two different methods can be adopted: (1) supervised training, where the known classification is used as target output; and (2) unsupervised training, which does not require a predetermined classification that produces a set of clusters. In the present study, two types of neural networks, that is, MLFFN with BPA and PNN, are used for the classification of eight types of ECG beats. In BPA, weight adjustment in the training phase consists of the following procedure (applied to the entire learning set):

An input vector is presented, and the output is computed.
The difference (error) between the output of the network and the weights are modified in order to minimize such error.

Table 7.2 Criteria for the Selection of Hidden Neurons

Number of Neurons in the Hidden Layer	MSE in Prediction
10	$3.2034\ e^{-004}$
20	$9.8389\ e^{-005}$
30	$1.2967\ e^{-005}$
40	$2.2967\ e^{-006}$

This procedure is repeated until the error in the entire set is acceptably low. To achieve this, the algorithm passes information from the output neurons backward to all hidden neurons to form error terms to update the weights of the multilayered network. The most important factor in multilayer network structure selection is the choice of the number of hidden neurons. A very low number of hidden neurons does not allow the reduction of the error function to a satisfactory value. A very high number of hidden neurons destroys the generalization ability of the network, and such numbers should also be avoided.

In the study, the *logsig* (logarithmic sigmoid) function is used as the activation function. The selection of hidden neurons is done by trial and error. The hidden neurons are chosen based on the MSE between target output and neural network output and are shown in Table 7.2. The number of hidden neurons corresponding to the minimum MSE in prediction is selected. Accordingly, the number of hidden neurons selected in the study was 40.

The performance goal selected was 1e-005, the number of neurons in the output layer was 8, to divide the output domain into 8 classes. A momentum term is used to achieve faster convergence with minimum oscillations in updating the network's weight. After some trials, the momentum factor was chosen for this study as 0.95, and the learning rate parameter selected was 0.9. Throughout the study, "trainrp" (resilient propagation) was used as the training function. It is the fastest algorithm for pattern recognition problems (Mathworks 1998). Figure 7.2 shows the architecture of MLFFN used for classification. It consists of 3 layers, input, hidden, and output layers, with the number of neurons as 5, 40, and 8, respectively. The fourth-order AR coefficients along with SE are given as the input vectors, and the eight types of abnormalities are taken as the outputs of the network.

The target vector selected for the training of MLFFN is shown in Table 7.3.

Since MLFFN with BPA applies the steepest descent method to update the weights, it suffers from a slow convergence rate and often yields suboptimal solutions (Brent 1991; Gori and Testi 1992). A variety of related algorithms has been introduced to address this problem. It was determined that RBFN is a better method, due to its higher speed in training and better classification performance compared to MLFFN. Among the various paradigms of RBFNs, PNNs are

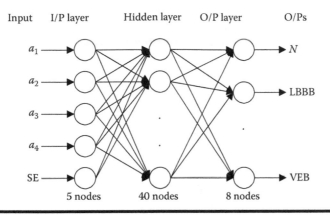

Figure 7.2 Architecture of multilayered feedforward network.

Table 7.3 Target Vector Selected for Neural Networks

Class	Type	Target Vector
1	N	0 0 0 0 0 0 0 1
2	LBBB	0 0 0 0 0 0 1 0
3	RBBB	0 0 0 0 0 1 0 0
4	PVC	0 0 0 0 1 0 0 0
5	APB	0 0 0 1 0 0 0 0
6	PB	0 0 1 0 0 0 0 0
7	VFB	0 1 0 0 0 0 0 0
8	VEB	1 0 0 0 0 0 0 0

particularly suitable for pattern classification tasks. In this study, recognition of the beats was then carried out by PNN.

7.3.5 Probabilistic Neural Network

The PNN developed by Donald Specht (Ubeyli 2008) is a network formulation of "probability density estimation." In PNN, the number of neurons in the hidden layer is usually the number of patterns in the training set because each pattern in the training set is represented by one neuron (Kanmani et al. 2007). It is a model based on competitive learning with a "winner takes all" attitude, and the core concept is based on multivariate probability. The PNN provides a general solution to

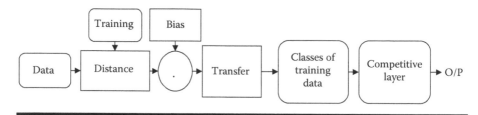

Figure 7.3 Schematic diagram of one layer of the probabilistic neural network.

pattern classification problems by following an approach developed in statistics, called "Bayesian classifiers." The PNN uses a supervised training set to develop distribution functions within a pattern layer. It is simpler to train PNN than multilayer neural network. However, the pattern layer can be huge if the distinction between categories is varied and at the same time similar in special areas (Yusubov, Gulbag, and Temurtas 2007).

A schematic diagram of one layer of PNN is shown in Figure 7.3. The new data is presented to the network. The distance between each data is computed, multiplied by a bias, and passed to the transfer function. Then all the amounts from the same class are added together and the class with the highest value is the winner.

For classification, consider objects drawn from K classes labeled as Θ_i, $i = \{1, 2, \ldots, K\}$. An object with p features is represented as a p vector $X = [x_1, x_2, \ldots,]^T$. The PNN adopts a Bayes decision rule, that is, classify an observation x into class $i \in \{1, 2, \ldots, K\}$ if

$$c_i m_i f_i(x) \rangle c_j m_j f_j(x), \ \forall j \neq i, \ j \in \{1, 2, \ldots, K\} \tag{7.8}$$

where c_k, m_k, and $f_k()$ are the cost of misclassifying a class K case, the prior probability of occurrence of class K, and the PDF of class K, respectively (Tsai 2000).

In this study, PNN with a two-layer structure has been used for the classification of ECG beats. It consists of a radial basis layer and a competitive layer. Each neuron in the radial basis layer calculates the probability that an input feature vector is associated with a specific class. The radial basis neuron compares the input vector p with the neuronal weight w and multiplies with a bias b to calculate the probability Y as given in Equation 7.9:

$$Y = \exp[-(\|w - p\|b)^2] \tag{7.9}$$

where $\|w - p\|$ denotes the Euclidean distance. Consequently, as the distance between k and p decreases the output increases and reaches the maximum value of unity while $w = p$. The bias b is set as follows (Ubeyli 2008):

$$b = \sqrt{-\ln(0.5)}/\text{spread} \tag{7.10}$$

such that the probability value crosses 0.5 when $w - p =$ spread. The sensitivity of radial basis neurons can be adjusted by varying the value of b through the coefficient spread. The competitive layer then assigns values unity to the class that demonstrates the maximum probability and null to the other classes. In this study, the value of spread factor chosen after experimentation was 0.1. It is reported in the literature (Yu and Chen 2007) that when spread factor increases, classification accuracy decreases. Variation of classification accuracy with different values of the smoothing factor is shown in Figure 7.4. The number of neurons in the competitive layer was seven, same as that of the desired classes.

Statistical parameters such as classification accuracy, sensitivity, and specificity were computed for all the classes to evaluate the performance of classifiers. Classification accuracy is defined as the ratio of the number of correctly classified beats to the total number of beats. True positive (TP) represents the pathological state ECG beat being classified as a pathological state, true negative (TN) represents a normal beat being classified as normal, false positive (FP) represents a normal beat being misclassified as a pathological state, and false negative (FN) represents a pathological beat being misclassified as normal. Sensitivity and specificity are defined as follows:

$$\text{Sensitivity} = 100 \times \frac{TP}{TP + FN} \tag{7.11}$$

$$\text{Specificity} = 100 \times \frac{TN}{TN + FP} \tag{7.12}$$

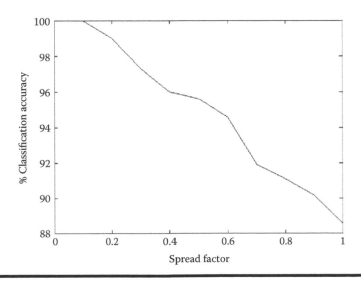

Figure 7.4 Variation of smoothing factor with percentage classification accuracy.

7.4 Results

The investigations in the methods that are addressed in this chapter show that proper mix of stochastic and soft computing techniques would render more reliable results in arrhythmia detection. The following sections reveal the fact more clearly.

7.4.1 Autoregressive Modeling Results

The parsimonious model order was determined by evaluating the FPE and NMSE of eight types of data under study. The NMSE, FPE, and AR coefficients were compared for different model orders for NSR data. The comparison is tabulated in Table 7.4.

From Table 7.4, it is confirmed that NMSE is reduced to a lower value after a model order of one, and following fourth order it remains almost constant, which shows that an AR model of order four is sufficient to model the ECG signals (Ge, Srinivasan, and Krishnan 2002). Comparison of AR coefficients after the third order gives a consistent value. The FPE value remains almost constant after the fourth order. It reemphasizes the fact that AR4 is enough to model the system. The same results were obtained after analyzing the other seven types of arrhythmia data under study. Thus, for the classification, fourth-order AR coefficients of different arrhythmia data were selected as one of the feature vectors. The original and simulated ECG signals with NSR, LBBB, RBBB, PVC, atrial premature (AP), PB, ventricular flutter (VF), and ventricular escape (VE) conditions are shown in Figures 7.5 through 7.12.

Table 7.4   Performance Measures for Parsimonious Model Order Selection

Model Order, NMSE, and FPE			AR Coefficients						
Model Order	NMSE	FPE	1	2	3	4	5	6	7
1	0.0648	0.023650	−0.920						
2	0.0112	0.000321	−1.652	0.898					
3	0.0119	0.000181	−2.243	2.108	−0.764				
4	0.0180	0.000101	−2.361	2.324	−1.541	−0.211			
5	0.0182	0.000102	−2.513	2.324	−1.552	−0.312	0.011		
6	0.0181	0.000100	−2.513	2.301	−1.116	−0.324	0.021	−0.227	
7	0.0182	0.000100	−2.501	2.351	−1.118	−0.192	0.765	−0.341	0.095

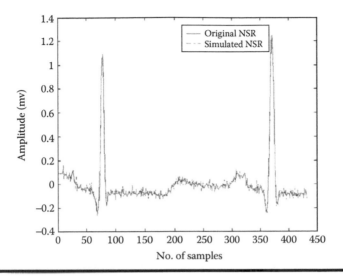

Figure 7.5   Original and simulated electrocardiogram with normal sinus rhythm.

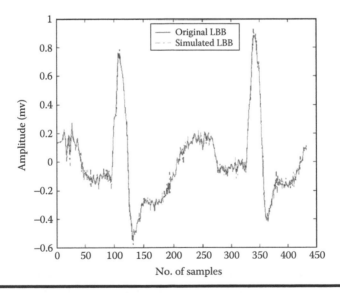

Figure 7.6   Original and simulated electrocardiogram with left bundle block condition.

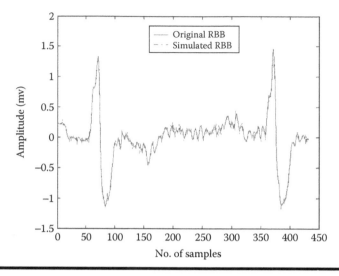

Figure 7.7 **Original and simulated electrocardiogram with right bundle block condition.**

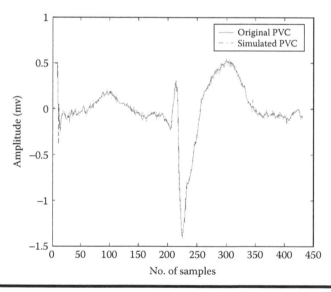

Figure 7.8 **Original and simulated electrocardiogram with premature ventricular contraction.**

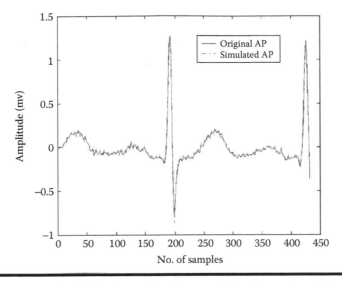

Figure 7.9 **Original and simulated electrocardiogram with atrial premature condition.**

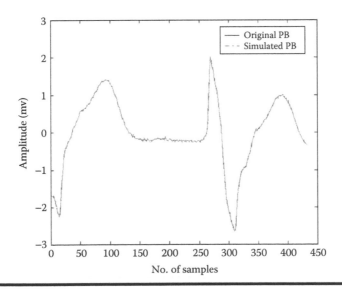

Figure 7.10 **Original and simulated electrocardiogram with paced beat.**

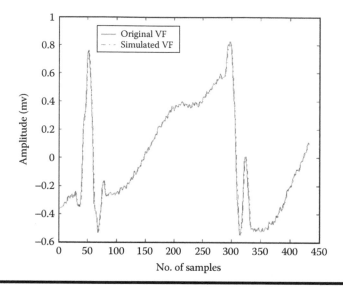

Figure 7.11 **Original and simulated electrocardiogram with ventricular flutter condition.**

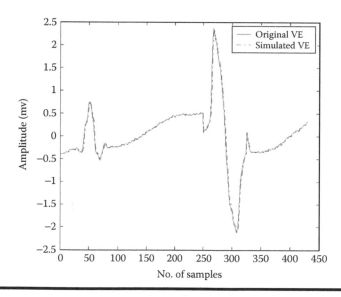

Figure 7.12 **Original and simulated electrocardiogram with ventricular escape condition.**

7.4.2 Validation Results

In the present study, it is found that fourth-order AR model results in a parsimonious model and $\dfrac{1.98}{\sqrt{432}} = 0.095263$. The fourth-order model is greater than the correlation coefficient of the sequence of residuals for lags up to 20. The correlation coefficients of the sequence of residuals for lags up to 10 for fourth-order NSR data are as follows:

$$\rho_1 = 0.0120, \quad \rho_2 = -0.0216, \quad \rho_3 = -0.0038, \quad \rho_4 = 0.0021,$$
$$\rho_5 = -0.0143, \quad \rho_6 = 0.0198, \quad \rho_7 = -0.0387, \quad \rho_8 = 0.0612,$$
$$\rho_9 = -0.0711, \quad \rho_{10} = -0.0651$$

The average values of all the selected parameters of the training data samples are tabulated in Table 7.5.

The selected parameters are then fed to the neural network classifiers for discriminating different ECG beats, and the results given by PNN are tabulated in Table 7.6.

From Table 7.6, it is calculated that the overall classification accuracy using PNN is 98.64%. The same procedure was applied for MLFFN and the accuracy was found to be 97.14%. A comparison between the two types of classifiers in terms of the performance index is shown in Table 7.7.

Table 7.5 Mean Feature Values of Selected Data for Training

ECG Beat Type	a1	a2	a3	a4	SE
N	−2.128	1.128	−0.129	−0.177	1.721
LBBB	−4.342	3.201	−2.108	0.867	1.189
RBBB	−3.114	3.428	−2.932	0.763	1.001
PVC	−2.389	2.191	−1.912	0.512	1.131
APB	−1.719	0.671	0.321	−0.312	1.414
PB	−2.632	2.531	−1.821	0.616	0.786
VFB	−1.781	0.491	0.521	−0.191	1.160
VEB	−1.496	0.392	0.167	−0.278	1.497

Table 7.6 Classification Result of Selected MIT–BIH Files by PNN

File	Sample No.	N	LBBB	RBBB	PVC	APB	PB	VFB	VEB
100	47919–61963	50/50	–	–	–	–	–	–	–
101	8414–23126	50/50	–	–	–	–	–	–	–
102	7857–22434	–	–	–	–	–	50/50	–	–
103	8417–23327	50/50	–	–	–	–	–	–	–
107	7946–22956	–	–	–	–	–	48/50	–	–
109	48044–63029	–	50/50	–	–	–	–	–	–
111	8195–23567	–	48/50	–	–	–	–	–	–
115	9164–26012	50/50	–	–	–	–	–	–	–
118	11687–26375	–	–	49/50	–	–	–	–	–
124	14528–36127	–	–	50/50	–	–	–	–	–
207	554826–560330	–	–	–	–	–	–	50/50	–
207	591405–603708	–	–	–	–	–	–	–	29/30
208	630057–641037	25/26	–	–	23/24	–	–	–	–
209	200210–206995	–	–	–	–	49/50	–	–	–

212	8068–19730	—	—	50/50	—	—	—	—	—
214	3694–17810	—	47/50	—	—	—	—	—	—
219	39235–53604	50/50	—	—	—	—	—	—	—
223	375669–385545	—	—	—	50/50	—	—	—	—
231	11224–27874	—	—	50/50	50/50	—	—	—	—
232	46639–64574	—	—	12/13	—	35/37	—	—	—
234	6271–17755	50/50	—	—	—	—	—	—	—

Note: "–" indicates the absence of that type of data. The numerators of fractions indicate the number of correctly identified beats, and the denominators indicate the total number of beats present.

Table 7.7 Performance Index Given by MLFFN and PNN

Performance Index	Type of Beat	By MLFFN	By PNN
Overall sensitivity (%)		95.10	96.78
Classification accuracy (%)	N	97.61	99.69
	LBBB	95.19	96.66
	RBBB	97.79	99.06
	PVC	96.41	98.65
	APB	94.53	96.55
	PB	96.86	98.00
	VFB	98.42	100.00
	VEB	95.49	96.66
Overall specificity (%)		96.56	98.21

7.5 Discussion

The objective of the present study is to model the ECG signals for extracting classifiable features using the AR model. The SE of the signal is also chosen as another parameter for the classification of ECG beats into normal and seven selected abnormal conditions. The results show that an AR order of four was sufficient to model the ECG signals to classify the selected abnormalities, and the normal system has a higher magnitude of SE than an abnormal system. A succinct evaluation of results of the proposed approach with those of the other methods presented in the literature is made (Finelli 1996; Zhang et al. 1999; Chen 2000; Melo, Caloba, and Nadal 2000; Yu and Chou 2008; Hu, Palreddy, and Tomkins 1997; Minami, Nakajima, and Toyoshima 1999; Prasad and Sahambi 2003).

Finelli (1996) proposes a time-sequenced adaptive filter to distinguish VT and VF alone. The complexity measure–based technique has been used to classify NSR, VT, and VF (Zhang et al. 1999). The total least squares–based Prony modeling technique produced an accuracy of 95.24%, 96%, and 97.78% for SVT, VT, and VF, respectively (Chen 2000). But the modeling technique does not include episodes from normal, atrial premature complex (APC), or PVC. Melo, Caloba, and Nadal (2000) achieved an overall accuracy of 93%–99% for the analysis of different types of arrhythmia data. The researchers proposed a QRS-feature-based algorithm for decimated ECG data using ANNs and including various types of beats such as APC and PVC. However, they did not include life-threatening conditions like VT and VF.

Yu and Chou (2008) took independent component (IC) analysis-based features, along with an RR interval, as the feature vectors for classifying 8 types of ECG beats with a classification accuracy of 98% by means of MLFFN and PNN classifiers. The discrimination power of PNN increased rapidly at smaller numbers of ICs. Hu, Palreddy, and Tomkins (1997) classified 4 types of ECG beats with an overall accuracy of 94%. In this method, the patient-specific classifier called "local expert" (LE) was trained specifically with the ECG record of the patient. The potential drawback of the method is that the development of an LE classifier for each patient is very costly.

A real-time discrimination algorithm with a Fourier-transform neural network (Minami, Nakajima, and Toyoshima 1999) has been proposed to distinguish between supraventricular and ventricular rhythms in cases where PVC and VT have been lumped together as belonging to a single class of ventricular rhythms. The systems obtained an accuracy of 98%. Each extracted QRS complex was converted to a Fourier spectrum. When the waveform of the QRS complex is directly applied to a neural network, the results depend on variations in the waveforms used during training. This is the limitation of this method when applied in arrhythmia detection. Prasad and Sahambi (2003) used combined wavelet network and ANN to classify normal and 12 different arrhythmia conditions with 96.77% accuracy. The ECG segments selected for analysis were 100 milliseconds of signal before and 150 milliseconds after the R point.

In the present study, a fixed sample size is used for AR modeling. The generalization capabilities of the AR model and the classification algorithms can be refined by applying the proposed approach to a larger data set. In addition to their utility in classification and diagnosis, AR coefficients can also be used for compression. It is found that AR modeling can lead to a low-cost, high-performance, simple-to-use, and portable telemedicine system for ECG, which offers a combination of diagnostic and compression capabilities.

In the proposed method, extracted features help to accommodate more data in the classifier instead of giving raw ECG data, and the approach improves accuracy. If raw data is directly applied, the results depend on variations in the waveforms; such problems are avoided in the study. In the present study, PNN outperforms the other methods with an impressive accuracy of 98.64% in discriminating 8 types of ECG beats, and MLFFN has a classification accuracy of 97.14%. Thus, it can be confirmed that the proposed scheme with PNN is an efficient and improved tool in the classification of clinical ECG beat types.

7.6 Conclusion

This chapter presents a performance study of MLFFNs and PNNs in the area of ECG arrhythmia discrimination. SE together with fourth-order AR parameters serve as input feature vectors for neural network classifiers. Among neural network classifiers, PNN performs better in terms of classification accuracy, overall sensitivity, and specificity. Comparison of the proposed method with the existing strategies reveals that PNN, with the extracted features, outperforms the other methods. The added advantage of PNN is that its classification speed is much higher than that of MLFFN.

References

http://www.physionet.org/ accessed June 13, 2008.

Afonoso, V. X., and W. J. Tompkins. 1995. Detecting ventricular fibrillation: Selecting the appropriate time-frequency analysis tool for the application. *IEEE Eng Med Biol* 14:152–9.

Al-Fahoum, A. S., and I. Howitt. 1999. Combined wavelet transformation and radial basis neural networks for classifying life-threatening cardiac arrhythmias. *Med Biol Eng Comput* 37:566–73.

Anuradha, V. C., and B. V. Reddy. 2008. ANN for classification of cardiac arrhythmias. *ARPN J Eng Appl Sci* 3:1–6.

Bishop, C. M. 1995. *Neural Networks for Pattern Recognition*. Oxford, UK: Oxford University Press.

Box, G., G. Genkins, and G. Reinsel. 1994. *Time Series Analysis: Forecasting and Control*. Englewood Cliffs, NJ: Prentice Hall.

Brent, R. P. 1991. Fast training algorithms for multi-layer neural nets. *IEEE Trans Neural Netw* 2:346–54.

Caswell, S. A., K. S. Kluge, C. M. J. Chiang, and J. M. Jenkins. 1993. Pattern recognition of cardiac arrhythmias using two intracardiac channels. In *Proceedings of Computers in Cardiology*, 181–4. London, UK: IEEE.

Ceylan, R., and Y. K. Ozbay. 2007. Comparison of FCM, PCA and WT techniques for classification of ECG arrhythmias using artificial neural network. *Expert Syst Appl* 33:286–95.

Chen, S. W. 2000. A two stage discrimination of cardiac arrhythmia using a total least square-based prony modelling algorithm. *IEEE Trans Biomed Eng* 47:1317–27.

Clifford, G. D., F. Azuaje, and P. E. McSharry. 2006. *Advanced Methods and Tools for ECG Data Analysis*. London: Artech House.

Coast, A. D., R. M. Stern, G. G. Cano, and S. A. Briller. 1990. An approach to cardiac arrhythmia analysis using hidden Markov models. *IEEE Trans Biomed Eng* 37:826–35.

Dokur, Z., and T. Olmez. 2001. ECG beat classification by a hybrid neural network. *Comput Methods Programs Biomed* 66:167–81.

Engin, M., and S. Demirag. 2003. Fuzzy-hybrid neural network based ECG beat recognition using three different types of feature sets. *Cardiovasc Eng Int J* 3:71–80.

Finelli, C. J. 1996. The time sequenced adaptive filter for analysis of cardiac arrhythmias in intraventricular electrocardiograms. *IEEE Trans Biomed Eng* 43:811–9.

Ge, D., N. Srinivasan, and S. M. Krishnan. 2002. Cardiac arrhythmia classification using autoregressive modelling. *Biomed Eng Online* 1:5 doi:10.1186/1475-925X-1-5.

Gori, M., and A. Tesi. 1992. On the problem of local minima in back propagation. *IEEE Trans Pattern Anal Mach Intell* 14:76–85.

Haykin, S. 1994. *Neural Networks: A Comprehensive Foundation*. New York: Macmillan College Publishing Company.

Hu, Y. H., S. Palreddy, and W. J. Tomkins. 1997. A patient adaptable ECG beat classifier using a mixture of experts approach. *IEEE Trans Biomed Eng* 44:891–900.

Hu, Y. H., W. J. Tompkins, J. L. Urrusti, and V. X. Alfonso. 1994. Applications of artificial neural networks for ECG signal detection and classification. *J Electrocardiol* 28:66–73.

Hudson, D. L., and M. E. Cohen. 1999. *Neural Networks and Artificial Intelligence for Biomedical Engineering*. New York: IEEE Press.

Kanmani, S., V. R. Uthariaraj, V. Sankaranarayanan, and P. Thambidurai. 2007. Object-oriented software fault prediction using neural networks. *Inform Softw Tech* 49:483–92.

Khadra, L., A. S. Al-Fahoum, and H. Al-Nashash. 1997. Detection of life threatening cardiac arrhythmias using the wavelet transformation. *Med Biol Eng Comput* 35:626–32.

Lagerholm, M., C. Peterson, G. Braccini, L. Edenbrandt, and L. Sornmo. 2000. Clustering ECG complexes using hermite functions and self-organizing maps. *IEEE Trans Biomed Eng* 47:838–48.

Mathworks, I. 1998. *Neural Networks Toolbox User's Guide*. version. 3 (Release 11). Natick, MA: MathWorks.

Melo, S. L., L. P. Caloba, and J. Nadal. 2000. Arrhythmia analysis using artificial neural networks and decimated electrocardiographic data. *Comput Cardiovasc* 27:73–6.

Minami, K., H. Nakajima, and T. Toyoshima. 1999. Real-time discrimination of ventricular tachyarrhythmia with Fourier-transform neural network. *IEEE Trans Biomed Eng* 46:179–85.

Niwas, S. I., R. S. S. Kumari, and V. Sadasivam. 2005. Artificial neural network based automatic cardiac abnormalities classification. In *Proceedings of Sixth International Conference on Computational Intelligence and Multimedia Applications (ICCIMA)*, 41–6.

Osowski, S., and T. H. Linh. 2001. ECG beat recognition using fuzzy hybrid neural network. *IEEE Trans Biomed Eng* 48:1265–71.

Owis, M. I., A. Abou-Zied, A. B. Youssef, and Y. M. Kadah. 2002. Study of features based on nonlinear dynamic modelling in ECG arrhythmia detection and classification. *IEEE Trans Biomed Eng* 49:733–6.

Ozbay, Y., R. Ceylan, and B. Karlik. 2006. A fuzzy clustering neural network architecture for classification of ECG arrhythmia. *Comput Biol Med* 36:376–88.

Prasad, G. K., and J. S. Sahambi. 2003. Classification of ECG arrhythmias using multiresolution analysis and neural networks. In *Proceeding of the IEEE Conference on Convergent Technologies (Tecon2003)*, vol. 1, 227–231. Bangalore, India: Pergamon Publishers.

Reddy, D. C. 2005. *Biomedical Signal Processing Principles and Techniques*. New Delhi: Tata McGraw-Hill.

Senhadji, L., G. Carrault, J. J. Bellanger, and G. Passariello. 1995. Comparing wavelet transforms for recognizing cardiac patterns. *IEEE Eng Med Biol* 14:167–73.

Silipo, R., and C. Marchesi. 1998. Artificial neural networks for automatic ECG analysis. *IEEE Trans Signal Process* 46:1417–25.

Sinha, N. K., and B. Kuszta. 1983. *Modelling and Identification of Dynamic Systems*. New York: Van Nostrand Reinhold Company.

Thakor, N. V., A. Natarajan, and G. Tomaselli. 1994. Multi way sequential hypothesis testing for tachyarrhythmia discrimination. *IEEE Trans Biomed Eng* 41:480–7.

Thayer, J. F. 2002. Progress in the analysis of heart rate variability. *IEEE Trans Eng Med Biol* 21:22–3.

Tian, L., and W. J. Tompinks. 1997. Time domain based algorithm for detection of ventricular fibrillation. In *Proceeding of EMBS 19th Annual International Conference*, vol. 3, 374–7. Chicago.

Tompkins, W. J. 1993. *Biomedical Digital Signal processing.* 253–61. Englewood Cliff, NJ: Prentice Hall.

Tsai, C. Y. 2000. An iterative feature reduction algorithm for probabilistic neural networks. *Omega* 28:513–24.

Ubeyli, E. D. 2008. Probabilistic neural networks employing Lyapunov exponents for analysis of Doppler ultrasound signals. *Comput Biol Med* 38:82–9.

Weigend, A., and N. Gershenfeld. 1993. *Time Series Prediction: Forecasting the Future and Understanding the Past,* 1–70. Redwood City, CA: Addison-Wesley.

Willems, J. L., and E. Lesaffre. 1987. Comparison of multigroup logistic and linear discriminant ECG and VCG classification. *J Electrocardiol* 20:83–92.

Yu, S. N., and Y. H. Chen. 2007. Electrocardiogram beat classification based on wavelet transformation and probabilistic neural network. *Pattern Recogn Lett* 28:1142–50.

Yu, S. N., and K. T. Chou. 2008. Integration of independent component analysis and neural networks for ECG beat classification. *Expert syst Appl* 34:2841–6.

Yusubov, I., A. Gulbag, and F. Temurtas. 2007. A study on mixture classification using neural network. *Electron Lett Sci Eng* 3:33–8.

Zhang, X. S., Y. S. Zhu, N. V. Thakor, and Z. Z. Wang. 1999. Detecting ventricular tachycardia and fibrillation by complexity measure. *IEEE Trans Biomed Eng* 46:548–55.

Chapter 8

Supervised and Unsupervised Metabonomic Techniques in Clinical Diagnosis: Classification of 677-*MTHFR* Mutations in Migraine Sufferers

Filippo Molinari, Pierangela Giustetto, William Liboni, Maria Cristina Valerio, Nicola Culeddu, Matilde Chessa, and Cesare Manetti

Contents

8.1 Migraine and Genetic, Hematochemical, and Instrumental Parameters

Migraine is a neurological disorder that correlates with an increased risk of subclinical cerebral vascular lesions (Kruit et al. 2004). Subjects suffering from migraines with aura (MwAs) are particularly exposed to such a vascular risk (Scher et al. 2005). There is a trend in considering a migraine as a systemic vasculopathy (Tietjen 2009a). The relationship between a MwA and cerebral lesions is not fully known; but it seems that the increased vascular risk of migraineurs is not caused by traditional risk factors (i.e., hypertension, hyperglycemia, hypercholesterolemia; Scher et al. 2005). However, these factors have implications on the possible drug therapies administered to migraine sufferers.

It has been demonstrated that migraines are associated with alterations in hematochemical parameters related to platelet aggregation (Kozubski et al. 1996; Tietjen and Khubchandani 2009b). The biological response to some agents acting on the platelet function has been found altered in women who suffer from migraines. Such alterations occur in the adenosine diphosphate (ADP), collagen, ristocetin, epinephrine, prostacyclin, arachidonic acid, and 5-hydro-triptamine (Allais et al. 1997). These alterations could increase the risk of vascular diseases on a thrombotic base.

Another important alteration that was correlated with MwA is increase in the concentration of homocysteine. This amino acid, which is linked to the metabolism of B_6 and B_{12} vitamins, represents a powerful risk factor for cardiovascular diseases when highly concentrated in blood serum (Bottini et al. 2006; Moschiano et al. 2008). The genetic mutation T/T677 in homozygosis of the genotype *677-MTHFR* (methylenetetrahydrofolate reductase) may cause increased levels of blood homocysteine. This mutation has been correlated with MwA (Scher et al. 2006) and with an increment in the angiopathies due to oxidative stress (Sucker et al. 2009). Recent studies prove that T/T677 mutation correlates with hyperhomocysteinemia, pathologies of peripheral arteries (Khandanpour et al. 2009), and cardiovascular events in the presence of hypertension (Durga et al. 2004). Therefore, due to the possible vascular implications they have, the study of genetic mutations in MwA is gaining in popularity.

Given the correlation between migraines and vascular disorders, migraine sufferers usually undergo an assessment of their cerebrovascular status. Transcranial Doppler sonography (TCD) is a noninvasive and reliable technique that measures the cerebral blood flow velocity (CBFV) in the arteries of the macrocirculation (Newell and Aaslid 1992). Some authors evidence that MwA patients present altered CBFV values compared with healthy controls both in resting conditions (Nowak and Kacinski 2009) and reactivity maneuvers (Anzola 2002). The assessment of cerebral vasomotor reactivity (i.e., the arteries' capacity of compensating systemic blood pressure alterations) is of primary importance in assessing the overall status of the artery bed. Despite the aforementioned assessments, the clinical utility of TCD in diagnosis and follow-up of migraine sufferers is still debated.

The aims of this study were to test the importance of TCD examinations in migraine sufferers by studying the correlations of CBFV with hematochemical and genetic parameters and to develop a classifier for the detection of the *677-MTHFR* genotype based on hematochemical parameters and instrumental data (i.e., TCD analysis).

8.2 Structure of the Database

In this study, we evaluated 50 subjects who suffer from migraines; they were diagnosed using the criteria given by the International Headache Society (1988), and were consecutively admitted to the Women's Headache Center, University of Torino, Italy. The mean age of the patients was 38.5 ± 1.6 years (range: 18–64 years). A total of 21 patients suffered from MwA. The patients were referred to the Neurology Division of the Gradenigo Hospital, Torino, where all the instrumental analyses (i.e., TCD examinations) were carried out. Exclusion criteria were the presence of headache other than migraine or of chronic metabolic diseases. All the subjects were informed about the purpose of the study and about data acquisition and treatment. All patients signed an informed consent before entering the study. The local Institutional Review Board approved this research.

In Section 8.2.1, we describe the database in detail. In the description, we separate instrumental data (Subsection 8.2.1.1), hematochemical data (Subsection 8.2.1.2), and genetic data (Subsection 8.2.1.3). To the best of our knowledge, this study is the only one that uses a mixed database consisting of metabolic, clinical, and instrumental variables.

8.2.1 Transcranial Doppler Sonography Data

All the subjects underwent TCD examination of the middle cerebral arteries (MCAs). No subjects were discarded due to poor bone window. We used a commercially available TCD device (X4 MultiDop; DWL, Germany) equipped with two 2-MHz ultrasound probes. The probes were placed on the temporal windows, and they were firmly held in place using an apposite holder. The same operator (a sonographer with more than 30 years of experience with vascular ultrasounds) performed all the examinations. We measured CBFV in the M1 tract of the subjects' MCAs.

The TCD recording took place in a quiet, dimmed room, with subjects lying in a comfortable supine position and breathing room air. All the TCD examinations were carried out in the interictal period, so that subjects were free of pain. After 5 minutes of baseline recording, subjects were asked to hold their breath. Breath holding (BH) is a powerful stimulus that induces vasodilation in cerebral arteries (Molinari et al. 2006). Vasodilation causes a drop in the resistance of the peripheral vessels, thus increasing cerebral blood flow. Figure 8.1 shows an example of BH as assessed by TCD.

The measurement of CBFV variation during BH is effective in assessing the cerebral vascular status of a subject (Silvestrini et al. 2004). We measured the vasomotor reactivity by using the BH index (BHI), which is defined as follows:

$$\text{BHI} = \frac{V_{\text{BH}} - V_{\text{BASE}}}{V_{\text{BASE}}} \times 100 \tag{8.1}$$

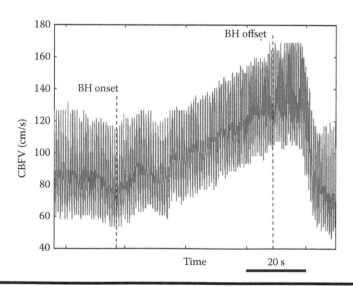

Figure 8.1 Cerebral blood flow velocity (CBFV) variation during breath holding (BH): The graph is relative to a healthy subject. The vertical dashed lines mark the onset and offset of voluntary BH. The CBFV linearly increases during BH due to peripheral vasodilation. (Reproduced from Molinari, F., W. Liboni, G. Grippi, and E. Negri, *J. Neuroeng. Rehabil.*, 3:16, 2006. With permission.)

where

- V_{BH} represents the CBFV measured at the end of the BH and averaged on a 60-second time window.
- V_{BASE} represents the CBFV measured at the beginning of the BH and averaged on a 60-second time window.

It has been shown that among MwA sufferers, an increased percentage suffers from cardiac atrial septal defects, including patent foramen ovale (PFO; Anzola et al. 2000; Rothrock 2008). The PFO has been indicated as one of the possible causes of cerebrovascular accidents affecting MwA sufferers. Therefore, we considered the presence of emboli in intracranial circulation as an index of cerebrovascular risk.

When a blood clot flows in the insonated cerebral artery, it produces a characteristic signal called a "high-intensity transient signal" (HITS). The TCD device we used was equipped with an automated HITS detection algorithm. We measured the number of detected HITSs in both the MCAs and inserted these numbers into the instrumental variables set.

Finally, a previous study conducted on a relatively large sample population of Italian women showed that CBFV decreases with age (Liboni et al. 2006). Therefore, we also considered age a variable in our database. Table 8.1 contains a summary of the six instrumental variables (plus age) that we used in this study.

8.2.2 Hematochemical Data

The same laboratory made all the hematological analyses. The subjects were asked to undergo blood analysis within 2 days of the TCD examination. Since our purpose was to develop a methodology based on data that are usually acquired during migraine treatment protocols, we kept a list of blood analyses that was similar to the

Table 8.1 TCD Data (with Unit) That Were Inserted in the Database

Variable	Unit
Age	year
Left CBFV	cm/s
Right CBFV	cm/s
Left BHI	%
Right BHI	%
Left HITS	a.u.
Right HITS	a.u.

Table 8.2 Data from Hematochemical Analysis and Born Test (Biochemical Data)

Hematochemical Variables	Agents for Platelet Aggregation Test (Born Test)
Antithrombin (%)	Arachidonic acid (%)
Activated partial thromboplastin time (APTT; second)	Adenosine diphosphate (ADP) 1 micromolar—ADP1 (%)
Hematocrit (%)	ADP 3 micromolar—ADP3 (%)
Hemoglobin (g/dl)	Collagen 2 micromolar (%)
Erythrocytes (×1000000/μ)	Ristocetin (%)
Fibrinogen (mg/dl)	
Leucocytes (×1000/μl)	
Platelets (×1000/μl)	
C protein (%)	
S protein (%)	
Prothrombin (%)	
PTINR (a.u.)	

Note: The first column shows the hematochemical data (with unit) that were inserted in the database. The second column shows the agents we considered for the platelet-aggregation response (Born test). The response is given in percentage.

one routinely used at the Women's Headache Center of Torino. Table 8.2 shows the complete list of variables that were considered in this study. The first column contains the hematochemical variables. Patients underwent the Born test for platelet aggregation under the effect of some agents (Born 1962). The agents we considered are reported in the second column of Table 8.2.

8.2.3 Genetic Data

All the patients were asked to undergo a genetic test to discover possible mutations of the *677-MTHFR* genotype. The C/T677-*MTHFR* variant (leading to an alanine-to-valine substitution) resulted in a thermolabile enzyme and a decreased production of folate, which is a cofactor required for homocysteine remethylation. Homozygosity for the *677-MTHFR* variant is associated with hyperhomocysteinemia up to 50% higher than normal. The testing methodology was "direct mutation analysis," an automated high-throughput system incorporating deoxyribonucleic acid (DNA) amplification (polymerase chain reaction) and primer extension, and allele resolution by matrix-assisted laser desorption ionization–time of flight (MALDI-TOF) mass spectrometry was used to test for variants in the *MTHFR* gene.

The genetic test revealed 14 subjects with T/T677-*MTHFR* mutation, 18 with C/T677-*MTHFR* mutation, and 18 with no mutation. There were also seven patients with T/T677 mutation, eight with C/T677 mutation, and six with no mutation resulting from MwA.

8.3 Statistical Analysis

Multivariate analysis was applied to the data set constituted by the variables described in Sections 8.2.1 through 8.2.3. The raw data set was a matrix having as rows (statistical units) 50 patients and as columns (variables) the 24 values relative to the hematochemical, instrumental, and genetic exams.

We used an analysis strategy combining multidimensional analysis of variance (ANOVA) analysis, unsupervised analysis, supervised analysis, and classification. The entire database was first analyzed using ANOVA and unsupervised techniques in order to extract the correlation structure that constitutes the inner texture of the data. We considered the mutation type and the presence or absence of aura as principal factors and independent variables. The right and left CBFVs were considered dependent variables. We used ANOVA to detect the most significant variables for the dependent variables, thus reducing the problem complexity and representation. Subsequent unsupervised and supervised techniques were then applied on a reduced set of variables, in order to better detail the inner correlation structure of the data set.

To build the classifier, the raw data set was then split into two parts in a procedure known as "cross-validation." We decided to use cross-validation in order to avoid possible overfitting of the data set given by the relatively small number of patients.

All statistical analyses were carried out using StatGraphics Centurion XV (Statpoint Technologies, Warrenton, Virginia, USA). In Section 8.4, we briefly explain the analysis techniques used in this research. Prior to performing unsupervised and supervised clustering, we performed a multivariate analysis of the raw data. Specifically, we aimed at evidencing the variables that are linked to genetic mutation in migraine.

In a previous study, we demonstrated that the left and right CBFVs of healthy women are highly correlated (Liboni et al. 2006). Right and left CBFVs, as well as right and left BHIs, were correlated for migraine subjects also (Pearson's correlation coefficient was always greater than 0.9; $p < .0001$).

We considered the left and right CBFVs as dependent variables. Multidimensional ANOVA tests were performed in order to extract the variables that influenced the dependent variables. Factors such as mutation and the presence of aura were considered. From the variables (biochemical data), we removed all the variables having a correlation coefficient greater than 0.6.

We performed an outlier analysis using Q and T^2 statistics. In processing the hematochemical and platelet-related data, we found six outliers. However, we decided to keep them since they could explain the pathology. When processing the instrumental data, we found four outliers relative to the right CBFV (one subject), left CBFV (two subjects), and right BHI (one subject). Again, we preferred to keep such subjects in the database and note them in subsequent statistical analysis.

8.4 Metabonomic Techniques

We applied metabonomic techniques to the database consisting of metabolic and hematochemical data, genetic data, and instrumental data collected from Doppler clinical analyses. Metabonomics, which belongs to the "omics" sciences, quantitatively measures living systems undergoing the effects of diseases caused by genetic mutations. Unlike genomics and proteomics, metabonomics focuses on the multiparameter evaluation of a complex living system by studying its overall physiological profile. In this study, we combined unsupervised and supervised analysis techniques. The result of unsupervised techniques is not guided by the maximization of externally imposed classification goals (i.e., discrimination between normal and pathologic, placebo versus drug administration, or different classes of disease). Rather, unsupervised techniques maximize the overall perception of data set features by minimizing the number of variables used to represent them. Therefore, unsupervised techniques allow for an unbiased description of the natural correlation structure in the data set. Conversely, supervised methods aim at maximizing an externally imposed task, such as the separation of a priori defined classes. Given the difficulty of the task and the biological variability that we expected in this study, we combined unsupervised and supervised techniques. The unsupervised approach was used to drive the supervised analysis. We used principal component (PC) analysis (PCA) as the unsupervised technique and partial least-squares discriminant analysis (PLS-DA) as the supervised approach.

8.4.1 Principal Component Analysis

PCA is an unsupervised technique that effectively represents the information embedded in multidimensional data sets (Benigni and Giuliani 1994). The raw data are represented in a transformed domain with lower dimensionality. The correlation structures of raw data generate few latent variables (or PCs). The PCs form an orthogonal basis and are sorted in the order of decreasing explained data variance. Each PC can be expressed as $PC = aX_1 + bX_2 + cX_3 + K$ where X_1, X_2, X_3, etc., represent the measured features and a, b, c, etc., represent numerical weights. Each statistical unit is assigned a score relative to each extracted component, whereas the correlation coefficients between original variables and extracted components (loadings) furnish the significance of a specific PC.

8.4.2 Partial Least Squares

The PLS is a supervised technique that aims at finding the best linear combination explaining experimental data. With reference to our study, we can define $\mathbf{X}$ as the matrix containing the measured variables and $\mathbf{Y}$ as the matrix of genetic mutation. Hence, $\mathbf{X}$ contains the independent variables, and $\mathbf{Y}$ contains the dependent variables. The purpose of using PLS is to find the linear combination of $\mathbf{X}$ that better explains $\mathbf{Y}$. The initial linear model can be expressed as follows:

$$\mathbf{Y} = \mathbf{XB} + \mathbf{E} \tag{8.2}$$

where $\mathbf{B}$ is the regression coefficient matrix and $\mathbf{E}$ the error matrix (same dimensions as $\mathbf{Y}$). Because of the difference in the numerical values of our data, we centered $\mathbf{X}$ and $\mathbf{Y}$ by subtracting their mean value and by normalizing with respect to their standard deviations.

When in the presence of many colinear variables, PCA may fail in properly extracting the covariance structure between the predictors. Conversely, PLS extracts only the covariance structure between the predictors and the response variables, thus leading to a more reliable system description. The PLS defines a matrix $\mathbf{T}$ so that $\mathbf{T} = \mathbf{XW}$, where $\mathbf{W}$ is an appropriate weight matrix and $\mathbf{T}$ is the factor score matrix. Then, the considered linear regression model becomes $\mathbf{Y} = \mathbf{TQ} + \mathbf{E}$, where $\mathbf{Q}$ is the matrix of regression coefficients (loadings) for $\mathbf{T}$ and $\mathbf{E}$ is the error matrix. Once $\mathbf{Q}$ is calculated, the overall system is equivalent to the one in Equation 8.2, where $\mathbf{B} = \mathbf{WQ}$. This equation can now be used as a linear predictive regression model. The PLS can be thought of as a PCA applied to the $\mathbf{X}$ matrix, a PCA applied to the $\mathbf{Y}$ matrix, and the correlation analysis of the two sets of obtained PCs. Using the standard nonlinear iterative partial least-squares (NIPALS) algorithm, we performed all PLS numerical computations.

8.4.3 Discriminant Analysis

The DA is used to determine which variables better discriminate between two or more groups in a sample population. In our study, there are three natural groups corresponding to genetic classifications: (1) no mutation, (2) C/T677 mutation, and (3) T/T677 mutation. Computationally, DA determines a set of weights by multiplying the $\mathbf{X}$ variables, so that the assignment error of each statistical unit to the correct $\mathbf{Y}$ class is minimized. We used PLS-DA to build a supervised model that could predict genetic mutation based on patients' instrumental and hematochemical data.

To build the classifier, the raw data set was split into two parts. Cross-validation was carried out in order to avoid possible overfitting of the data set given the relatively small number of patients. We randomly selected 40 patients to build the classifier, and we used the remaining 10 subjects for validation. We tested 100 different possible combinations of 40 (classifier set) and 10 (validation set) subjects, with the condition that in the validation subset there were at least 2 subjects with C/T677 mutation, 2 with T/T677, and 2 with no mutation. We forced this condition since a preliminary study (results are not reported in this chapter) revealed that the classifier did not perform correctly when built on a heterogeneous subset and validated on a homogeneous one. Therefore, we only considered the random combinations leading to subsets in which all three classes of subjects (no mutation, C/T677, and T/T677 mutation) were present.

8.5 Migraine Subjects Classification

Table 8.3 shows the result of the multidimensional ANOVA analysis performed on the hematochemical and platelet-related data presented in Table 8.2 when considering the left and right CBFVs as dependent variables. The principal factors were considered in the presence of mutation and aura. The CBFVs are influenced by response to ADP 3 micromolar (ADP3) and collagen ($p < .001$). The left CBFV

Table 8.3 Covariable Significance as Computed by ANOVA

Dependent Variable	Covariable	p Value
Left CBFV	ADP3	.0040
	Collagen	.0207
	Ristocetin	.0340
	Mutation	.0424
	Aura	.5307
Right CBFV	ADP3	.0021
	Collagen	.0057
	Mutation	.0230
	Aura	.1553

Note: Principal factors are mutation and aura (presented in italics as the last row of each dependent variable). The dependent variables are left and right CBFVs (listed in the first column). The second column contains the significant covariables, and the third column contains the associated *p* values.

is also related to the response to ristocetin ($p < .001$), whereas the right CBFV is not. Among the principal factors, only mutation is related to blood velocity, whereas the presence of aura is not. The significant variables were used in the following unsupervised step to confirm relations between instrumental and biochemical data:

The PCA was performed on a data set consisting of all the 50 subjects (rows) and the following 7 observations: (1) ADP3, (2) collagen, (3) ristocetin, (4) right CBFV, (5) left CBFV, (6) right BHI, and (7) left BHI. All the variables were standardized. We extracted three PCs; their combination explained 79.9% of the total variance of the data. Figure 8.2 contains the graph of the seven eigenvalues. Eigenvalues are proportional to the percent of variability in the data they explain. We chose eigenvalues greater than one (horizontal line in Figure 8.2), since eigenvalues lower than one explain the same variance of the original variables and are not useful for data analysis in a reduced domain.

Table 8.4 presents the weights of the seven variables on the three components. Both CBFVs have a high weight in the first component, whereas the BHIs are higher in the second and third components. In addition, ristocetin, collagen, and ADP3 responses have a negative high weight in the third component.

Figure 8.3 shows the distribution of subjects on the hyperplanes formed by components 1 and 2 (uppermost graph), components 2 and 3 (central graph), and components 1 and 3 (bottom graph). Black circles represent subjects in the T/T677 group, gray circles represent those in the C/T677 group, and white circles represent the ones without mutation. The continuous black lines represent the projection of original variables on hyperplanes. The first component separates the subjects with T/T677

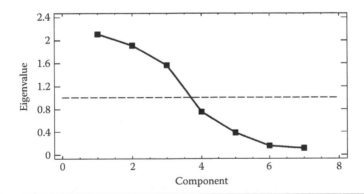

Figure 8.2 Eigenvalues of the principal component analysis applied to the data set consisting of seven variables (ADP3, collagen, ristocetin, right and left cerebral blood flow velocities and right and left breath-holding indexes): The first three eigenvalues are higher than one (horizontal dashed line). Therefore, these three eigenvalues were used to represent the subjects in a reduced domain of three principal components.

Table 8.4 Weights of the Seven Variables with Respect to the First Three Principal Components of PCA

Variable	Component 1	Component 2	Component 3
ADP3	−0.27	0.40	−0.45
Collagen	0.13	0.18	−0.57
Ristocetin	−0.04	0.37	−0.41
Left CBFV	0.60	−0.18	−0.21
Right CBFV	0.59	−0.21	−0.22
Right BHI	0.34	0.54	0.28
Left BHI	0.27	0.55	0.35

mutation from the subjects with C/T677 mutation or no mutation (Figure 8.3, upper panel in the top-right quadrant and bottom panel in the upper part of the chart). Component 2 separates T/T677 mutation in the upper part of the chart (Figure 8.3, middle panel). Table 8.5 presents the results of the t-test performed on the mutation subgroups. It confirms that component 1 differentiates T/T677 from C/T677, whereas component 2 separates T/T677 from C/T677 and from the subjects with no mutation.

The overall result of the PCA was that subjects with T/T677 mutation could be distinguished from the other subjects because of their PC values. However, such PCs are mainly correlated with CBFVs (first component), BHIs (second and third components), collagen, ADP3, and ristocetin (third component). This result confirms the importance of the vascular parameter alterations carried out by genetic mutations in migraine.

The analysis of subjects with respect to mutation was refined by using PLS as the supervised approach. Independent variables were the same as those used with PCA; dependent variables were mutation and aura. Table 8.6 summarizes the weights of the dependent and independent components. Figure 8.4 shows the PLS discrimination of the subjects in the hyperplane (factor 2, factor 4). Factor 2, in particular, is a discriminant of the T/T677 (black circles: left part of the graph) and C/T677 mutations (gray circles: right part of the graph). In particular, collagen response is higher in subjects without mutation with respect to T/T677 (t-test, $p < .001$) and in subjects with C/T677 mutation with respect to T/T677 (t-test, $p < .002$).

The supervised PLS-DA classifier was built by coding the dummy variable **Y** by a number (0 = no mutation, 1 = C/T677 mutation, 2 = T/T677 mutation). To build the PLS-DA classifier, we relied on the previously performed ANOVA analysis and removed all the variables having a correlation coefficient higher than 0.6.

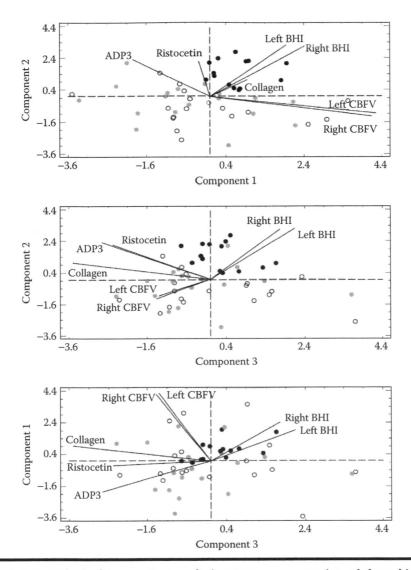

Figure 8.3 Principal component analysis (PCA) representation of the subjects in three hyperplanes: Black circles represent the subjects in the T/T677 group, gray circles those in the C/T677 group, and white circles the subjects without mutation. The dashed lines represent the zero level of principal components (PCs). The black lines represent projection of the original variables on the hyperplanes. The upper panel shows PC1 versus PC2, central panel shows PC3 versus PC2, and the lower panel shows PC3 versus PC1.

Table 8.5 Results of the *t*-Test Comparing the Three Mutation Subgroups

Patient Groups	Component 1 (30.13)	Component 2 (27.31)	Component 3 (22.46)
No mutation versus C/T677	0.805	−0.946	0.783
No mutation versus T/T677	−1.459	−6.623[a]	−5.075
T/T677 versus C/T677	2.581[a]	−5.075[a]	−1.387

Note: The significance threshold is 0.05. In parenthesis, the explained variance by the specific PCA component is given.

[a] Statistically significant difference.

Table 8.6 Weights for the Dependent and Independent Variables for the PLS Analysis

Independent Variables	Factor 1	Factor 2	Factor 3	Factor 4
Aura	0.36	−0.15	0.41	0.07
ADP3	−0.72	−0.41	−0.28	−0.18
Collagen	0.51	0.54	−0.03	−0.29
Ristocetin	−0.04	0.40	0.40	−0.05
Left BHI	0.01	−0.60	−0.29	−0.75
Right BHI	0.32	0.02	0.72	0.55
Dependent variables	Factor 1	Factor 2	Factor 3	Factor 4
Left CBFV	0.50	0.29	0.24	0.22
Right CBFV	0.52	0.28	0.10	0.22

The following four variables were removed for the **X** matrix: (1) antithrombin, (2) prothrombin time in international normalized ratio (PTINR), (3) hematocrit, and (4) erythrocytes. The classifier was developed on a set of 40 subjects (corpus set). Figure 8.5 shows a sample of the discrimination of a corpus set performed by the first two functions. The black "+" (plus sign) marks the centroids of the clusters. In this specific case, sensitivity was 100% and specificity was 100% (on the corpus set); the Wilks' lambda parameter was equal to 0.071, leading to $p < .003$.

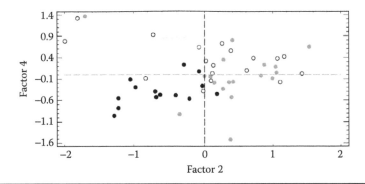

Figure 8.4 **Sample partial least-squares (PLS) representation of the subjects in the hyperplane defined by factor 2 and factor 4 (factors are numerically given in Table 8.6): Black circles represent the subjects in the T/T677 group, gray circles those in the C/T677 group, and white circles the subjects without mutation. The dashed lines represent the zero level of the principal components. It can be observed that factor 2 separates the genetic mutation T/T677 (left part of the chart—black circles) from C/T677 (right part of the chart—gray circles).**

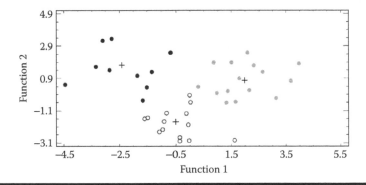

Figure 8.5 **Sample partial least-squares discriminant analysis representation of the 40 subjects of a corpus set in the hyperplane defined by the discriminant functions 1 and 2: Black circles represent the subjects in the T/T677 group, gray circles those in the C/T677 group, and white circles the subjects without mutation. The black "+" (plus sign) marks the centroids of the clusters.**

The PLS-DA classifier was then used to cluster the remaining 10 subjects (validation set). Corpus and validation sets were randomly selected. We tested 100 combinations leading to corpus and validation sets containing at least 2 subjects for each group. Hence, we classified 1000 subjects of which 363 belonged to the T/T677 group, 364 to the C/T677 group, and 273 to the no mutation group.

Each element of the validation set was assigned to a group based on its distance from the centroids. For all the considered classifiers the discriminant function p

value was always lower than .01. The classification performance was as follows (we indicate the range of variation in parentheses):

- 100% sensitivity (95.6%–100%) and 97.6% specificity (94.1%–98.3%) for the T/T677 mutation
- 91.8% sensitivity (87.5%–95.5%) and 97.6% specificity (92.3%–99.0%) for the C/T677 mutation
- 89.0% sensitivity (84.7%–92.4%) and 95.9% specificity (90.2%–98.8%) for the group without mutation

Overall classifier performance was 95.9% sensitivity, 89.0% specificity, and 94.0% efficiency.

8.6 Migraine and Metabonomic Techniques

A migraine is a complex neurological disorder involving a wide range of biochemical problems, including metabolic imbalances and neuronal and vascular anomalies. In this study, we profiled different groups of women suffering from migraines. The independent observation we considered in this study was the genetic mutation of the *677-MTHFR* gene. We believe that, though a preliminary study, this study could be important in the clinical management of migraines.

We built a mixed database incorporating instrumental data (the CBFVs in the middle cerebral arteries and the cerebral reactivity index to BH), genetic data (the mutation type C/T677 or T/T677, or the absence of mutation), and some biochemical parameters (including hematochemical variables and data related to platelet aggregation). The complete list is presented in Table 8.2.

The analyses PCA and multivariate ANOVA showed that in our sample population of migraine sufferers, *677-MTHFR* mutations are associated with altered platelet response to collagen, ristocetin, and ADP3. However, we also documented that *MTHFR* mutations in migraines correlate with the instrumental measures of CBFVs and BHIs. The PLS results confirmed the importance of CBFVs and BHI patterns in different subgroups of genetic mutations (the last four rows of Table 8.6). When compared with the C/T677 subgroup, T/T677 subjects showed increased BHIs (t-test, $p < 5 \times 10^{-6}$ for left and $p < 6 \times 10^{-7}$ for the right side).

The PLS-DA offered satisfactory classification performance when classifying a population of migraineurs based on their genetic mutation. The overall classification performance was 95.9% sensitivity and 89.0% specificity. The most important variables for classification were the left and right CBFVs; the BHI; and, again, the platelet response to ADP3, collagen, and ristocetin. This finding confirms the importance of assessing vasculature-related variables in migraine. Therefore, we believe that TCD could be of primary importance in the follow-up of migraine sufferers.

We did not perform outlier rejection. Outliers for the biochemical variables and for the instrumental variables were correctly classified in all the tests performed. Therefore, we assumed that the outliers might be representative of the variability of the data that can be obtained when working with a complex pathology, like migraine.

8.7 Migraine and Transcranial Doppler Sonography Variables

We found that BHI is altered in subjects with genetic mutations (t-test, $p < 5 \times 10^{-6}$). This result confirms that vascular reactivity may be impaired in migraineurs (Liboni et al. 2007). In 2008, Vernieri et al. showed that subjects suffering from MwA presented an increased BHI in the predominant migraine side with respect to nonpredominant subjects and to control subjects, suggesting a possible deficit in the cerebrovascular autonomic control.

Our results confirm previously published studies. The novelty of our study is in two results that we obtained using the applied multidimensional approach: First, we found that aura is not a major factor when related to mutation. In other words, the presence of aura was not correlated to the overall profile of the genetic mutation. This result is new, since many studies have pointed out that migraines with and without aura are characterized by different pathogenesis. Also, some pathologies have a greater prevalence in MwA than in a migraine without an aura (i.e., the patent foramen ovale (Diener, Kurth, and Dodick 2007). We did not find any significant correlation between aura and the instrumental or biochemical variables. The PCA and PLS techniques showed that the metabolic profile of subjects with C/T677 and T/T677 mutations is significantly different from that of subjects without mutation, even if aura is discarded among the classification variables.

The second important result of our study is the finding that biochemical data alone are insufficient to perform classification. We tested a PLS-DA classifier based only on the variables of Table 8.2. We removed the instrumental data from the data matrix $\mathbf{X}$. Performance dropped to 54% sensitivity and 58% specificity ($p > .05$). We determined, therefore, that mutation induces a global alteration of the metabolic profile of the subject that is better described by the joint analysis of hematochemical and vascular parameters.

8.8 Innovation, Limitations, and Future Perspectives

The principal limitation of this study is that we relied on the relative paucity of the sample database. However, the results seem promising since even the first multivariate analysis we performed (unsupervised PCA; Figure 8.3) showed a natural tendency in clustering T/T677 mutations. The application of this method results in a sort of "natural normalization" of the studied data set, given that PCs correspond to the eigenvectors of the correlation matrix, which in turn correspond by definition to the covariance matrix of the standardized variables. This method is convenient when dealing with heterogeneous variables defined by different measurement units, ruling out all questionable a priori defined standardization processes. Another limitation of the study is the small number of hematochemical data inserted into the database. We decided to develop a study based on exams that are routinely performed. Therefore, we did not specifically test any hematochemical variable other than the ones listed in Table 8.2 (first column), which reflects the list of blood examinations requested by the institution where the patients were followed up.

To the best of our knowledge, this is the first attempt to apply metabonomic statistical techniques to a clinical data set consisting of metabolic and instrumental data. We believe that this approach is a pilot experience to bring metabonomic (and, in general, data mining) techniques into clinical practice, when tuned for specific and well-defined groups of patients.

By using a metabonomic approach, we developed a classifier to detect genetic mutations of the *677-MTHFR* gene in a population of women suffering from migraine. Our database consisted of biochemical and instrumental data. Transcranial Doppler ultrasonography was used to measure the baseline values of cerebral blood flow and their variations during voluntary BH. Unsupervised and supervised approaches were used to extract the variables correlating with mutation. Our classifier showed an overall satisfactory performance: 95.9% sensitivity and 89.0% specificity. We found that cerebral blood flow velocities and response to BH are important variables in characterizing mutations in migraine.

This pilot study adapting metabonomic techniques to the profiling of pathologic subjects could be a basis for the development of a clinically applicable methodology devoted to the profiling, classification, and evolutionary modeling of migraine patients.

References

Allais, G., G. Facco, D. Ciochetto, C. De Lorenzo, M. Fiore, and C. Benedetto. 1997. Patterns of platelet aggregation in menstrual migraine. *Cephalalgia* 17(Suppl. 20):39–41.

Anzola, G. P. 2002. Clinical impact of patent foramen ovale diagnosis with transcranial Doppler. *Eur J Ultrasound* 16:11–20.

Anzola, G. P., M. Del Sette, L. Rozzini, P. Zavarise, E. Morandi, C. Gandolfo, S. Angeli, and C. Finocchi. 2000. The migraine-PFO connection is independent of sex. *Cerebrovasc Dis* 10:163.

Benigni, R., and A. Giuliani. 1994. Quantitative modeling and biology: The multivariate approach. *Am J Physiol* 266:R1697–704.

Born, G. V. 1962. Aggregation of blood platelets by adenosine diphosphate and its reversal. *Nature* 194:927–9.

Bottini, F., M. E. Celle, M. G. Calevo, S. Amato, G. Minniti, L. Montaldi, D. Di Pasquale, R. Cerone, E. Veneselli, and A. C. Molinari. 2006. Metabolic and genetic risk factors for migraine in children. *Cephalalgia* 26:731–7.

Diener, H. C., T. Kurth, and D. Dodick. 2007. Patent foramen ovale and migraine. *Curr Pain Headache Rep* 11:236–40.

Durga, J., P. Verhoef, M. L. Bots, and E. Schouten. 2004. Homocysteine and carotid intima-media thickness: A critical appraisal of the evidence. *Atherosclerosis* 176:1–19.

International Headache Society. 1988. Classification and diagnostic criteria for headache disorders, cranial neuralgias and facial pain. Headache Classification Committee of the International Headache Society. *Cephalalgia* 8(Suppl. 7):1–96.

Khandanpour, N., G. Willis, F. J. Meyer, M. P. Armon, Y. K. Loke, A. J. Wright, P. M. Finglas, and B. A. Jennings. 2009. Peripheral arterial disease and methylenetetrahydrofolate reductase (MTHFR) C677T mutations: A case-control study and meta-analysis. *J Vasc Surg* 49:711–8.

Kozubski, W., B. Walkowiak, Z. Pawlowska, A. Prusinski, and C. S. Cierniewski. 1996. Blood platelet fibrinogen receptors in migraine and related headaches. *Neurol Neurochir Pol* 30(Suppl. 2):25–33.

Kruit, M. C., M. A. van Buchem, P. A. Hofman, J. T. Bakkers, G. M. Terwindt, M. D. Ferrari, and L. J. Launer. 2004. Migraine as a risk factor for subclinical brain lesions. *JAMA* 291:427–34.

Liboni, W., G. Allais, O. Mana, F. Molinari, G. Grippi, E. Negri, and C. Benedetto. 2006. Transcranial Doppler for monitoring the cerebral blood flow dynamics: Normal ranges in the Italian female population. *Panminerva Med* 48:187–91.

Liboni, W., F. Molinari, G. Allais, O. Mana, E. Negri, G. Grippi, C. Benedetto, G. D'Andrea, and G. Bussone. 2007. Why do we need NIRS in migraine? *Neurol Sci* 28(Suppl. 2):S222–4.

Molinari, F., W. Liboni, G. Grippi, and E. Negri. 2006. Relationship between oxygen supply and cerebral blood flow assessed by transcranial Doppler and near-infrared spectroscopy in healthy subjects during breath-holding. *J Neuroeng Rehabil* 3:16.

Moschiano, F., D. D'Amico, S. Usai, L. Grazzi, M. Di Stefano, E. Ciusani, N. Erba, and G. Bussone. 2008. Homocysteine plasma levels in patients with migraine with aura. *Neurol Sci* 29(Suppl. 1):S173–5.

Newell, D. W., and R. Aaslid. 1992. Transcranial Doppler: Clinical and experimental uses. *Cerebrovasc Brain Metab Rev* 4:122–43.

Nowak, A., and M. Kacinski. 2009. Transcranial Doppler evaluation in migraineurs. *Neurol Neurochir Pol* 43:162–72.

Rothrock, J. F. 2008. Patent foramen ovale (PFO) and migraine. *Headache* 48:1153.

Scher, A. I., G. M. Terwindt, H. S. Picavet, W. M. Verschuren, M. D. Ferrari, and L. J. Launer. 2005. Cardiovascular risk factors and migraine: The GEM population-based study. *Neurology* 64:614–20.

Scher, A. I., G. M. Terwindt, W. M. Verschuren, M. C. Kruit, H. J. Blom, H. Kowa, R. R. Frants, et al. 2006. Migraine and MTHFR C677T genotype in a population-based sample. *Ann Neurol* 59:372–5.

Silvestrini, M., R. Baruffal, M. Bartolini, F. Vernieri, C. Lanciotti, M. Matteis, E. Troisi, and L. Provinciali. 2004. Basilar and middle cerebral artery reactivity in patients with migraine. *Headache* 44:29–34.

Sucker, C., C. Kurschat, F. Farokhzad, G. R. Hetzel, B. Grabensee, B. Maruhn-Debowski, R. Loncar, R. E. Scharf, and R. B. Zotz. 2009. The TT genotype of the C677T polymorphism in the methylentetrahydrofolate reductase as a risk factor in thrombotic microangiopathies: Results from a pilot study. *Clin Appl Thromb Hemost* 15:283–8.

Tietjen, G. E. 2009a. Migraine as a systemic vasculopathy. *Cephalalgia* 29:987–96.

Tietjen, G. E., and J. Khubchandani. 2009b. Platelet dysfunction and stroke in the female migraineur. *Curr Pain Headache Rep* 13:386–91.

Vernieri, F., F. Tibuzzi, P. Pasqualetti, C. Altamura, P. Palazzo, P. M. Rossini, and M. Silvestrini. 2008 Increased cerebral vasomotor reactivity in migraine with aura: An autoregulation disorder? A transcranial Doppler and near-infrared spectroscopy study. *Cephalalgia* 28:689–95.

Chapter 9

Automatic Grading of Adult Depression Using a Backpropagation Neural Network Classifier

Subhagata Chattopadhyay, Preetisha Kaur, Fethi Rabhi, and Rajendra Acharya U.

Contents

9.1 Introduction to Grading of Adult Depression

Depression is one of the most serious psychological disorders affecting approximately 100 million people across the world (Gotlib and Hammen 1992). The World Health Organization (WHO) has highlighted depression as the third most common disease in the world. According to WHO, depression is defined as "a mental disorder characterized by sadness, loss of interest, pleasure, feeling of guilt or low self-worth, disturbed sleep or appetite, low energy, and poor concentration" (WHO, http://www.who.int/topics/depression/en/). Moreover, approximately 2.6%–12.0% males and 7%–21% females suffer from diagnosable depression states globally (Keller 1994). Often, depression has a recurrent course (Kaplan, Saddock, and Greb 1994) and, thereby, tends to enhance the morbidity, mortality, and economic loss of individuals who suffer from the disease (Kaplan, Saddock, and Greb 1994).

There exist several depression rating tools, for example, Hamilton's scale (Hamilton 1960) and Zung's rating scale (Zung 1972). These scales are made for manual screening and grading of adult depression. Hence, the diagnosis is often biased and lacks clear boundaries among the grades (Bagby et al. 2004), for example, "mild to moderate" and "moderate to severe" rather than "mild," "moderate," and "severe." Doctors (especially novices) often suffer from the obvious dilemma of how to determine the stage of a patient's depression. For instance, in a mild-to-moderate diagnosis, it is often difficult for doctors to determine how close the disease progression is to mild or moderate. This problem is very significant since its consequences can be critical; there is a chance that a wrong inference-making leads to either "under" diagnosis or "over" diagnosis that, in turn, results in erratic management plans. Having stated this practical issue of manual depression grading, the present study attempts to clearly differentiate the grades of depression into mild, moderate, and severe for a given set of adult depression patients using a backpropagation neural network (BPNN) technique.

A neural network (NN), a connected graph of many nodes (i.e., neurons) having weighted links, attempts to achieve humanlike performance by modeling the human nervous system (Chen et al. 2005). An NN can usually be one of two types: (1) feedforward and (2) backpropagation. Other types include Kohonen's (1995) self-organizing map (SOM) and the Hopfield (1982) network. The BPNNs are fully connected and layered feedforward networks in which activations flow from an input layer through a hidden layer to an output layer (Rumelhart, Hilton, and McCelland 1986). The network usually starts with a set of random weights according to each learning example, each of which is passed through the network to activate the nodes. Finally, the network's output (i.e., calculated output) is compared with the actual output (i.e., target output) and the error estimates are propagated back to the hidden layer and then to the input layer. The network updates these error estimates iteratively until it becomes stable.

The BPNN approach is widely used in various medical domains, for example, pathology and laboratory medicine (Astion and Wilding 1992), diagnosis of breast and ovarian cancers (Wilding et al. 1994), chromosome classifications (Cho 2000), detection of ophthalmic artery stenosis (Guler and Ubeyli 2003), psychiatric diagnosis (Aruna, Puviarasan, and Palaniappan 2005), and treatment of gastrointestinal disorders (Aruna, Puviarasan, and Palaniappan 2003). There are obvious good reasons due to which BPNN is used as a popular classification technique in the medical domain (Tu 1996).

9.2 Literature Review

Medical decision support systems (MDSSs) play an important role in efficient and accurate patient management as patients show variability in their preferences for doctors who make decisions for them, which may be due to a number of reasons (Robinson and Thomas 2001).

9.2.1 Medical Decision Support Systems as an Emerging Tool

Studies also show that the use of MDSSs considerably enhances physician performance (Johnston et al. 2001; Kawamoto et al. 2005). Therefore, artificial intelligence and knowledge engineering techniques are widely used in health care due to their high efficiency in dealing with large volumes of heterogeneous, subjective, and nonlinear (together called as raw or ad hoc) medical data (Prather et al. 1997; Cios and Moore 2002). In this study, we focus on developing an NN-based classifier; hence, useful clustering and classification methods, such as MDSSs, are discussed in Section 9.2.2.

9.2.2 Clustering and Classification

Clustering, which is an unsupervised learning method in which class labels are unknown, is a potentially useful technique for determining common outcomes and facilitating the assessment of preferences (Lin et al. 1995). Hence, a clustering search is more data driven and concise (Lee and Siau 2001). Classification, on the other hand, is a popular form of supervised learning (i.e., the class labels are known) used in health care, and it includes methods such as artificial NNs, support vector machines, Bayesian networks, fuzzy logic, and genetic algorithms (Lucas 2004). Classifiers are robust at handling uncertainties, missing and noisy data, integration of data sources, and data without much domain knowledge support (Cho et al. 2008). They also have high accuracy in terms of interpretability (Sordo and Zeng 2005) and show remarkable stability in performance as the data set increases in size (Curiac et al. 2009).

9.2.3 Clustering and Classification in Mental Health

Clustering and classification techniques are widely used to study mental health issues in humans. They can be used to study various mental diseases and their correlations (Cougnard et al. 2004). Such approaches also help in early detection of psychosis (Chattopadhyay, Pratihar, and De Sarkar 2007). It is reported that fuzzy classifiers are used to classify and predict adult psychotic disorders (Chattopadhyay, Pratihar, and De Sarkar 2008; Gross et al. 1990).

9.2.4 Neural Networks in Medicine

Artificial NNs are used as an efficient approach in handling diverse medical problems, some of which are in the fields of radiological diagnosis (Bhatikar, DeGroff, and Mahajan 2004), cardiac analysis (Das et al. 2008), gastrointestinal studies (Lisoba and Taktak 2006), cancer studies (Li et al. 2000), and brain studies (Magnotta et al. 1999). A detailed discussion of these studies is beyond the scope of this chapter.

9.2.5 Neural Networks in Mental Health

In the mental health domain, some applications of artificial NNs include the measurement of brain structures (Pradhan, Sadasivan, and Arunodaya 1999), detection of seizure activity in electroencephalogram (EEG; Zou et al. 1996), and assistance in psychiatric diagnosis and prediction of length of stay in psychiatric hospitals (Davis, Lowell, and Davis 1993).

Hence, due to the high success rates of data mining classifiers in the health-care domain, efficiency of artificial NNs for handling large heterogeneous data volumes, and the diverse and complex nature of mental health data, we use BPNN in our study.

9.3 Materials and Methods

As mentioned in Section 9.1, depression is often manually graded as mild to moderate and moderate to severe in practice. This grading leads to confusion among doctors: How should mild, moderate, and severe be defined? Therefore, the objective of this study is to clearly distinguish depression levels into mild, moderate, and severe using a BPNN classifier in a set of given depression states. The processes, tools, and techniques used in the study are discussed here. Our processes include the following:

- Capturing the biological predictors of depression from the available literature
- Generating a questionnaire with the help of domain experts (e.g., psychiatrists and psychologists)
- Collecting anonymous data from various hospital sources after taking appropriate ethical measures

Our tools include the following:

- Iris data (available in MATLAB® by default) and adult depression data
- SPSS 15 statistical software for data analysis
- MATLAB for BPNN algorithm development, BPNN algorithm testing, and relevant numeric computations

Our techniques include the following:

- Statistical data mining using SPSS: Checking data integrity by Cronbach's (1951) alpha (α) and regressions (Han and Kamber 2006) for checking the fidelity/fit of the proposed data model
- Using the BPNN classifier in MATLAB

9.3.1 Processes Adopted

We first develop a "hierarchical tree" (HT) of major constructs, which is an orchestrated effect of various biological, environmental, and socioeconomic factors that finally leads to the onset of the illness (level 0; see Figure 9.1). Based on the available literature, several biological factors (level I) have been identified; such biological factors include "emotional" (E; Ball, McGuffin, and Farmer 2008), "cognitive" (C; Austin, Mitchell, and Goodwin 2001; Forsell, Jorm, and Winblad 1994), "motivational" (M; DiMatteo, Lepper, and Croghan 2000; Aneshensel, Frerichs,

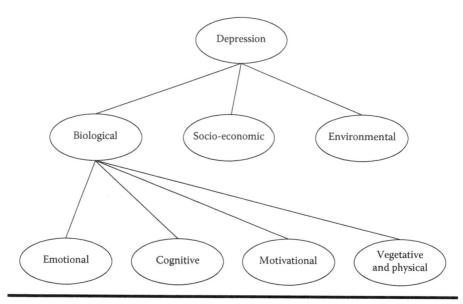

Figure 9.1 **Hierarchical tree structure of biological constructs and indicators.**

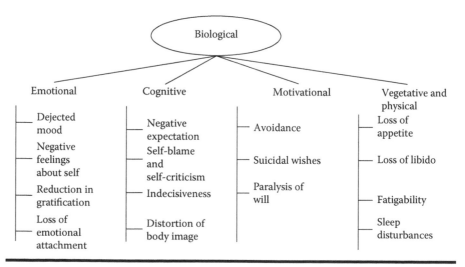

Figure 9.2 **Various biological indicators.**

and Huba 1984), and "vegetative and physical" (V; Weiss, Longhurst, and Mazure 1999; level II). Finally, a set of indicators (Is) addresses the signs/symptoms of depression (Figure 9.2), which are captured by relevant questions (Table 9.1). Such an HT structure is important for accurately constructing the depression model. The questions are close-ended ones, and the answer to each question is measured under a 3-point scale [0, 0.5, 1.0] denoting "disagree," "neutral," and "agree" (Table 9.2).

Table 9.1 Identified Biological Indicators Captured by Relevant Questions

Indicators	Question Number	Ticked Answer
Dejected mood	Q1	Agree; neutral; disagree
Negative expression	Q2, 9	-do-
Reduction in gratification	Q3, 4, 5	-do-
Loss of emotional attachment	Q6, 7	-do-
Negative feeling about self	Q8, 10, 26	-do-
Sleep disturbances	Q11, 12	-do-
Fatigability	Q13, 14	-do-
Loss of libido	Q15	-do-
Loss of appetite	Q16, 17	-do-
Paralysis of will	Q18, 19	-do-
Suicidal intents	Q20, 21	-do-
Avoidance	Q22	-do-
Distortion of body image	Q23	-do-
Individualism	Q24	-do-
Self-blame and self-criticism	Q25	-do-

Table 9.2 Questionnaire to Address the Signs/Symptoms of Depression

Name			
Address			
Age			
Questions	Agree	Disagree	Neutral
1. I often feel sad and downhearted (DEJECTED MOOD).			
2. I don't feel good about my career (NEGATIVE EXPECTATIONS).			

(Continued)

**Table 9.2 Questionnaire to Address the Signs/Symptoms
of Depression (*Continued*)**

Questions	Agree	Disagree	Neutral
3. I feel unwanted (REDUCTION IN GRATIFICATION).			
4. I still enjoy the things I used to do (REDUCTION IN GRATIFICATION).			
5. I get irritated/agitated easily (REDUCTION IN GRATIFICATION).			
6. The pleasure and joy has gone out of my life (LOSS OF EMOTIONAL ATTACHMENT).			
7. I have lost interest in the aspects of life that are important to me (LOSS OF EMOTIONAL ATTACHMENT).			
8. I feel others would be better off if I were dead (NEGATIVE FEELINGS ABOUT SELF).			
9. I feel hopeless about the future (NEGATIVE EXPECTATIONS).			
10. I think of myself as a failure (NEGATIVE FEELINGS ABOUT SELF).			
11. I have trouble sleeping at night (SLEEP DISTURBANCES).			
12. My sleep is disturbed—too little, too much, or broken sleep (SLEEP DISTURBANCES).			
13. I feel fatigued (FATIGABILITY).			
14. I feel tired very often (FATIGABILITY).			
15. I still enjoy sex as before (LOSS OF LIBIDO).			
16. I eat as much as I used to (LOSS OF APETITE).			
17. I notice that I am losing weight (LOSS OF APETITE).			
18. I do things slowly, slower than I used to (PARALYSIS OF WILL).			

Table 9.2 Questionnaire to Address the Signs/Symptoms of Depression (*Continued*)

Questions	Agree	Disagree	Neutral
19. I find it easy to do things that I used to (PARALYSIS OF WILL).			
20. I spend time thinking how I might kill myself (SUICIDAL WISHES).			
21. I feel killing myself would solve many problems in my life (SUICIDAL WISHES).			
22. I am restless and can't keep still (AVOIDANCE).			
23. I don't feel good about myself when I look in the mirror (DISTORTION OF BODY IMAGE).			
24. I find it easy to make decisions (INDECISIVE).			
25. I feel I am a guilty person and deserve to be punished (SELF-BLAME AND SELF-CRITICISM).			
26. My mind is as clear as it used to be (NEGATIVE FEELINGS ABOUT SELF).			

Note: Block letters show indicators.

Considering 26 questions as the attributes and 124 depression cases (tuples), a matrix is generated. In this study, 62 cases are used as the training data and the remaining 62 are kept for testing the performance of the developed BPNN classifier.

9.3.2 Tools and Techniques Used

Sections 9.3.2.1 through 9.3.2.7 describe the various tools and techniques used in this study.

9.3.2.1 Checking Internal Consistency (α) in Collected Depression Data

In data mining, before actual experiments it is mandatory for all data (whether primary or secondary) to be checked for its internal consistency as inconsistent data can yield fallacious results. Hence, the internal consistency of adult depression data has been checked by measuring Cronbach's α (Pradhan, Sadasivan, and Arunodaya 1999) that may be expressed as follows:

$$\alpha = \frac{N}{N-1}\left(1-\frac{\sum_{i=1}^{N}\sigma_{Y_i}^2}{\sigma_X^2}\right) \tag{9.1}$$

where N is the number of item sets, σ_X^2 is the variance of the observed score, and $\sigma_{Y_i}^2$ is the variance of the component i.

9.3.2.2 Iris Data (for Testing Backpropagation Neural Network Initial Code)

Iris data (Fisher 1936) has been obtained from MATLAB (by default, it is there in MATLAB). The data set is well acclaimed and consists of 150×4 matrices of several species of iris flowers that all have 3 class labels, *Iris setosa, Iris virginica*, and *Iris versicolor*, where each class has equal number of data in the matrix. The data set is divided into four attributes: (1) petal length, (2) petal width, (3) septal length, and (4) septal width. In this study, 75 out of 150 data are chosen for the training, and the rest are chosen for testing the performance of the BPNN classifier. The reason for using iris data is to test the BPNN code and the performance of the BPNN algorithm developed in MATLAB. Hence, iris data serves not only merely as another data set but also as a tool in this study.

9.3.2.3 Development of the Proposed Backpropagation Neural Network Classifier

This study proposes a BPNN-based classifier for clearly differentiating depression levels (i.e., mild, moderate, and severe). Since BPNN involves a gradient-based search for a set of weights that fit the training data to minimize the mean-squared distance (MSD) between a network's class prediction and the known target values of tuples, it can be a useful classification technique in health sciences (Tu 1996; Han and Kamber 2006). Furthermore, the assigned learning rate helps to avoid getting stuck at a local minima in the given decision space and, hence, it is appropriately chosen for such a classification problem in which precision is the desired output and the degree of precision depends on the maturity gained by the network with iterations. Other important advantages of NN are that it has a high tolerance for adapting to noisy data and that it can be applied to classify patterns on which it has not been trained. Hence, BPNN is largely used on medical data that often suffer from these issues. Figure 9.3 shows a BPNN topology (Tu 1996; Han and Kamber 2006), while Figure 9.4 describes the steps of BPNN algorithm.

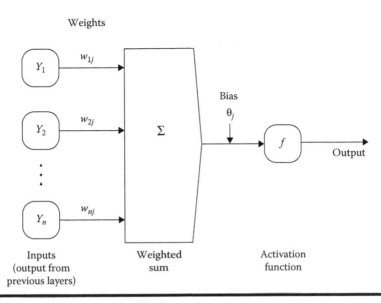

Figure 9.3   The topology of a backpropagation neural network.

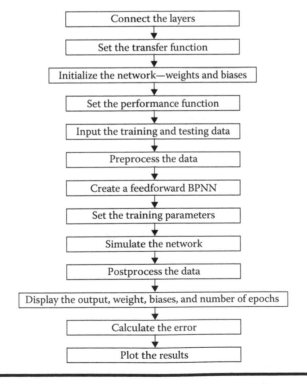

Figure 9.4   General algorithm for a backpropagation neural network.

9.3.2.4 k-Times Holdout Random Subsampling

It is also worth mentioning here that a *k*-times holdout random subsampling is performed to test the classifiers' best accuracy as well as its robustness. Therefore, collected depression data is subdivided into two segments: (1) Two-thirds is used for training the classifier, and (2) one-third is kept for testing. This procedure is repeated with the same proportion 10 times.

9.3.2.5 Errors in Prediction

Errors in prediction by the developed BPNN classifier are calculated using a percentage deviation ∂ as follows:

$$\partial = \frac{(T_i - C_i)}{C_i} \times 100 \tag{9.2}$$

where T_i is the target value and C_i the calculated value of the component i.

9.3.2.6 Multiple Linear Regressions

A regression study is then performed to find the best correlation coefficient for ascertaining the goodness of fit of the model:

$$A = b \times T + m \tag{9.3}$$

where A is the output vector, b is the *y*-axis intercept, T is the target vector, and m is the slope. This regression model is integrated by default in MATLAB and is automatically calculated for ascertaining the goodness of fit of a model.

9.3.2.7 Accuracy Measure

The accuracy of the classifier (denoted as AC) is measured as follows:

$$AC = \frac{CCT}{TT} \times 100 \tag{9.4}$$

where CCT denotes correctly classified tuples and TT denotes total tuples. Here, it is important to note that as only one class label is predicted at a time, sensitivity and specificity are not measured. Instead, accuracy is estimated using this equation.

9.4 Results and Discussions

Sections 9.4.1 through 9.4.3 discuss the experimental results of the study.

9.4.1 Results of Internal Consistency (α) Test (Depression Data)

As mentioned in Section 9.3.2.1, internal consistency testing is a useful technique for understanding the reliability of data. Hence, it is a mandatory check performed prior to the actual experiment. As iris data comprises a well-acclaimed data set, its internal consistency checking was not performed. However, adult depression data is mixed in nature. Only 10% is primary data (voluntary information provided by subjects); the remaining 90% is secondary data (collected from hospital records). Thus, data reliability in this case must be rechecked (see Section 9.3.2.2). The measure of α for the depression data is 0.77, which indicates a good/acceptable reliability of data according to a study by Nunnaly (1978).

9.4.2 Results of the Study on Iris Data

As mentioned in Section 9.3.2.2, the iris data matrix is 150×4; 75 cases each are used to train and test the developed BPNN classifier. The topology of BPNN includes the following:

- An input layer consisting of four neurons
- A hidden layer consisting of two neurons
- An output layer consisting of two neurons

It is important to mention here that the optimized values for learning rate and momentum or gradient are 0.3 and 0.6, respectively, based on the most converged values, obtained through iterations (which are 5 per each increment made in learning rate and momentum). Since we know the target outputs, the calculated outputs are matched and in this case, all the class labels of the iris data can be predicted with 100% accuracy. Tables 9.3 through 9.6 show the network convergence/optimization results with varied learning rates. While, Table 9.7 shows to the classification results.

9.4.3 Results of the Study on Depression Data

There are 124 adult depression cases registered in this study. Out of them, 62 cases are used for training the BPNN classifier and 62 are used for testing the classifier. For this data, the topology of BPNN includes

- An input layer consisting of 26 neurons
- A hidden layer consisting of 10 neurons
- An output layer consisting of 2 neurons

Table 9.3 Optimization of Learning Rate Keeping the Momentum Constant (for Iris Data)

Sl. No.	Learning Rate When Momentum = 0.6	Minimum Performance	Maximum Performance	Minimum Number of Epochs	Remarks
1	0.0	0.0520	0.0520	1	Performance goal not met
		0.8874	0.8874	1	Performance goal not met
		0.7158	0.7158	1	Performance goal not met
		0.1127	0.1127	1	Performance goal not met
		0.2136	0.2136	1	Performance goal not met
	Average value	0.3963	0.3963	1	Performance goal not met
2	0.1	0.0247	0.7712	191	Performance goal met
		0.0247	0.8693	129	Performance goal met
		0.0247	0.7712	191	Performance goal met
		0.0246	0.1444	220	Performance goal met
		0.0246	1.8464	240	Performance goal met
	Average value	0.0247	0.8805	194	Performance goal met
3	0.2	0.0155	0.1842	206	Performance goal met
		0.0242	0.1444	115	Performance goal met
		0.0247	0.0557	267	Performance goal met
		0.0247	0.0557	267	Performance goal met
		0.0241	0.5264	144	Performance goal met
	Average value	0.0226	0.1932	200	Performance goal met

4	0.3	0.0246	0.1842	49	Performance goal met
		0.0217	0.5042	36	Performance goal met
		0.0137	0.9387	20	Performance goal met
		0.0140	0.9689	26	Performance goal met
		0.0242	0.5960	24	Performance goal met
	Average value	0.0196	0.6384	31	Performance goal met
5	0.4	0.0115	2.0045	23	Performance goal met
		0.0138	1.2463	23	Performance goal met
		0.0239	1.7195	22	Performance goal met
		0.0245	1.5156	21	Performance goal met
		0.0170	2.9528	22	Performance goal met
	Average value	0.0181	1.887	22	Performance goal met
6	0.5	0.0067	1.8766	26	Performance goal met
		0.0231	1.8560	24	Performance goal met
		0.0150	1.3230	22	Performance goal met
		0.0245	0.2737	26	Performance goal met
		0.0210	1.8273	26	Performance goal met
	Average value	0.0180	1.4313	25	Performance goal met

(Continued)

Table 9.3 Optimization of Learning Rate Keeping the Momentum Constant (for Iris Data) (*Continued*)

Sl. No.	Learning Rate When Momentum = 0.6	Minimum Performance	Maximum Performance	Minimum Number of Epochs	Remarks
7	0.6	0.0087	2.8010	25	Performance goal met
		0.0134	0.3286	22	Performance goal met
		0.0084	2.6361	26	Performance goal met
		0.0089	1.1397	25	Performance goal met
		0.0102	2.2443	25	Performance goal met
	Average value	0.0099	1.829	25	Performance goal met
8	0.7	0.0192	2.3179	22	Performance goal met
		0.0178	4.3600	27	Performance goal met
		0.0194	1.3546	24	Performance goal met
		0.0138	5.2588	27	Performance goal met
		0.0171	1.2952	22	Performance goal met
	Average value	0.0174	2.9173	24	Performance goal met
9	0.8	0.0114	2.0511	25	Performance goal met
		0.0113	0.6796	21	Performance goal met
		0.0118	3.1791	22	Performance goal met
		0.0068	3.6815	26	Performance goal met
		0.0093	1.6435	23	Performance goal met
	Average value	0.010	2.246	23	Performance goal met

10	0.9	0.0234	2.3156	20	Performance goal met
		0.0170	2.4940	24	Performance goal met
		0.0120	7.7820	26	Performance goal met
		0.0117	0.8218	21	Performance goal met
		0.0155	1.8304	24	Performance goal met
	Average value	0.0159	3.048	23	Performance goal met
11	1.0	0.0180	0.8580	22	Performance goal met
		0.0185	0.3675	23	Performance goal met
		0.0120	7.7820	26	Performance goal met
		0.0128	7.2598	25	Performance goal met
		0.0233	6.0768	27	Performance goal met
	Average value	0.0169	4.562	25	Performance goal met

Table 9.4 Optimization of Momentum Keeping the Learning Rate Constant (for Iris Data)

Sl. No.	Momentum When Learning Rate = 0.3	Minimum Performance	Maximum Performance	Minimum Number of Epochs	Remarks
1	0.0	0.0246	0.3913	80	Performance goal met
		0.0246	0.1949	86	Performance goal met
		0.0240	0.0217	65	Performance goal met
		0.0245	0.7186	83	Performance goal met
		0.0246	0.0157	70	Performance goal met
	Average value	0.0247	0.2684	77	Performance goal met
2	0.1	0.0247	0.0023	105	Performance goal met
		0.0247	0.3251	91	Performance goal met
		0.0247	0.8556	95	Performance goal met
		0.0238	0.3945	101	Performance goal met
		0.0244	0.3272	112	Performance goal met
	Average value	0.0289	0.3809	101	Performance goal met
3	0.2	0.0237	3.8819	42	Performance goal met
		0.0236	0.0662	30	Performance goal met
		0.0217	0.7302	32	Performance goal met
		0.0206	1.1236	39	Performance goal met
		0.0243	1.1166	36	Performance goal met
	Average value	0.0227	1.3837	36	Performance goal met

4	0.3	0.0222	0.0845	42	Performance goal met
		0.0241	0.0782	55	Performance goal met
		0.0245	1.0117	56	Performance goal met
		0.0242	0.8118	47	Performance goal met
		0.0243	1.2237	54	Performance goal met
	Average value	0.0190	0.641	51	Performance goal met
5	0.4	0.0235	0.7546	17	Performance goal met
		0.0121	4.3081	16	Performance goal met
		0.0234	0.7241	15	Performance goal met
		0.0179	1.2606	13	Performance goal met
		0.0119	0.8961	15	Performance goal met
	Average value	0.0177	1.588	15	Performance goal met
6	0.5	0.0230	0.6198	26	Performance goal met
		0.0229	1.8464	20	Performance goal met
		0.0229	1.8464	20	Performance goal met
		0.0226	2.7406	21	Performance goal met
		0.017	0.3512	18	Performance goal met
	Average value	0.0216	1.4808	21	Performance goal met

(Continued)

Table 9.4 Optimization of Momentum Keeping the Learning Rate Constant (for Iris Data) (Continued)

Sl. No.	Momentum When Learning Rate = 0.3	Minimum Performance	Maximum Performance	Minimum Number of Epochs	Remarks
7	0.6	0.0220	0.6832	24	Performance goal met
		0.0205	0.5068	23	Performance goal met
		0.0242	0.5960	24	Performance goal met
		0.0193	0.2205	21	Performance goal met
		0.0205	0.5773	22	Performance goal met
	Average value	0.0213	0.5167	23	Performance goal met
8	0.7	0.0240	0.0278	31	Performance goal met
		0.0215	0.6556	38	Performance goal met
		0.0183	1.1166	36	Performance goal met
		0.0244	0.1404	33	Performance goal met
		0.0222	0.8073	39	Performance goal met
	Average value	0.0220	0.5495	35	Performance goal met
9	0.8	0.0210	0.3710	47	Performance goal met
		0.0173	0.2815	46	Performance goal met
		0.0149	0.1426	44	Performance goal met
		0.0173	1.2237	44	Performance goal met
		0.0106	1.0117	47	Performance goal met
	Average value	0.0162	0.6061	46	Performance goal met

10	0.9	0.0204	0.3068	78	Performance goal met
		0.0212	0.3341	78	Performance goal met
		0.0188	3.9142	77	Performance goal met
		0.0219	0.0473	73	Performance goal met
		0.0179	0.0473	74	Performance goal met
	Average value	0.0200	0.9299	76	Performance goal met
11	1.0	0.1194	3193.6149	2	Performance goal not met
		0.5454	1074.8956	2	Performance goal not met
		0.4905	1421.3462	2	Performance goal not met
		0.1424	1736.3575	2	Performance goal not met
		0.0590	426.5452	2	Performance goal not met
	Average value	0.2713	1570.5527	2	Performance goal not met

Table 9.5 Optimization of Learning Rate Keeping the Momentum Constant (for Depression Data)

Sl. No.	Learning Rate When Momentum = 0.4	Minimum Performance	Maximum Performance	Error Testing	Remarks
1	0.0	0.4489	0.4489	2.4749e-061	Performance goal not met
		1.4467	1.4467	5.1882e-063	Performance goal not met
		0.4271	0.4271	2.5644e-061	Performance goal not met
		0.6878	0.6878	4.8163e-061	Performance goal not met
		0.3964	0.3964	2.6617e-061	Performance goal not met
		0.6813	0.6813	2.6617e-061	Performance goal not met
	Average value	0.6813	0.6813	8.0978×10^{-27}	Performance goal not met
2	0.1	9.9976e-004	0.9785	1.1429e-062	Performance goal met
		9.9954e-004	2.1143	8.8865e-063	Performance goal met
		9.9943e-004	5.5067	3.1671e-063	Performance goal met
		9.9985e-004	1.4616	1.0753e-061	Performance goal met
		9.9996e-004	1.0139	1.2028e-062	Performance goal met
	Average value	0.1831	2.2150	3.3902×10^{-27}	Performance goal met
3	0.2	9.9914e-004	2.8155	5.6196e-063	Performance goal met
		9.9918e-004	1.3865	5.6604e-062	Performance goal met
		9.9887e-004	0.3869	3.5649e-062	Performance goal met
		9.9950e-004	0.7079	5.9817e-062	Performance goal met
		9.9914e-004	2.8155	5.6196e-063	Performance goal met
	Average value	0.1830	1.622	3.8706×10^{-27}	Performance goal met

4	0.3	9.9875e-004	0.5993	1.2433e-062	Performance goal met
		9.9840e-004	2.3079	2.9399e-062	Performance goal met
		9.9929e-004	0.5191	2.2065e-063	Performance goal met
		9.9907e-004	0.7297	2.1221e-062	Performance goal met
		9.9782e-004	0.4555	2.2690e-062	Performance goal met
	Average value	0.1829	0.9211	2.0845×10^{-27}	Performance goal met
5	0.4	9.9951e-004	0.5425	1.0404e-062	Performance goal met
		9.9903e-004	1.4912	1.6629e-062	Performance goal met
		9.9955e-004	0.5891	1.3934e-062	Performance goal met
		9.9937e-004	0.8292	1.1058e-062	Performance goal met
		9.9893e-004	1.6279	1.6404e-062	Performance goal met
	Average value	0.1830	1.0159	1.6218×10^{-27}	Performance goal met
6	0.5	9.9835e-004	1.5435	1.1686e-061	Performance goal met
		9.9885e-004	0.2811	1.7677e-061	Performance goal met
		9.9880e-004	1.1583	2.5881e-062	Performance goal met
		9.9990e-004	0.6335	2.8300e-062	Performance goal met
		9.9935e-004	0.9229	1.5757e-062	Performance goal met
	Average value	0.1829	0.9078	8.6170×10^{-27}	Performance goal met

(Continued)

Table 9.5 Optimization of Learning Rate Keeping the Momentum Constant (for Depression Data) (*Continued*)

Sl. No.	Learning Rate When Momentum = 0.4	Minimum Performance	Maximum Performance	Error Testing	Remarks
7	0.6	9.9623e-004	0.5394	1.8825e-061	Performance goal met
		9.9787e-004	0.8526	4.5554e-062	Performance goal met
		9.9709e-004	0.8909	1.8940e-062	Performance goal met
		9.9908e-004	0.8070	1.4197e-061	Performance goal met
		9.9915e-004	0.5316	1.6485e-062	Performance goal met
Average value		0.1827	0.7243	9.7459×10^{-27}	Performance goal met
8	0.7	9.9873e-004	0.4496	1.2992e-061	Performance goal met
		9.9972e-004	2.5152	1.458e-062	Performance goal met
		9.9968e-004	1.7797	8.7738e-063	Performance goal met
		9.9846e-004	3.8065	4.9368e-062	Performance goal met
		9.9963e-004	4.9775	3.2425e-063	Performance goal met
Average value		0.1830	2.7057	4.8798×10^{-27}	Performance goal met
9	0.8	9.9924e-004	0.5667	3.6378e-063	Performance goal met
		9.9969e-004	2.0912	4.8638e-063	Performance goal met
		9.9605e-004	0.7954	3.6837e-063	Performance goal met
		9.9992e-004	0.8831	5.3082e-063	Performance goal met
		9.9890e-004	1.8241	3.0262e-062	Performance goal met
Average value		0.1829	1.2321	4.1639×10^{-27}	Performance goal met

10	0.9	9.9994e-004	0.6731	3.3303e-062	Performance goal met
		9.9993e-004	2.1923	2.6447e-062	Performance goal met
		9.9973e-004	4.6183	3.9460e-063	Performance goal met
		9.9845e-004	3.4364	4.7455e-063	Performance goal met
		9.9245e-004	2.3466	2.6665e-062	Performance goal met
	Average value	0.1828	3.3406	2.2541×10^{-27}	Performance goal met
11	1.0	9.9987e-004	7.9361	3.4888e-063	Performance goal met
		9.9904e-004	4.1768	1.1744e-062	Performance goal met
		9.9988e-004	8.5171	2.6478e-061	Performance goal met
		9.9981e-004	4.2981	7.1292e-063	Performance goal met
		9.9899e-004	0.2967	6.1693e-062	Performance goal met
	Average value	0.1830	5.0449	8.2678×10^{-27}	Performance goal met

Table 9.6 Optimization of Momentum Keeping the Learning Rate Constant (for Depression Data)

Momentum When Learning Rate = 0.3	Minimum Performance	Maximum Performance	Error Testing	Remarks
0.0	9.9976e-004	0.6672	7.0528e-063	Performance goal met
	9.9773e-004	0.5993	2.1833e-064	Performance goal met
	9.9982e-004	0.9551	2.4003e-062	Performance goal met
	9.9966e-004	0.7982	1.3190e-061	Performance goal met
	9.9972e-004	0.7297	2.3246e-062	Performance goal met
Average value	0.1830	0.7499	4.4183×10^{-27}	Performance goal met
0.1	9.9975e-004	0.5891	1.6880e-062	Performance goal met
	9.9949e-004	0.8292	7.3215e-064	Performance goal met
	9.9987e-004	1.6279	1.1120e-062	Performance goal met
	9.9958e-004	0.2811	1.7882e-061	Performance goal met
	9.9999e-004	0.9470	1.8034e-061	Performance goal met
Average value	0.1831	0.8548	9.1934×10^{-27}	Performance goal met
0.2	9.9904e-004	0.6335	2.8097e-062	Performance goal met
	9.9889e-004	1.6363	3.6651e-063	Performance goal met
	9.9946e-004	0.9229	1.2369e-062	Performance goal met
	9.9985e-004	0.5537	1.4934e-061	Performance goal met
	9.9858e-004	0.5836	1.5129e-062	Performance goal met
Average value	0.1830	0.8660	4.9440×10^{-27}	Performance goal met

0.3	9.9945e-004	0.5341	1.1685e-062	Performance goal met
	9.9990e-004	0.3964	2.7386e-063	Performance goal met
	9.9995e-004	0.4641	3.9749e-063	Performance goal met
	9.9928e-004	0.6878	3.9790e-063	Performance goal met
	9.9964e-004	0.8431	1.4050e-062	Performance goal met
Average value	0.1830	0.5851	3.1761×10^{-27}	Performance goal met
0.4	9.9857e-004	0.6731	3.6605e-062	Performance goal met
	9.9803e-004	0.7280	3.4757e-063	Performance goal met
	9.9981e-004	0.7632	4.5409e-063	Performance goal met
	9.9956e-004	0.7299	2.1378e-062	Performance goal met
	9.9905e-004	0.5291	3.3617e-063	Performance goal met
Average value	0.1829	0.6846	1.6439×10^{-27}	Performance goal met
0.5	9.9892e-004	1.0080	2.4070e-062	Performance goal met
	9.9878e-004	0.8861	3.0865e-062	Performance goal met
	9.9928e-004	1.5038	4.6227e-063	Performance goal met
	9.9952e-004	1.9135	1.0587e-062	Performance goal met
	9.9806e-004	1.5241	3.9215e-062	Performance goal met
Average value	0.1829	1.1671	2.5919×10^{-27}	Performance goal met

(Continued)

Table 9.6 Optimization of Momentum Keeping the Learning Rate Constant (for Depression Data) (Continued)

Momentum When Learning Rate = 0.3	Minimum Performance	Maximum Performance	Error Testing	Remarks
0.6	9.9995e-004	1.6164	4.8782e-063	Performance goal met
	9.9998e-004	1.3822	2.2786e-061	Performance goal met
	9.9949e-004	1.2719	2.8905e-063	Performance goal met
	9.9894e-004	1.0263	3.7466e-062	Performance goal met
	9.9929e-004	0.7808	2.3585e-063	Performance goal met
Average value	0.1830	1.2155	6.5286×10^{-27}	Performance goal met
0.7	9.9987e-004	2.5993	1.0493e-062	Performance goal met
	9.9922e-004	0.9527	2.3569e-062	Performance goal met
	9.9708e-004	0.5307	4.8139e-063	Performance goal met
	9.9826e-004	1.1002	3.2426e-062	Performance goal met
	9.9978e-004	2.4447	2.6054e-062	Performance goal met
Average value	0.1829	1.5255	2.3074×10^{-27}	Performance goal met
0.8	9.9994e-004	0.5189	2.6255e-063	Performance goal met
	9.9845e-004	0.4282	2.6489e-062	Performance goal met
	9.9903e-004	0.6108	3.5646e-062	Performance goal met
	9.9872e-004	0.7619	3.0564e-062	Performance goal met
	9.9983e-004	0.8981	2.2712e-061	Performance goal met
Average value	0.1830	1.0841	7.6423×10^{-27}	Performance goal met

0.9	0.0062	3.8480	2.2954e-062	Performance goal not met
	0.0016	1.1225	7.0545e-063	Performance goal not met
	0.0036	5.8506	3.5743e-062	Performance goal not met
	0.0040	4.9478	2.0352e-062	Performance goal not met
	0.0012	7.8749	1.9254e-062	Performance goal not met
Average value	0.0026	4.7287	2.4971×10^{-27}	Performance goal not met
1.0	0.4567	3.6911e + 006	7.4440e-058	Performance goal not met
	0.2772	2.1036e + 005	3.9057e-059	Performance goal not met
	0.4851	1.0858e + 007	9.1807e-058	Performance goal not met
	0.2872	4.8472e + 006	3.6958e-058	Performance goal not met
	0.4292	3.3759e + 006	2.2876e-058	Performance goal not met
Average value	0.3870	4596512	2.9761×10^{-25}	Performance goal not met

Table 9.7 Classification Accuracy of the BPNN Classifier (See Section 9.3.2.7)

Tuples/Cases	Grades of Depression		
	Mild	Moderate	Severe
Correctly classified	2	7	46
Incorrectly classified	0	2	5
Accuracy (%)	100	77	90

In this classification task, the optimum learning rate and the momentum/gradient through iterations (as done in iris data) are checked, which are 0.4 for each. Finally, the accuracy of classification is measured for each class label (mild, moderate, and severe).

9.4.3.1 Errors (δ) in Terms of Prediction

Errors in terms of prediction (δ) by the developed classifier are also calculated, and the plots are shown below (refer to Figures 9.5, 9.6 and 9.7). The calculated value of error over 10 iterations of training and testing was found to be negligible (see Section 9.3.2.5). The plot shows improvement in precision as the iterations advance further. In this plot, it may be noted that from the sixth iteration there is improvement in predictions until the tenth iteration, and there is no further improvement afterward (Figures 9.1 and 9.2, respectively).

9.4.3.2 k-Times Holdout

The calculated value of error for 10 iterations of *k*-times holdout (see Section 9.3.2.4) was found to be negligible.

9.4.3.3 Regression

A regression study for 10 iterations was performed and a comparison of regression coefficients done for ascertaining the best fit for the proposed classifier model for 10 iterations (see Section 9.3.2.6). It was found that the average correlation coefficient for 10 iterations was 94%, which denotes a good linear fit as it is very close to 1.0 (the best fit), which may be seen in figure 9.8.

9.5 Conclusion

This study proposes a BPNN-based method that can clearly classify depression grades as mild, moderate, and severe. This automatic grading system is an improvement over the practice of manual grading using mixture terms (e.g., mild-to-moderate and moderate-to-severe). In this study, depression parameters

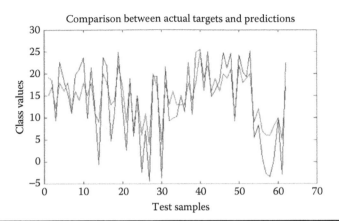

Figure 9.5 Target versus predicted plots for 62 depression cases.

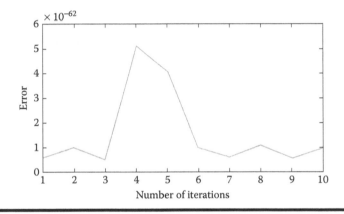

Figure 9.6 Error plot for each of 10 iterations.

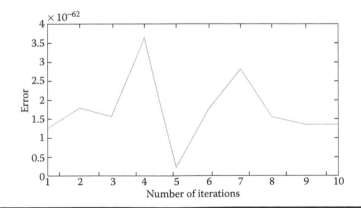

Figure 9.7 Error plot for each of 10 iterations during *k*-times holdout.

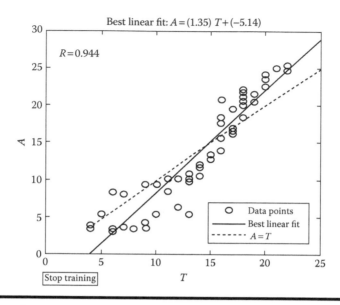

Figure 9.8 The best-fit regression curve.

were collected from the available literature to define an HT that, in turn, helped to generate a questionnaire for data collection and develop a neural classifier model. It was observed that the collected depression data is reliable and consistent ($\alpha = 0.77$) according to the study by Nunnaly (1978).

The proposed BPNN classifier predicted mild depression with 100% accuracy, whereas for moderate and severe depression it showed 77% and 90% accuracies, respectively, which can be further increased if tested on a larger data set (due to the classifier's incremental learning property). Calculated error rates are negligibly low even for k-times holdout. The regression model shows that the correlation coefficient (R) is 94%, which is closer to the best linear fit.

One potential application of this work is to assist general practitioners located at rural health centers that seriously suffer from a lack of specialist services. The proposed method can be integrated in a medical support system for screening, grading, and referring depression cases to specialty centers for treatment.

References

Aneshensel, C. S., R. R. Frerichs, and G. J. Huba. 1984. Depression and physical illness: A multiwave, nonrecursive causal model. *J Health Soc Behav* 25:350–371.

Aruna, P., N. Puviarasan, and B. Palaniappan. 2005. Diagnosis of gastrointestinal disorders using DIAGNET. *Expert Syst Appl* 32(2):329–35.

Aruna, P., N. Puviarasan, and B. Palaniappan. 2005. An investigation of neurofuzzy system in psychosomatic disorders. *Expert Syst Appl* 28:673–79.

Astion, M. L., and P. Wilding. 1992. The application of backpropagation neural network to problems in pathology and laboratory medicine. *Arch Pathol Lab Med* 116:995–1001.

Austin, M. P., P. Mitchell, and G. M. Goodwin. 2001. Cognitive defects in depression: Possible implications for functional neuropathology. *Br J Psychiatry* 178:200–06.

Bagby, R. M., G. R. Andrew, R. S. Deborah, and M. B. Marshall. 2004. The Hamilton Depression rating scale. *Am J Psychiatry* 161:2163–77.

Ball, H. A., P. McGuffin, and A. E. Farmer. 2008. Attributional style and depression. *Br J Psychiatry* 192:275–78.

Bhatikar, S. R., C. DeGroff, and R. L. Mahajan. 2004. A classifier based on the artificial neural networks for the cardiologic auscultation in pediatrics. *Artif Intell Med* 33:251–60.

Chattopadhyay, S., D. K. Pratihar, and S. C. De Sarkar. 2007. Some study on fuzzy clustering of psychosis data. *Int J Bus Intell Data Min* 2:143–59.

Chattopadhyay, S., D. K. Pratihar, and S. C. De Sarkar. 2008. Developing fuzzy classifier to predict the chance of occurrence of adult psychosis. *Knowl based Syst* 21:479–97.

Chen H., Fuller S.S., Friedman C., Hersh W. 2005. Knowledge management, data mining and text mining in medical informatics. In *Medical Informatics Knowledge Management and Data Mining in Biomedicine*, eds. H. Chen, S. S. Fuller, C. Friedman, and W. Hersh, 4–30. New York: Springer's Integrated Series in Information Systems.

Cho, J. M. 2000. Chromosome classification using backpropagation neural networks. *IEEE Eng Med Biol Mag* 19:28–33.

Cho, B. H., H. Yu, J. Lee, Y. J. Chee, S. I. Kim, and I. Y. Kim. 2008. Nonlinear support vector machine visualization for risk factor analysis using nomograms and localized radial basis function kernels. *IEEE Trans Inf Technol Biomed* 12920:247–56.

Cios, K. J., and G. W. Moore. 2002. Uniqueness of medical data mining. *Artif Intell Med* 26:1–24.

Cougnard, A., L. R. Salmi, R. Salamon, and H. Verdoux. 2004. A decision analysis and model to assess the feasibility of early detection of psychosis in the general population. *Schizophrenia Res* 74:27–36.

Cronbach, L. J. 1951. Coefficient alpha and the internal structures of tests. *Psychometrika* 16:297–334.

Curiac, D., I. Vasile, G. Banias, O. Volosencu, C. Albu A. 2009. Bayesian network model for diagnosis of psychiatric diseases. In *IEEE Information Technology Interfaces, Proceedings of the ITI-09*, eds. V. Luzar-Stier, I. Jarec, Z. Bekic, 61–6. Dubrovnik, Croatia.

Das, A., T. BenMenachem, F. T. Farooq, G. S. Cooper, A. Chak, M. V. Sivak Jr, and R. C. K. Wong. 2008. Artificial neural network as a predictive instrument in patients with acute nonvariceal upper gastrointestinal hemorrhage. *Gastroenterology* 134:65–74.

Davis, G. E., W. E. Lowell, and G. L. Davis. 1993. A neural network that predicts psychiatric length of stay. *MD Comput* 10:87–92.

DiMatteo, M. R., H. S. Lepper, and T. W. Croghan. 2000. Depression is a risk factor for noncompliance with medical treatment. *Arch Intern Med* 160:2101–07.

Fisher, R. A. 1936. The use of multiple measurements in axonomic problems. *Ann Eugen* 7:179–88.

Forsell, Y., A. F. Jorm, and B. Winblad. 1994. Association of age, sex, cognitive dysfunction and disability with major depressive symptoms in an elderly sample. *Am J Psychiatry* 151:1600–04.

Gotlib, I. H., and C. L. Hammen. 1992. Psychological aspects of depression: Toward a cognitive-interpersonal integration. In *The Wiley Series in Clinical Psychology*, vol. xi, 330. New York: John Wiley & Sons.

Gross, G. W., J. M. Boone, V. Grecohunt, and B. Greenberg. 1990. Neural network in radiologic diagnosis. *Invest Radiol* 25:165–224.

Guler, I., and E. D. Ubeyli. 2003. Detection of ophthalmic artery stenosis by least-mean squares backpropagation neural network. *Eng Appl Artif Intell* 18:413–22.

Hamilton, M. 1960. A rating scale for depression. *J Neurol Neurosurg Psychiatry* 23:56–62.

Han, J., and M. Kamber. 2006. *Data Mining Concepts and Techniques.* San Francisco, CA: Morgan Kaufmann Publishers.

Hopfield, J. J. 1982. Neural network and physical systems with collective computational abilities. *Proc Natl Acad Sci USA* 79(8):2554–58.

http://www.who.int/topics/depression/en/ is a web reference. (Accessed December 18, 2010.)

Johnston, M. E., K. B. Langton, R. B. Haynes, and A. Mathieu. 2001. Effects of computer decision support systems on clinician performance and patient outcome: A critical appraisal of research. *Ann Intern Med* 120:2135–42.

Kaplan, H. I., B. J. Saddock, and J. A. Greb. 1994. Synopsis of psychiatry. In *Behavioural Sciences and Clinical Psychiatry*, eds. H. I. Kaplan, B. J. Saddock, J. A. Greb, 803–23. New Delhi, India: B.I. Waverly Pvt Ltd. New Delhi, India.

Kawamoto, K., C. A. Houlihan, E. A. Balas, and D. F. Lobach. 2005. Improving clinical practice using clinical decision support systems: A systematic review of trials to identify features critical to success. *Br Med J* 330(7494):765.

Keller, M. B. 1994. Depression: A long term illness. *Br J Psychiatry* 26:9–15.

Kohonen, T. 1995. *Self Organizing Maps.* Berlin: Springer Verlag.

Lee, S. J., and K. Siau. 2001. A review of data mining techniques. *Ind Manage Data Syst* 101:41–6.

Li, Y., L. Liu, W. Chiu, and W. Jian. 2000. Neural network modeling for surgical decisions in traumatic brain injury patients. *Int J Med Inform* 57:1–9.

Lin, A., L. A. Lenert, M. A. Hlatky, K. M. McDonald, R. A. Olshen, and J. Homberger. 1995. Clustering and the design of the preference-assessment surveys in healthcare. *Health Serv Res* 34:1033–45.

Lisoba, P. J., and A. F. G. Taktak. 2006. The use of artificial neural network in decisive support in cancer: A systematic review. *Neural Netw* 19:408–15.

Lucas, P. 2004. Bayesian analysis, pattern analysis and data mining in healthcare. *Curr Opin Crit Care* 10:399–403.

Magnotta, V. A., N. C. Andreasen, D. Heckel, T. Cizadlo, P. W. Corson, J. C. Ehrhardt, and W. T. C. Yuh. 1999. Measurement of brain structures with artificial neural networks: Two and three dimensional applications. *Radiology* 211:781–90.

Nunnaly, J. 1978. *Psychometric Theory.* New York: McGraw-Hill.

Pradhan, N., P. K. Sadasivan, and G. R. Arunodaya. 1999. Detection of seizure activity EEG by an artificial neural network: A preliminary study. *Comput Biomed Res* 29:303–13.

Prather, J. C., D. F. Lobach, L. K. Goodwin, J. W. Hales, M. L. Hage, and W. E. Hammond. 1997. Medical data mining: Knowledge discovery in a clinical data warehouse. In *American Medical Informatics Association Annual Fall Symposium Proceedings Archive*, ed. D. R Masys, 101–05. Nashville, TN: Hanley and Befus Inc.

Robinson, A., and R. Thomas. 2001. Variability in patient preferences for participating in medical decision making: implication for the use of decision support tools. *Qual Saf Health Care* 10:34–8.

Rumelhart, D. E., G. E. Hilton, and J. L. McCelland. 1986. A general framework for parallel distributed processing. In *Parallel Distributed Processing*, ed. PDP research group, 45–76. Cambridge MA: The MIT Press.

Sordo, M., and Q. Zeng. 2005. On sample size and classification accuracy: A performance comparison. In *Biological and Medical Data Analysis, Lecture Notes in Computer Science,* eds. J. L. Oliveira, M. S. Fernando, V. Maojo, vol. 3745, 193–201. 6th International Symposium ISBMDA 2005. Aveiro, Portugal: Springer Berlin/Heidelberg.

Tu, J. V. 1996. Advantages and disadvantages of using artificial neural networks versus logistic regression for predicting medical outcomes. *J Clin Epidemiol* 49:1225–31.

Weiss, E. L., J. G. Longhurst, and C. M. Mazure. 1999. Childhood sexual abuse as a risk factor for depression in women: Psychosocial and neurobiological correlate. *Am J Psychiatry* 156:816–28.

Wilding, P., M. A. Morgan, A. E. Grygotis, M. A. Shoffner, and E. F. Rosato. 1994. Application of backpropagation neural networks for diagnosis of breast and ovarian cancer. *U.S. National Library of Medicine Cancer Lett* 77:145–53.

Zou, Y., Y. Shan, L. Shu, Y. Wang, F. Feng, K. Xu, Y. Song et al. 1996. Artificial neural network to assist psychiatric diagnosis. *Br J Psychiatrists* 169:64–7.

Zung, W. W. K. 1972. The depression status inventory: An adjunct to the self rating depression scale. *J Clin Psychol* 28:539–43.

Chapter 10

Alignment-Based Clustering of Gene Expression Time-Series Data

Numanul Subhani, Luis Rueda,
Alioune Ngom, and Conrad Burden

Contents

10.1 Introduction

Clustering is a multivariate analysis technique used to discover unknown patterns or groups in data and is an appropriate method to use when there is no a priori knowledge about the data. Clustering can be done on any data including genes, samples, time points in a time series, etc. The type of input makes no difference to the performance of a clustering algorithm. The algorithm treats all inputs as n-dimensional feature vectors. To group objects that are similar, we need a very precise definition of the measure of similarity. There are many different ways in which such a measure of similarity can be calculated depending on the representation of gene expression profiles.

We discuss the clustering problem of microarray time-series gene expression profiles. The time-series clustering problem is formally stated in order to discuss these approaches. A data set $D = \{x_1(t), x_2(t), \ldots, x_s(t)\}$ is given. Then, $x_i = [x_{i1}, x_{i2}, \ldots, x_{in}]^t$ is an n-dimensional feature vector that represents the expression level of gene i at n different time points, $t = [t_1, \ldots, t_n]^t$. We want to partition a set of s profiles, D, into k disjoint clusters, $C_1, \ldots, C_k, 1 \le k \le s$, such that $C_i \ne \phi$, $i = 1, \ldots, k$, $U_{i=1}^k C_i = D$, and $C_i \cap C_j = \phi$; $i \ne j$; $i, j = 1, \ldots, k$. Each profile is assigned to the cluster that is closest to it. We consider the specific case of time-series clustering, in which the order of time points cannot be changed because different permutations give different results that are biologically meaningless.

In this chapter, a comprehensive review of clustering approaches for analyzing microarray time-series data is presented. The key features of each method are highlighted, a methodological comparison is provided, the advantages and drawbacks of the methods are examined, and an experimental comparison of some of these methods is performed.

10.2 Current Clustering Methods for Gene Expression Time-Series Profiles

Many clustering methods for time-series gene expression data have been developed. Most of these approaches are either directly adopted or somewhat modified from the standard clustering methodology available in the fields of classification and pattern recognition. These methods, in general, apply some conventional metrics and algorithms to gene temporal expression profiles for clustering and do not perform any profile alignment before applying a similarity measure.

A partitional clustering method based on the k-means approach was applied to cluster gene expression temporal data in the study by Cho et al. (1998). In k-means clustering, a data set is partitioned into k predefined clusters by iteratively reallocating cluster members such that the overall within-cluster dispersion is minimized. For the similarity measure, the authors used Euclidean distance to compute the dissimilarity between each pair of genes in the feature space. The application of k-means clustering to this data set revealed new sets of coregulated genes and their putative cis-regulatory elements. For this approach, only the value of k must be known a priori; no other assumptions about the dynamics of the expression profile are required before application. The lack of structure in clustering may result in inconsistent clusters, since the clustering may reinforce local groups (Tamayo et al. 1999).

In a study conducted by Tamayo et al. (1999), the researchers applied self-organizing maps (SOMs) to visualize and interpret patterns of gene temporal expression profiles. The SOM, a type of mathematical cluster analysis, works well with exploratory analysis of data and reveals relevant patterns in large, high-dimensional data sets. This method consists of maintaining a set of nodes, a topology, and a distance function (on the nodes). The algorithm iteratively maps the nodes into the feature space of the genes. Deciding on the number of nodes is important, since if the number is small the patterns cannot be distinguished properly (due to a large within-cluster scatter). On the other hand, addition of nodes exceeding a certain value (that produces distinct patterns) fails to produce any new patterns.

A Bayesian approach for improving the clustering results of a gene expression time series using rough knowledge regarding the general shapes of the classes was proposed (Brehelin 2005). Knowledge about the general shapes of classes can be elementary regarding the change of the mean expression level over time. Information regarding the shape of a class is directly integrated into the model so that classes with the desired profiles are favored. However, if no such information is available then classical clustering using a Bayesian approach is performed. The Bayesian approach intuitively deals with the temporal nature of data. The effectiveness of this method for recovering a particular class of genes, regarding which there was prior knowledge, was demonstrated using a data set composed of a mixture of real and synthetic data. It was constructed by injecting some synthetic data into the original fibroblast data set (Iyer et al. 1999). The experiments show that the method can recover the synthetic data with very high accuracy.

A Bayesian method was also applied for model-based clustering where the models were autoregressive curves of a fixed order (Ramoni, Sebastiani, and Kohane 2002). To search for the most likely set of clusters out of the given temporal expression data, an agglomerative procedure was used. During clustering, the approach explicitly took into account the dynamic nature of gene expression time-series data and detailed a way that identifies the number of distinct clusters. This method can be viewed as a specialized version of Bayesian clustering by dynamics (BCD), where the concept of similarity is defined in such a way that two time series are considered similar if they are generated by the same stochastic process. The proposed method models temporal gene expression profiles by autoregressive equations that derive a posterior probability, and agglomerates models having maximum posterior probabilities. This method is also able to identify the optimal number of clusters based on the well-known Akaike information criterion.

A hidden Markov model (HMM) that accounts for horizontal dependencies in gene expression time-series data was proposed in a study by Schliep and Torney (2003). The approach proposed by these authors focuses on univariate emission probability densities. Usually, model-based clustering approaches are effective when grammatical or structural properties of the data can be explicitly expressed, and the HMM can be used by these types of approaches to partition gene expression time-series data into clusters. The process starts with an initial collection of HMMs that encompass typical qualitative behavior, and the process iteratively finds cluster models and assigns data points to these models in such a way that the joint likelihood of the clustering is maximized. The approach also provides a method for partially supervised learning, which allows users to add groups of labeled data to the initial collection of clusters. The method can cope effectively with unlabeled data, as well as with additional labeled data. The information contained in the labeled data is used to produce high-quality clusters. This method is also robust with respect to noisy and frequently missing data.

In a study by Kruglyak, Heyer, and Yooseph (1999), the authors proposed a new correlation coefficient–based method for clustering gene expression time-series profiles. They introduced a similarity measure that gives high scores to gene pairs that exhibit similar behavior throughout the time points and is robust to outliers. It can also reduce false positives, that is, giving a high score to a pair of dissimilar profiles by the similarity measure.

In the work by Peddada et al. (2003), an order-restricted inference method was used to select and cluster gene expression profiles for time-series or dose-response data. This approach uses known inequalities among parameters and applies the ideas of order-restricted inference. The entire process is carried out in multiple steps. Potential candidate profiles of interest are defined and expressed in terms of inequalities between the expected gene expression levels at various time points, and then the mean expression level of each gene is estimated by using the method proposed in the work by Hwang and Peddada (1994). The gene is assigned to a profile based on the goodness-of-fit criterion and the bootstrap test procedure for a given gene (Peddada, Prescott, and Conaway 2001), and the process continues for each gene in the data set under consideration. This method makes use of the ordering in a time-series study and detects genes more sensitively using their temporal ordering and ability of finding consistent patterns over time.

Conesa, Nueda, and Ferrer (2006) proposed a statistical procedure that identifies genes with different expression profiles across analytical groups in time-series experiments. This statistical procedure is a general regression-based approach suitable for analyzing single or multiple microarray temporal data. This method, referred to as maSigPro (microarray significant profiles), is a two-step regression strategy in which experimental groups are identified by dummy variables. The procedure adjusts a global regression model with all the defined variables to identify differentially expressed genes. Profiles with statistically significant difference are then found by applying a variable selection strategy that studies the differences between the groups. The model parameters are adjusted based on the data under study and the specific interests of the researcher. The proposed method does not require multiple pairwise comparisons, although it can detect significant profile differences. It also allows for unbalanced designs and heterogeneous sampling times. This approach can be used to find genes with significant temporal expression changes between experimental groups and to analyze the magnitude of these differences.

Ernst, Nau, and Bar-Joseph (2005) proposed an algorithm that specifically addresses the challenges inherent to short time-series expression data sets for clustering such expression data. The algorithm first selects a set of distinct patterns that can be expected from the experiment, and then assigns genes to the profile that best represents them among the preselected profiles. This method assigns genes to profiles and determines the significance of each profile. Significant profiles are retained for further analysis and can be combined to form clusters. This approach uses the correlation coefficient to measure the similarity between expression profiles, and identifies a more coherent set of genes than contemporary clustering methods based on the *k*-means approach, by grouping together the temporal profiles of relevant functional categories. This method can work with a data set with no repeats by leveraging the statistical power obtained from the large number of genes being profiled simultaneously.

Bar-Joseph et al. (2003) focused on the analysis of gene temporal expression profiles that can cope with the problems of missing values (unobserved time points) and nonuniformly sampled data. Gene temporal expression profiles are represented as continuous curves using statistical spline estimation. In this method, a model-based clustering algorithm (using a modified expectation-maximization [EM] algorithm) operates directly on the continuous representations of gene expression profiles, thus permitting interpolation of missing values, and an alignment algorithm uses the spline representation to warp the time scale of one realization into another.

The algorithm attempts to produce an optimal alignment (that maximizes the similarity between the two sets of expression profiles) by adjusting the function parameters. This approach reconstructs unobserved time points with 10%–15% fewer errors than other methods. Further, the clustering approach was effective with nonequidistant sampled time-series data, and the continuous alignment algorithm, which allows the user to control the number of degrees of freedom of the warp, helps to avoid overfitting and the difficulties faced by discrete methods. Although this method is suitable for the analysis of relatively long time series (10 time points or more), the suitability of the approach for shorter time-series experiments is yet to be demonstrated.

In a study by Déjean et al. (2007), the researchers also attempted to obtain relevant clusterings of gene expression temporal profiles by identifying homogeneous clusters of genes. This method focuses on the shapes of the curves and not on the absolute levels of expression (or expression ratios). It combines spline smoothing and first-derivative computation with hierarchical and partitional clustering. To smooth the temporal profiles, it is necessary to obtain regular and differentiable functions. Therefore, the approach is based on the framework of functional data analysis (Ramsay and Silverman 2005) with focus on the first derivative of curves by means of a priori spline smoothing. Spline smoothing is performed, and clustering of derivatives of the continuous smoothed curves results from the smoothing. Both hierarchical clustering and the k-means approach are applied to obtain the clusters. The smoothing parameter is tuned in the first step by using a heuristic approach that takes into account both statistical and biological considerations.

Cho et al. (2005) proposed a similarity measure for coexpressed genes based on the expression-level rate of change across time points. For this method, profiles are considered as piecewise linear (PL) functions, and the similarity is calculated by measuring the difference of slopes between the functions. Unequal time intervals are considered and viewed as weights. The proposed algorithm, motivated by the advantages of fuzzy clustering, is referred to as "fuzzy short time-series" (FSTS) clustering, and it incorporates the distance measure proposed in this work in the fuzzy c-means clustering scheme. The performance of the proposed method was shown to be better than that of conventional approaches (fuzzy c-means, k-means, average linkage hierarchical clustering, and random clustering) for unevenly distributed time points. The proposed method benefits from the fuzzy approach that inherently accounts for noise in the data and allows genes to belong to more than one group. The method is designed to cope with short and unevenly sampled time-series data.

Area-based profile alignment proposed by Rueda, Bari, and Ngom (2008) makes use of two features vectors and produces two new vectors in such a way that the area between "aligned" vectors is minimized. A profile alignment method that takes the length of intervals between the time points into consideration was proposed by Rueda, Bari, and Ngom (2008). This approach considers the weights of all intervals equally, irrespective of the actual size of the interval of the measurement. The profile-alignment algorithm takes two feature vectors from the original space as input, and outputs two feature vectors in the transformed space after aligning them in such way that the sum of squared errors is minimized. The profiles are aligned using an area-based distance function rather than conventional distance functions. The area-based distance function is defined by computing the integral distance between the two aligned profiles. In the studies of Rueda, Bari, and Ngom (2008), and Rueda and Bari (2006), hierarchical agglomerative clustering is used, where the decision rule is based on the furthest-neighbor distance or complete linkage distance between two clusters. The complete linkage, or furthest-neighbor, approach calculates the distance between the furthest pair of points for each pair of clusters and merges the two clusters that have the minimum distance among all such distances between all pairs of clusters under consideration. This clustering approach performs pairwise alignment before measuring the distance between pairs of given profiles, in each iteration. This step slows down the computational process.

Yin and Chiang (2008) proposed a variation-based clustering method. In this approach, gene expressions are translated into gene variation vectors, and the cosine values of these vectors are used to evaluate their similarity over time. The proposed algorithm is designated as the "variation-based coexpression detection algorithm" (VCD). The algorithm has two main advantages: First, it is unnecessary to determine the number of clusters in advance since the algorithm automatically detects those genes having profiles that are grouped together and creates patterns for these groups. Second, the algorithm features a new measurement criterion for calculating the degree of change of expressions between adjacent time points and evaluating their trend similarities. The use of the cosine measure based on variation vectors not only enables the algorithm to evaluate trend similarities in time-series expressions but also allows the accurate evaluation of the co-occurring time of the peaks within the expressions. The performance of the proposed algorithm has been verified via application to three real-world microarray data sets. The experimental results confirm the enhanced grouping performance of the proposed algorithm.

10.3 Profile Alignment of Gene Expression Time Series

The aim of clustering gene expression temporal data is to group together profiles with similar patterns. Deciding on the similarity often involves pairwise distance measurements of coexpressions. Conventional distance measures include correlation, rank correlation, Euclidean distance, and angle between vectors of observations, among others. However, clustering algorithms that apply a conventional distance function (e.g., Euclidian distance, correlation coefficient) on a pair of profiles can

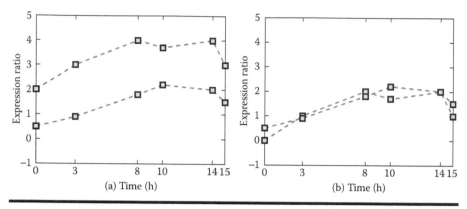

Figure 10.1 Typical examples for profile alignment: (a) Unaligned profiles and (b) aligned profiles.

fail to reflect the temporal information embedded in the underlying expression profiles. Some alignment techniques resolve the issue effectively prior to applying the distance function. The basic idea and the effect of profile alignment when computing the distance between two profiles can be visualized in the following example.

Typical examples for profile alignment are depicted in Figure 10.1 with a pair of genes. Figure 10.1a shows the genes prior to alignment. Using the Pearson correlation distance (Peddada et al. 2003), genes in Figure 10.1a are not most likely to be clustered together. If the prime interest is to cluster genes according to the variation of their expression level at different time points, then the genes from Figure 10.1b (after alignment) make more accurate clusters.

10.3.1 Pairwise Gene Expression Profile Alignment

In this section, we describe the concept of alignment of pairs of gene expression profiles. We will first discuss the alignment of piecewise linear profiles as introduced in Rueda and Bari (2006) and then describe a generalization to continuous integrable profiles.

10.3.1.1 Piecewise Linear Profiles

The motivation for profile alignment based on mean-squared error, as proposed by Rueda and Bari (2006), is the consideration of two feature vectors and production of two new vectors in such a way that the sum of squared-error differences between the aligned expression ratios at each time point is minimized. This approach treats weights of the intervals between the measurements of an experiment equally, irrespective of actual size of the interval of the measurement. In the study by Rueda and Bari (2006), the profile alignment scheme works as follows: Let $x_1, x_2 \varepsilon D$ be two *m*-dimensional feature vectors. The aim is to find two new vectors x_1 and $x_2' = x_2 - a$ (e.g., to find a scalar a). The algorithm takes two feature vectors from the original space as inputs (which are two temporal gene expression data in this case) and outputs two feature vectors in the transformed space after aligning them in such a way that the sum of squared errors is minimized.

10.3.1.2 Continuous and Integrable Profiles

Given two profiles, $x(t)$ and $y(t)$ (either PL or continuously integrable functions), where $y(t)$ is to be aligned to $x(t)$; the basic idea of alignment is to vertically shift $y(t)$ toward $x(t)$ in such a way that the squared errors between the two profiles is minimal. Let $\hat{y}(t)$ be the result of shifting $y(t)$. Here, the error is defined in terms of the areas between $x(t)$ and $\hat{y}(t)$ in the interval $[0, T]$. The functions $x(t)$ and $\hat{y}(t)$ may cross each other many times, but we want the sum of all the areas where $x(t)$ is above $\hat{y}(t)$ minus the sum of those areas where $\hat{y}(t)$ is above $x(t)$ to be minimal (see Figure 10.2). Let a denote the amount of vertical shifting of $y(t)$. Then, we want to find the value a_{min} of a that minimizes the integrated squared error between $x(t)$ and $\hat{y}(t)$. Once we obtain a_{min}, the alignment process consists of performing the shift on $y(t)$ as $\hat{y}(t) = y(t) - a_{min}$.

The pairwise alignment results of the study by Rueda, Bari, and Ngom (2008) are generalized from the case of PL profiles to profiles that are integrable functions on a finite interval. Suppose that we have two profiles, $x(t)$ and $y(t)$ defined on the time interval $[0, T]$. The alignment process consists of finding the value a that minimizes

$$f^a(x(t), y(t)) = \int_0^T [x(t) - y(\hat{t})]^2 \, dt = \int_0^T [x(t) - [y(t) - a]]^2 \, dt \qquad (10.1)$$

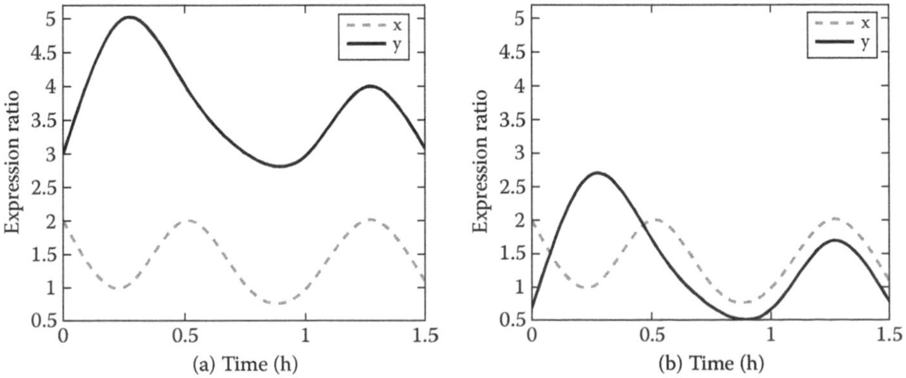

Figure 10.2 Profiles before and after applying shift: (a) Unaligned profiles and (b) aligned profiles $x(t)$ and $y(t)$ after applying $y(t) \leftarrow y(t) - a_{min}$.

Differentiating with respect to a yields

$$\frac{d}{da} f_a(x(t), y(t)) = 2\int_0^T [x(t) + a - y(t)] dt = 2\int_0^T [x(t) - y(t)] dt + 2aT \quad (10.2)$$

Setting $\dfrac{d}{da} f_a(x(t), y(t)) = 0$ and solving for a gives

$$a^{\min} = -\frac{1}{T} \int_0^T [x(t) - y(t)] dt \quad (10.3)$$

Since $\dfrac{d^2}{da^2} f_a(x(t), y(t)) = 2T > 0$, $a_{\min}$ is a minimum. The integrated error between $x(t)$ and the shifted $\hat{y}(t) = y(t) - a_{\min}$ is then

$$\int_0^T [x(t) - \hat{y}(t)] dt = \int_0^T [x(t) - y(t)] dt = \cdots + a_{\min} T = 0 \quad (10.4)$$

In terms of Figure 10.2, Equation 10.4 means that the sum of all the areas where $x(t)$ is above $y(t)$ minus the sum of those areas where $y(t)$ is above $x(t)$ is zero. Given an original profile $x(t) = [e_1, e_2, ..., e_n]$ (with n expression values taken at n time points $t_1, t_2, ..., t_n$), we use natural cubic spline (NCS) interpolation, with n knots $(t_1, e_1), ..., (t_n, e_n)$, to represent $x(t)$ as a continuously integrable function.

$$x(t) = x_j(t) \quad \text{if} \quad t_j \leq t \leq t_{j+1} \quad (10.5)$$

where $x^j(t) = x^{j3}(t - t^j)^3 + x^{j2}(t - t^j)^2 + x^{j1}(t - t^j)^1 + x^{j0}(t - t^j)^0$ interpolates $x(t)$ in the interval $[t^j, t^{j+1}]$, with spline coefficients $x^{jk} \in R$, for $1 \leq j \leq n - 1$ and $0 \leq k \leq 3$.

For practical purposes, given the coefficients $x_{jk} \in R$ associated with $x(t) = [e_1, e_2, ..., e_n] \in R^n$, we need only to transform $x(t)$ into a new space as follows: $x(t) = [x_{13}, x_{12}, x_{11}, x_{10}, ..., x_{j3}, x_{j2}, x_{j0}, ..., x_{(n-1)3}, x_{(n-1)2}, x_{(n-1)1}, x_{(n-1)0}] \in R^{4(n-1)}$. We can add or subtract polynomials given their coefficients, and the polynomials are continuously differentiable. This yields an analytical solution for $a_{\min}$ in Equation 10.3 as follows:

$$a^{\min} = -\frac{1}{T} \sum_{j=1}^{n-1} \int_{t_j}^{t_{j+1}} [x_j(t) - y_j(t)] dt = -\frac{1}{T} \sum_{j=1}^{n-1} \sum_{k=0}^{3} \frac{(x_{jk} - y_{jk})(t_{j+1} - t_j)^{k+1}}{k+1} \quad (10.6)$$

Figure 10.2b shows a pairwise alignment of the two initial profiles in Figure 10.2a, after the vertical shift $y(t) \leftarrow y(t) - a_{\min}$ has been applied. The two aligned profiles cross each other many times, but the integrated error, Equation 10.4, is zero.

Using Equation 10.4, the horizontal t axis will bisect a profile $x(t)$ into two halves with equal areas, when $x(t)$ is aligned to the t axis. In Section 10.3.1.3, we use this property of Equation 10.4 to define the multiple alignment of a set of profiles.

10.3.1.3 Multiple Gene Expression Profile Alignment

Given a set $D = \{x_1(t) \dots x_s(t)\}$, we want to align the profiles such that the integrated squared error between any two vertically shifted profiles is minimal. Thus for any $x_i(t)$ and $x_j(t)$, we want to find the values of a_i and a_j that minimize

$$f^{ai,aj}(x_i(t), x^j(t)) = \int_0^T [\hat{x}_i(t) - \hat{x}^j(t)]^2 \, dt = \int_0^T \Big[[x_i(t) - a_i] - [x_j(t) - a_j]\Big]^2 dt \quad (10.7)$$

where both $x_i(t)$ and $x_j(t)$ are shifted vertically by the amounts a_i and a_j, respectively, in possibly different directions, whereas in the pairwise alignment of Equation 10.1 the profile $y(t)$ is shifted toward a fixed profile $x(t)$. The multiple alignment process then consists of finding the values of $a_1, \dots, a_s$ that minimize

$$F_{a_1 \dots a_S}(x_1(t), \dots, x_s(t)) = \sum_{1 \le i \le j \le s} f_{a_i, a_j}(x_i(t), x_j(t)) \quad (10.8)$$

We use Lemma 10.3.1 to find the values a_i and a_j, $1 \le i \le j \le s$, that minimize $F_{a_1 \dots a_s}$.

Lemma 10.3.1. If $x_i(t)$ and $x_j(t)$ are pairwise aligned each to a fixed profile, $z(t)$, then the integrated error $\int_0^T [\hat{x}_i(t) - \hat{x}_j(t)] dt = 0$.

Proof. If $x_i(t)$ and $x_j(t)$ are pairwise aligned each to $z(t)$, then from Equation 10.3, we have $a_{\min} = -\frac{1}{T} \int_0^T [z(t) - x_i(t)] dt$ and $a_{\min} = -\frac{1}{T} \int_0^T [z(t) - x_j(t)] dt$.

Then,

$$\int_0^T \Big[\hat{x}_i(t) - \hat{x}_j(t)\Big] dt = \int_0^T \Big[x_i(t) - a_{\min_i}\Big] - \Big[x_j(t) - a_{\min_j}\Big] dt$$

$$= \int_0^T x_i(t) dt + \int_0^T \Big[z(t) - x_i(t)\Big] dt - \int_0^T x_j(t) dt - \int_0^T \Big[z(t) - x_j(t)\Big] dt = 0 \quad \blacksquare$$

Corollary 10.3.2. If $x_i(t)$ and $x_j(t)$ are pairwise aligned each to a fixed profile, $z(t)$, then $f_{a_{\min_i}, a_{\min_j}}(x_i(t), x_j(t))$ is minimal.

Proof. From Lemma 10.3.1, $\int_0^T [\hat{x}_i(t) - \hat{x}_j(t)] dt = 0 \Rightarrow \int_0^T \Big[[x_i(t) - a_{\min_i}] - [x_j(t) - a_{\min_{ji}}]\Big]^2 dt$ is minimal. $\blacksquare$

Lemma 10.3.3. If profiles $x_1(t), \ldots, x_s(t)$ are pairwise aligned each to a fixed profile, $z(t)$, then $F_{a_{min_1}, \ldots, a_{min_s}}(x_1(t), \ldots, x_s(t))$ is minimal.

Proof. From Corollary 10.3.2, $f_{a_i, a_j}(x_i(t), x_j(t)) \geq f_{a_{min_i}, a_{min_j}}(x_i(t), x_j(t))$ with equality holding when $a_k = a_{min_k}$, which is attained by aligning each $x_k(t)$ independently with $z(t)$, $1 \leq k \leq s$. From the definition of Equation 10.8, it follows that $F_{a_1, \ldots, a_s}(x_1(t), \ldots, x_s(t)) \geq \sum_{1 \leq i \leq j \leq s} f_{a_{min_i}, a_{min_j}}(x_i(t), x_j(t))$, with equality holding when $a_k = a_{min_k}, 1 \leq k \leq s$.

Thus, for a fixed profile $z(t)$, the application of Corollary 10.3.2 to all pairs of profiles minimizes $F_{a_{min_1}, \ldots, a_{min_s}}(x_1(t), \ldots, x_s(t))$ in Equation 10.8. ▪

Theorem 10.3.4. For a fixed profile, $z(t)$, and a set of profiles, $X = \{x_1(t), \ldots, x_s(t)\}$, there always exists a multiple alignment, $\hat{x} = \{\hat{x}_1(t), \ldots, \hat{x}_s(t)\}$, such that

$$\hat{x}_i(t) = x_i(t) - a_{min_i}, \quad \text{where} \quad a_{min_i} = -\frac{1}{T} \int_0^T [z(t) - x_i(t)] \, dt \qquad (10.9)$$

and, in particular, for profile $z(t) = 0$, defined by the horizontal t axis, we have

$$\hat{x}_i(t) = x_i(t) - a_{min_i}, \quad \text{where} \quad a_{min_i} = -\frac{1}{T} \int_0^T [x_i(t)] \, dt \qquad (10.10)$$

We use the multiple alignment of Equation 10.10 in all subsequent discussions. Using spline interpolations, each profile $x_i(t)$, $1 \leq i \leq s$, is a continuously integrable profile.

$$x^{i(t)} = x_{i,j}(t) \quad \text{if} \quad t_j \leq t \leq t_{j+1} \qquad (10.11)$$

where $x_{i,j}(t) = x_{ij3}(t - t^j)^3 + x_{ij2}(t - t^j)^2 + x_{ij1}(t - t^j)^1 + x_{ij0}(t - t^j)^0$ represents $x_i(t)$ in the interval $[t_j, t_{j+1}]$, with spline coefficients x_{ijk} for $1 \leq i \leq s, 1 \leq j \leq n - 1$, and $0 \leq k \leq 3$. Thus, the analytical solution for a_{min_i} in Equation 10.10 is

$$a_{min} = \frac{1}{T} \sum_{j=1}^{n-1} \sum_{k=0}^{3} \frac{x_{ijk}(t_{j+1} - t_j)^{k+1}}{k+1} \qquad (10.12) \; ▪$$

All the aforementioned theoretical results on NCS representations including lemmas and theorems also apply to PL representations of time-series profiles. The distance function and centroid of a set that is presented in Section 10.4.1 also apply to PL representations of time-series profiles.

10.4 Alignment-Based Clustering Approaches

In this section, we define the distance between two gene profiles based on our concept of profile alignment. Clustering approaches are then proposed using our new distance measure and an appropriate definition of the centroid of a set.

10.4.1 Distance Function

The distance between any two PL profiles is defined as $f(a_{min})$ in the work by Rueda, Bari, and Ngom (2008). Here, for convenience we slightly change the definition to

$$d(x,y) = \frac{1}{T} f^{(a_{min})} = \frac{1}{T} \int_0^T [x(t) + a_{min} - y(t)]^2 \, dt \qquad (10.13)$$

For any function $\varphi(t)$ defined on $[0; T]$, we also define

$$\langle \hat{\phi} \rangle = \frac{1}{T} \int_0^T \varphi(t) \, dt \qquad (10.14)$$

Then, from Equations 10.1 and 10.3,

$$d(x,y) = \frac{1}{T} \int_0^T \Big[[x(t) - y(t)]^2 + 2a_{min}[x(t) - y(t)] + a_{min}^2 \Big] \, dt$$

$$= \frac{1}{T} \int_0^T [x(t) - y(t)]^2 \, dt - 2a_{min}^2 + a_{min}^2 \qquad (10.15)$$

$$= \langle [x(t) - y(t)]^2 \rangle - \langle x(t) - y(t) \rangle^2$$

Apart from the factor $\dfrac{1}{T}$, this is precisely the distance $d_{PA}(x, y, t)$ in the work by Rueda, Bari, and Ngom (2008). By performing the multiple alignment of Equation 10.10 to obtain the new profiles $\hat{x}(t)$ and $\hat{y}(t)$, we have

$$d(x,y) = \langle [\hat{x}(t) - \hat{y}(t)]^2 \rangle = \frac{1}{T} \int_0^T [\hat{x}(t) - \hat{y}(t)]^2 \, dt \qquad (10.16)$$

Thus, $d(x,y)^{\frac{1}{2}}$ is the 2-norm, satisfying all the properties we might want for a metric. On the other hand, $d(x, y)$ in Equation 10.16 does not satisfy the triangle inequality and, hence, it is not a metric. We, however, use $d(x, y)$ in Equation 10.16 as our distance function, since this expression is algebraically easier to work with than the metric $d(x, y)^{\frac{1}{2}}$. Equation 10.16 is closer to the spirit of regression analysis and, thus, we can dispense with the requirement for the triangle inequality. Also the distance as defined in Equation 10.16 is unchanged by an additive shift and, hence, is order-preserving, that is, $d(u, v) \le d(x, y)$ if and only if $d(\hat{u}, \hat{v}) \le d(\hat{x}, \hat{y})$.

This property has important implications for distance-based clustering methods that rely on pairwise alignments of profiles, as discussed in Section 10.4.4.

With the spline interpolations of Equation 10.5, the analytical solution for d(x, y) in Equation 10.16 was derived using the symbolic computational package, *Maple*, as follows:

$$
\begin{aligned}
d(x, y) = {} & \frac{P^2(n^7 - m^7)}{7} + \frac{(2PQ - 6p^2m)(n^6 - m^6)}{6} + \frac{(2PR - 10PQM + Q^2 + 15P^2m^2)(n^5 - m^5)}{5} \\
& + \frac{(-8PRm - 4Q^2m + 2PS + 20PQm^2 + 2QR - 20P^2m^3)(n^4 - m^4)}{4} \\
& + \frac{(-6QRm - 20Pm^3Q + R^2 + 6Q^2m^2 + 12Pm^2R - 6PmS + 15P^2m^4 + 2QS)(n^3 - m^3)}{3} \\
& + \left\{ \frac{(10Pm^4Q + 6Qm^2R + 2RS - 8Pm^3R - 2R^2m - 6P^2m^5 + 6Pm^2S - 4QmS - 4Q^2m^3)}{2} \right. \\
& \left. \times (n^2 - m^2) \right\} - 2RmS(n - m) + S^2(n - m) + p^2m^6(n - m) + Q^2m^4(n - m) \\
& + R^2m^2(n - m) - 2Qm^3R(n - m) - 2Pm^5Q(n - m) - 2Pm^3S(n - m) \\
& + 2Pm^4R(n - m) + 2Qm^2S(n - m)
\end{aligned}
\tag{10.17}
$$

where $P = (x_{j3} - y_{j3})$, $Q = (x_{j2} - y_{j2})$, $R = (x_{j1} - y_{j1})$, $S = (x_{j0} - y_{j0})$, $m = t^j$, and $n = t^{j+1}$.

10.4.2 Centroid of a Cluster

For a set of profiles $D = \{(x_1(t), \dots, x_s(t)\}$, we aim to find a centroid profile $\mu(t)$ that represents D well. An obvious choice is the function that minimizes

$$
\Delta[\mu] = \sum_{\mu=1}^{s} d(x_i, \mu)
\tag{10.18}
$$

where Δ plays the role of the within-cluster scatter defined in the work by Rueda, Bari, and Ngom (2008). Since $d(.,.)$ is unchanged by an additive shift $x(t) \rightarrow x(t) - a$ in either of its arguments, we have

$$
\Delta[\mu] = \sum_{\mu=1}^{s} d(\hat{x}_i, \mu) = \frac{1}{T} \int_{0}^{T} \sum_{i=1}^{s} [\hat{x}_i(t) - \mu(t)]^2 \, dt
\tag{10.19}
$$

where $\hat{X} = \{\hat{x}_i(t), \dots, \hat{x}_s(t)\}$ is the multiple alignment of Equation 10.10. This alignment is a functional of μ, that is, a mapping from the set of real-valued functions defined on $[0, T]$ to the set of real numbers. To minimize with respect to μ, we set the functional derivative to zero*. This functional is of the following form:

$$
F[\phi] = \int L(\phi(t)) dt
\tag{10.20}
$$

for some function L, for which the functional derivative is simply $\dfrac{\delta F[\phi]}{\delta \phi(t)} = \dfrac{d L(\phi(t))}{d \phi(t)}$.

---

* For a function $F[\phi]$, the functional derivative is defined as $\dfrac{\delta F[\phi]}{\delta \phi} = \lim_{\epsilon \to 0} \dfrac{(F[\phi + \epsilon \, \delta_t] - F[\phi])}{\epsilon}$, where $\delta_t(T) = \delta(T - t)$ is the Dirac delta function centered at t.

In our case, we have

$$\frac{\delta \Delta[\mu]}{\delta \mu(t)} = -\frac{2}{T} \sum_{i=1}^{s} [\hat{x}_i(t) - \mu(t)] = -\frac{2}{T} \left(\sum_{i=1}^{s} \hat{x}_i(t) - s\mu(t) \right) \qquad (10.21)$$

Setting $\dfrac{\delta \Delta[\mu]}{\delta \mu(t)} = 0$ gives

$$\mu(t) = \frac{1}{S} \sum_{i=1}^{s} \hat{x}_i(t) \qquad (10.22)$$

With the spline coefficients, x_{ijk}, of each $x_i(t)$ interpolated as in Equation 10.11, the analytical solution for $\mu(t)$ in Equation 10.22 is

$$\mu_j(t) = \frac{1}{S} \sum_{i=1}^{s} \left[\sum_{k=0}^{3} x_{ijk}(t - t_j)^k \right] - a_{\min_i}, \text{ in each interval } [t_j, t_{j+1}] \qquad (10.23)$$

Equation 10.22 applies to aligned profiles, whereas Equation 10.23 applies to unaligned profiles.

10.4.3 Clustering via Profile Alignment

In the studies of Rueda, Bari, and Ngom (2008) and Rueda and Bari (2006), the profiles were represented as PL functions. Area-based profile alignment takes two features vectors and produces two new vectors in such a way that the area between aligned vectors is minimized.

In the study by Rueda, Bari, and Ngom (2008), hierarchical agglomerative clustering is used where the decision rule is based on the furthest-neighbor distance or complete linkage distance between two clusters. The complete linkage or furthest neighbors approach calculates the distance between the furthest pair of points for each pair of clusters and merges the two clusters that have the minimum distance among all such distances between all pairs of clusters under consideration. The proposed clustering algorithms are discussed in this chapter. Validity indices to determine the accuracies of the proposed approaches and to determine the number of clusters are also discussed.

10.4.4 k-Means Clustering via Multiple Alignment

The k-means algorithm is one of the simplest and fastest clustering algorithms. It takes the number of clusters, k, as an input parameter. The program starts by randomly choosing k points as the centers of the clusters. These points can be random points from more densely populated volumes of input space or just randomly chosen patterns from the data itself. Once some cluster centers are chosen, the algorithm takes each profile and calculates the distance from it to all cluster centers. The second step starts by considering all the profiles associated with one cluster center and calculating a new position for this cluster center. The coordinates of this new center are usually obtained by calculating the mean of the coordinates of the points belonging to that cluster. Since the centers have moved, profile memberships need to be updated by recalculating the distance from each profile to the new cluster centers. The algorithm continues to update the cluster centers based on the new membership and update the membership of each profile until the cluster centers are such that no profile moves from one cluster to another. Since no profile has changed its membership, the centers remain the same and the algorithm terminates. A more formal definition of *k*-means clustering is stated as follows:

In *k*-means clustering (Xu and Wunsch 2009), we want to partition a set of *s* profiles, $D = \{x_1(t), \ldots, x_s(t)\}$, into *k* disjoint clusters $C_1, \ldots, C_s, 1 \le k \le s$, such that $C_i \ne \varphi$, $i = 1, \ldots, k$; $U_{i=1}^k C_i = D$; and $C_i \cap C_j = \phi$, $i \ne j$, $i, j = 1, \ldots, k$. Further, each profile is assigned to the cluster with the closest mean. The *k*-means clustering is similar to EM for mixtures of Gaussians in the sense that both methods attempt to find the centers of natural clusters in data. It assumes that object features form a vector space. Let $U = \{u_{ij}\}$ be the membership matrix defined as follows:

$$u^{ij} = \{1 \text{ if } d(x_i, \mu_j) = \min^{l=1,\ldots,k} d(x_i, \mu_l) \quad \text{where} \quad i = 1, \ldots, s \quad (10.24)$$
$$0 \text{ otherwise}$$

The aim of *k*-means is to minimize the sum of squared distances:

$$J(\theta, U) = \sum_{i=1}^n \sum_{j=1}^k u_{id} \, d(x_i, \mu_j) \quad (10.25)$$

where $\theta = \mu_1, \mu_2, \ldots, \mu_n$

In *k*-MCMA (*k*-means clustering with multiple alignment; see Figure 10.3), we first multiple-align the set of profiles *D*, using Equation 10.10, and then cluster the multiple-aligned $\hat{D}$ with *k*-means. Recall that the process of Equation 10.10 is to pairwise align each profile with the *t* axis. The *k* initial centroids are found by randomly selecting *k* pairs of profiles in $\hat{D}$ and then taking the centroid of each pair. In step (4.a) of *k*-MCMA, see Subhani et al. (2009), we do not use pairwise alignment to find the centroid $\hat{\mu}_i(t)$ closest to $\hat{x}_j(t)$ since by Lemma 10.3.1 they are automatically aligned relative to each other. When profiles are multiple-aligned, any arbitrary distance function other than Equation 10.16 can be used in step (4.a), including the Euclidean distance. Also, by Theorem 10.4.1, there is no need to multiple-align $\hat{C}_{\hat{\mu}_i}$ in step (4.b), to update its centroid $\hat{\mu}_i(t)$.

Algorithm 1 *k-MCMA: k-Means Clustering with Multiple Alignment*

Input: Set of profiles, $D = \{x_1(t), \ldots, x_s(t)\}$, and desired number of clusters, k

Output: Clusters $\hat{C}_{\hat{\mu}_1}, \ldots, \hat{C}_{\hat{\mu}_k}$

1. Apply natural cubic spline is interpolation on $x_i(t) \in D$, for $1 \leq i \leq k$ (see Section 13.3)

2. Multiple-align transformed D to obtain $\hat{D} = \{\hat{x}_1(t), \ldots, \hat{x}_s(t)\}$, using Eq. (13.10)

3. Randomly initialize centroid $\hat{\mu}_i(t)$, for $1 \leq i \leq k$ repeat

4. a. Assign $\hat{x}_j(t)$ to cluster $\hat{C}_{\hat{\mu}_i}$ with minimal $d(\hat{x}_j, \hat{\mu}_i)$, for $1 \leq j \leq s$ and $1 \leq i \leq k$

   b. Update $\hat{\mu}_i(t)$ of $\hat{C}_{\hat{\mu}_i}$, for $1 \leq i \leq k$

until Convergence: that is, no change in $\hat{\mu}_i(t)$, for $1 \leq i \leq k$

return Clusters $\hat{C}_{\hat{\mu}_1}, \ldots, \hat{C}_{\hat{\mu}_k}$

Figure 10.3 *k*-means clustering with multiple-alignment (*k*-MCMA) algorithm.

Theorem 10.4.1. Let $\bar{\mu}(t)$ be the centroid of a cluster of m multiple-aligned profiles. Then $\hat{\bar{\mu}}(t) = \bar{\mu}(t)$.

Proof. We have $\hat{\bar{\mu}}(t) = \bar{\mu}(t) - a_{\min_\mu}$. However, $a_{\min_\mu} = \dfrac{1}{T}\displaystyle\int_0^T \bar{\mu}(t)\,dt = \dfrac{1}{T}\displaystyle\int_0^T \dfrac{1}{m}\sum_{i=1}^m \hat{x}_i$ $(t) = 0$, since each $\hat{x}_i(t)$ is aligned with the t axis.

Thus, Lemma 10.3.1 and Theorem 10.4.1 make k-MCMA perform much faster than the application of k-means directly on the nonaligned data set D and when the Euclidean distance is used to assign a profile to a cluster. An important implication of Equation 10.16 is that applying k-means on the nonaligned data set D (i.e., clustering on D), without any multiple alignment, is equivalent to k-MCMA (i.e., clustering on $\hat{D}$). That is, if a profile $x_i(t)$ is assigned to a cluster C_{μ_i} by k-means on D, its shifted profile $\hat{x}_i(t)$ will be assigned to cluster $\hat{C}_{\mu_i}$ by k-MCMA (k-means on $\hat{D}$). This clustering assignment can be easily shown by the fact that multiple-alignment is order-preserving, as pointed out in Section 10.4. In k-means on D, step (4.a) requires O(sk) pairwise alignments to assign s profiles to k clusters, whereas no pairwise alignment is needed in k-MCMA. In other words, we show that we can multiple-align once, and obtain the same k-means clustering results, provided that we initialize the means in the same manner. This demonstration reinforces a known fact presented in the study by Roth et al. (2003) that a dissimilarity function that is not a metric can be made a metric by using a shift operation (in our case, any metric can be used in step (4.a) such as the Euclidean distance). In this case, the objective function of k-means does not change, and convergence is assured. Thus, this method saves computational time and opens the door for applications of multiple-alignment methods to many distance-based clustering methods. ■

10.4.5 Expectation-Maximization Clustering via Multiple Alignment

In the work by Subhani et al. (2009), we devised a clustering approach, k-MCMA, in which we combined the multiple-alignment of Equation 10.10 and the k-means clustering method with a distance function based on the pairwise alignment of Equation 10.3. In this section, we use the EM clustering algorithm and combine it with alignment methods. The EM algorithm is used for clustering in the context of mixture models (Laird, Dempster, and Rubin 1977). The goal of EM clustering is to estimate the means and standard deviations for each cluster so as to maximize the likelihood of the observed data (distribution). The EM algorithm attempts to approximate the observed distributions of values based on mixtures of different distributions in different clusters. A mixture of Gaussians is a set of k probability distributions, in which each distribution represents a cluster. With an initial approximation of the cluster parameters, the algorithm iteratively performs two steps: First, the expectation step computes the values expected for the cluster probabilities; and second, the maximization step computes the distribution and its likelihood. It iterates until the log-likelihood reaches a (possibly local) maximum. The algorithm is similar to k-means in the sense that the centers of natural clusters in data are recomputed until a desired convergence is achieved.

In EM (Xu and Wunsch 2009), we want to partition a set of s profiles, $D = \{x_1(t), \ldots, x_s(t)\}$, into k disjoint clusters $C_1, \ldots, C_s, 1 \le k \le s$, such that $C_i \ne \varphi$, $i = 1, \ldots, k$; $U_{i=1}^k C_i = D$; and $C_i \cap C_j = \phi$, $i, j = 1, \ldots, k$ and $i \ne j$. Let D be the complete data space drawn independently from the mixture density:

$$\text{E—step: } p(x/\theta) = \sum_{i=1}^k p(x/C_i, \theta_i) P(C_i) \tag{10.26}$$

where parameter $\theta = [\theta_1, \ldots, \theta_k]^t$ is fixed but unknown, and $P(C_i)$ is the known posterior probability of class C_i. The aim is to maximize the likelihood:

$$\text{M—step: } p(D/\theta) = \prod_{e=1}^s p(x_e/\theta) \tag{10.27}$$

To maximize the likelihood function, log-likelihood is used in the normal distribution of component densities given by $p(x_k/C_i, \theta_i) \sim N(\mu_i, \sum_i)$, where $\theta_i = [\mu_i, \sum_i]^t$; μ_i and $\sum_i$ are the means and the covariances of the classes, respectively. Both steps iterate until the log-likelihood reaches a maximum. Thus, EM assigns profiles to multiple clusters, similar to fuzzy clustering. In addition, unlike in k-means, each profile is assigned to the cluster that finds the maximum posterior probability.

In EMMA (EM with multiple-alignment; see Figure 10.4), we first multiple-align the set of profiles D, using Equation 10.10, and then cluster the multiple-aligned $\hat{D}$ with EM. Recall that the process of Equation 10.10 is to pairwise align each profile with the t axis. The k centroids can be initialized randomly in step (3) of EMMA,

---

Algorithm 2 *EMMA: EM Clustering with Multiple Alignment*

Input: Set of profiles, $D = \{x_1(t), \ldots, x_s(t)\}$, and desired number of clusters, k

Output: Clusters $\hat{C}_{\hat{\mu}_1}, \ldots, \hat{C}_{\hat{\mu}_k}$

1. Apply natural cubic spline is interpolation on $x_i(t) \in D$, for $1 \leq i \leq k$ (see Section 13.3)

2. Multiple-align transformed D to obtain $\hat{D} = \{\hat{x}_1(t), \ldots, \hat{x}_s(t)\}$, using Eq. (13.10)

3. Initialize centroid $\hat{\mu}_i(t)$, for $1 \leq i \leq k$

4. Compute the initial log-likelihood (see Eq. (13.27)) repeat

1. **E-step:** $p(x \mid \theta) = \sum_{i=1}^{k} p(x \mid \hat{C}_{\hat{\mu}_i}, \theta_i)\, P\,(\hat{C}_{\hat{\mu}_i})$

2. Assign $\hat{x}_j(t)$ to cluster $\hat{C}_{\hat{\mu}i}$ with maximum log-likelihood, for $1 \leq j \leq s$ and $1 \leq i \leq k$

3. **M-step:** $p(D \mid \theta) = \prod_{e=1}^{s} p(x_e \mid \theta)$ until the log-likelihood reaches its maximum return Clusters $\hat{C}_{\hat{\mu}_1}, \ldots, \hat{C}_{\hat{\mu}_k}$

Figure 10.4 Algorithm for expectation-maximization (EM) clustering with multiple-alignment.

or by any initialization approach. However, to obtain better clustering results with EMMA, it is necessary to start with near-optimal centroids; thus, we applied the k-MCMA algorithm detailed in the work by Subhani et al. (2009) to generate the k initial centroids in step (3).

For distance-based clustering algorithms such as the k-means method, it has been shown that any arbitrary distance function (including Euclidean) can be used in the cluster assignment step of the algorithm, when the profiles are multiple-aligned first (Subhani et al. 2009). Moreover, Theorem 10.4.1 shows that for distance-based methods there is no need to multiple-align $\hat{C}_{\hat{\mu}_i}$ to update its centroid $\hat{\mu}_i(t)$. By Theorem 10.4.1, there is also no need to multiple-align a cluster $\hat{C}_{\hat{\mu}_i}$ in step (1) of EMMA, for updating its centroid $\hat{\mu}_i(t)$. Likewise, any arbitrary distance function can be used in step (1) for computing the centroids. Thus, Theorem 10.4.1 makes EMMA run much faster than direct application of EM on the nonaligned data set D. The EMMA is not a distance-based clustering method; nevertheless, the quantities $p(x/\theta)$, $p\left(x/\hat{C}_{\hat{\mu}_i}, \theta_i\right)$, $P(\hat{C}_{\hat{\mu}_i})$, and $p(D/\theta)$ are also preserved when the distances are preserved. In other words, we show that we can multiple-align once and obtain the same EM clustering results if we initialize the means in the same manner.

10.5 Assessment of Clustering Quality

The two fundamental properties, number of clusters and goodness of the clustering, need to be determined in any typical clustering system. To determine the appropriate number of clusters and also the goodness or validity of the resulting clusters, we run our k-MCMA algorithm in conjunction with four cluster validity indices (Maulik and Bandyopadhyay 2002), that is, the (1) Davies–Bouldin (DB) index, (2) Dunn index, (3) Calinski–Harabasz (CH) index, and (4) I index. Once the appropriate number of clusters is determined, k-MCMA is used for proper partitioning of the data into the predetermined number of clusters. Let K be the number of clusters.

The DB index is a function of the ratio of the sum of within-cluster scatters to between-cluster separation. The within-cluster scatter for the ith cluster is S_i, and the distance between clusters is $d_{ij} = \| \mu_i - \mu_j \|$.

Then, the DB index is defined as follows:

$$
DB = \frac{1}{K} \sum_{i=1}^{k} \max_{j, j \neq i} \frac{S_{i,q} + S_{j,q}}{d_{ij}}
\tag{10.28}
$$

The optimal number of clusters is the one that minimizes this index.

The Dunn index is defined as follows: Let S and T be two nonempty subsets of R^N. The diameter Δ of S is defined as $\Delta(S) = \max_{x,y \in S}\{d(x,y)\}$ and the distance between S and T is defined as $\delta(S,T) = \min_{x \in S, y \in T}\{d(x,y)\}$, where $d(x, y)$ denotes the distance between x and y. Then, the Dunn index is defined as follows:

$$
Dunn = \min_{1 \leq i \leq k} \left\{ \min_{1 \leq k \leq K, j \neq i} \left\{ \frac{\delta(C_i, C_j)}{\max_{1 \leq k \leq K} \{\Delta(C_k)\}} \right\} \right\}
\tag{10.29}
$$

The optimal number of clusters is the one that maximizes this index.

The CH index is defined as follows:

$$
CH = \frac{[traceB/(K-1)]}{[traceW/(n-k)]}
\tag{10.30}
$$

where B is the between-cluster matrix and W is the within-cluster scatter matrix. The optimal number of clusters is the one that maximizes this index.

The I index is defined as follows:

$$
I(K) = \left(\frac{1}{k} \frac{E_1}{E_k} D_k \right)^p
\tag{10.31}
$$

where $E_k = \sum_{k=1}^{k} \sum_{j=1}^{n} \mu_{kj} \| x_j - z_k \|$, $D_k = \max_{i,j=1}^{k} \| z_i - z_j \|$. $U(X) = [u_{kj}]_{kXn}$ is a partition matrix for the data, and z_k is the centroid of the kth cluster. The optimal number of clusters is the one that maximizes this index. We have taken $p = 2$, which is used to control the contrast between different cluster configurations. This index is typically used in many applications. To find the optimal number of clusters, we apply k-MCMA and EMMA on a data set in conjunction with the four aforementioned validity indices for $k = 1,\dots,\sqrt{s}$ (where s is the number of profiles) clusters. Among the four validity indices, we note which number is the most frequent and that number is chosen as the best number of clusters for that data set.

10.6 Computational Experiments

To evaluate the performances of our alignment-based clustering approaches, we performed experiments on six well-known data sets. We performed three types of experiments to illustrate the different capabilities of our approaches. In the first type, we compared our k-MCMA and EMMA algorithms on two sets of data sets: (1) a preclustered data set in which the number k of clusters is known, which allows us to measure the abilities of k-MCMA and EMMA to find the correct "known" classes; and (2) data sets with unknown number of clusters, which allow us to measure the abilities of our methods to obtain clustering results having the correct number of clusters. In the second type of experiments, we investigated the effect of two profile representations, PL representations and NCS representations of profiles, on the clustering results of k-MCMA when it is applied to the two types of data sets. In the third type of experiments, we compared our alignment-based clustering approaches with a variation-based clustering algorithm, that is, the "variation-based coexpression detection" (VCD) algorithm. The VCD algorithm, which was recently introduced by Yin and Chiang (2008), is also based on first applying a transformation to the data and then, like our methods, clustering the resulting transformed data with any clustering method. Our clustering methods were implemented in MATLAB® (version 7.7.0.471 (R2008b)), and all tests were run on an Intel Core 2 Duo CPU 1.83 GHz with 2 GB of RAM under Windows Vista Home Premium with Service Pack 1.

10.6.1 Data Sets

Six groups of data have been used in the experiments: (1) *Saccharomyces cerevisiae* budding yeast data set, (2) *Pseudomonas aeruginosa transcriptomes* data set, (3) serum data set obtained from the work of Iyer et al. (1999), (4) *Micrococcus luteus* infection challenge to *Anopheles gambiae* data set, (5) *Escherichia coli* infection challenge to *Anopheles gambiae* data set, and (6) *Schizosaccharomyces pombe* data set of *cdc25* mutant cell gene expression profiles.

The degree to which a gene is expressed to produce a protein, under certain environmental conditions within the cell, is called the "expression level" of the gene. This expression is measured in the microarray and reflected in the quantification process. The expression ratio is the ratio of two expression levels of a gene at different cellular environments, that is, "normal" versus "abnormal," or "control" versus "experiment." In our experiments, the *Saccharomyces cerevisiae* data set and the *Schizosaccharomyces pombe* data set contained expression levels, and other data sets contained expression ratios. More precisely, the typical microarray experiment involves five processing stages: (1) target preparation; (2) design of deoxyribonucleic acid (DNA) chips for targeted genes; (3) hybridization; (4) expression level, or ratio measurement; and (5) data preparation for analysis.

The data preparation stage itself involves certain steps like image processing, quantification, data preprocessing, normalizations, and data representation for analysis. The normalization process helps to identify the abnormalities of gene expression levels. To increase the reliability of the experiments, the *Saccharomyces cerevisiae* and serum data sets were made in replicates, whereas the *Schizosaccharomyces pombe* data set contains *cdc25* mutant cell gene raw expression levels. The remaining three data sets were made in triplicate.

The detailed description and the normalization procedure of each data set are described in Section 10.6.2 and Sections 10.6.2.1 through 10.6.2.7.

10.6.2 Saccharomyces cerevisiae *Data Set*

The data set contains time-series gene expression profiles of the complete characterization of messenger ribonucleic acid (mRNA) transcript levels during the yeast cell cycle. These experiments measure the expression levels of 6220 yeast genes during the cell cycle at 17 different points, from 0 to 160 minutes, at every 10-minute time interval. From these gene profiles, 221 profiles were analyzed. We normalized each expression profile as described by Cho et al. (1998); that is, we divided each transcript level by the mean value of each profile with respect to each other.* The data set contains five known clusters called "phases": (1) the early G1 phase (32 genes), (2) the late G1 phase (84 genes), (3) the S phase (46 genes), (4) the G2 phase (28 genes), and (5) the M phase (31 genes).

10.6.2.1 Pseudomonas aeruginosa *Data Set*

The experiments mentioned in Section 10.6.2 measured the expression ratios of 3315 *P. aeruginosa* genes (Waite et al. 2006) in the planktonic cultures at 0, 4, 8, 14, 24, and 48 hours. The resulting expressions are averaged values of the three replicates for each condition, which are then normalized to zero mean and unit variance.

10.6.2.2 Serum Data Set

The serum data set contains data on the transcriptional response of cell cycle–synchronized human fibroblasts to serum. These experiments measure the expression levels of 8613 human genes after a serum stimulation at 12 different time

---

* http://genomics.stanford.edu/yeast_cell_cycle/cellcycle.html

points, at 0 hour, 15 minutes, 30 minutes, 1 hour, 2 hours, 3 hours, 4 hours, 8 hours, 16 hours, 20 hours, and 24 hours. From these 8613 gene profiles, 517 profiles were separately analyzed, as their expression ratios had changed substantially at two or more time points. The experiments and analysis focused on this data set, which is the same group of 517 genes used in the study by Iyer et al. (1999).

10.6.2.3 Micrococcus luteus *Data Set*

Immune responses of the malaria vector mosquito *A. gambiae* were monitored systematically by the induced expression of five RNA markers after the *M. luteus* infection challenge. Bacterial infection of third and fourth instar larvae and adult female mosquitoes were performed by pricking with a needle dipped in a concentrated solution of *M. luteus* at seven time points: 1, 4, 8, 12, 18, and 24 hours. Expression values were log-2-transformed, normalized ratios of medians, as described by Muller et al. (1997).

10.6.2.4 Escherichia coli *Data Set*

Immune responses of the malaria vector mosquito *A. gambiae* were monitored systematically by the induced expression of five RNA markers after the *E. coli* infection challenge. Bacterial infection of third and fourth instar larvae and adult female mosquitoes were performed by pricking with a needle dipped in a concentrated solution of *E. coli* at seven time points: 1, 4, 8, 12, 18, and 24 hours. Expression values were log-2-transformed, normalized ratios of medians, as described by Muller et al. (1997).

10.6.2.5 Schizosaccharomyces pombe *Data Set*

A data set containing the cell cycle progressions of the fission yeast *Schizosaccharomyces pombe* was prepared by Peng et al. (2005). This data set contains 747 genes and 2 types of cells: (1) wild-type cells and (2) *cdc25* mutant cells. We have used the *cdc25*-type mutant cells genes.

10.6.2.6 Experiments

Experiments were carried out using our methods in conjunction with four validity indices, described in Section 10.5, on all five data sets to determine the number of clusters. After determining the number of clusters for each data set, clustering was performed using our methods. Table 10.1 shows the results of the indices for finding the number of clusters. After analyzing all the tables, the most frequent value for a data set was chosen as the best number of clusters. Interestingly, we found $k = 5$ as the best number of clusters for *Saccharomyces cerevisiae*, which is exactly the number of known yeast phases.

Table 10.1   Best Number of Clusters for All Data Sets

Data Sets	Number of Clusters
Saccharomyces cerevisiae 5 (DB, CH, I)	5 (DB, CH, I)
P. aeruginosa 7 (CH, I)	7 (CH, I)
Serum 15 (CH, I)	15 (CH, I)
M. luteus 8 (CH, I)	8 (CH, I)
E. coli 7 (CH, Dunn)	7 (CH, Dunn)
Schizosaccharomyces pombe 8 (DB, CH)	8 (DB, CH)

10.6.2.7 Results and Discussions

After setting $k = 5$, we ran both the proposed approaches on the *Saccharomyces cerevisiae* data set. Once the clusters were found, to compare the k-MCMA and EMMA clustering with the preclustered data set of Cho et al. (1998), the next step was to label the clusters, where the labels were the phases in the preclustered data set. Although this can be done in many different ways, we adopted the following approach: To compare the clusters with preclustered yeast phases, we faced a combinatorial assignment problem. We assigned each k-MCMA and EMMA cluster to a yeast phase using the Hungarian algorithm (Kuhn 1955). The Hungarian method is a combinatorial optimization algorithm that solves the assignment problem in polynomial time. Our phase assignment problem is formulated using a complete bipartite graph $G = (C, P, E)$ with k cluster vertices (C) and k phase vertices (P), and each edge in E has a nonnegative cost $c(\hat{C}_{\hat{\mu}_i}, \hat{P}_{\hat{v}_j})$, $\hat{C}_{\hat{\mu}_i} \in C$ and $\hat{P}_{\hat{v}_j} \in P$. We wanted to find a perfect match with the minimum cost. We determined the cost of an edge between a cluster vertex to be $\hat{C}_{\hat{\mu}_i}$ and a phase vertex $\hat{P}_{\hat{\mu}_i}$ to be the distance between their centroids $\hat{\mu}_i$, $\hat{v}_j$, that is, $c(\hat{C}_{\hat{\mu}_i}, \hat{P}_{\hat{v}_j}) = d(\hat{C}_{\hat{\mu}_i}, \hat{P}_{\hat{v}_j})$, and the distances were computed using Equation 10.16. In terms of such a bipartite graph, the Hungarian method selected the k perfect, matching pairs $(\hat{C}_{\hat{\mu}_i}, \hat{P}_{\hat{v}_j})$ with minimum cost.

In Figure 10.5, the cluster and the phase of each of the five selected pairs, found by the Hungarian algorithm, are shown at the same level; for example, cluster C5 of k-MCMA is assigned to the late G1 phase mentioned by Cho et al. (1998) by our phase assignment approach. Hence, they are at the same level in the figure. The five clusters found by EMMA are shown in Figure 10.5a and those found by k-MCMA are shown in Figure 10.5c, whereas the corresponding phases given by Cho et al. (1998) after the phase assignment are shown in Figure 10.5b. The horizontal axis represents the time points in minutes, and the vertical axis represents the expression values. Each cluster is vertically shifted by six units up, in order to distinguish them visually. The dashed black lines are the cluster centroids learned by EMMA

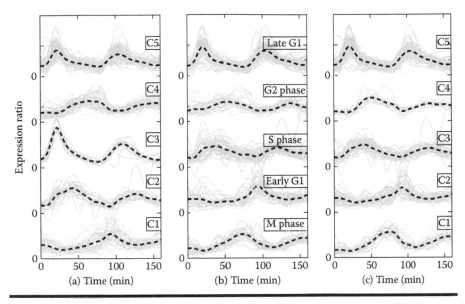

Figure 10.5 (a) Expectation-maximization with multiple-alignment (EMMA) clusters; (b) *Saccharomyce cerevisiae* phases (From Cho, R., M. Campbell, E. Winzeler, L. Steinmetz, A. Conway, L. Wodicka, T. Wolfsberg, A. Gabriellan, D. Landsman, D. Lockhart, and R. Davis, *Mol Cell*, 2:65–73, 1998. With permission.); and (c) *k*-means clustering with multiple-alignment (*k*-MCMA) clusters, with centroids shown.

(Figure 10.5a) and the known phase centroids of the yeast data (Figure 10.5b). In the figure, each cluster and phase is multiple-aligned using Equation 10.10 to enhance visualization.

After setting $k = 7$, we applied k-MCMA and EMMA on the *P aeruginosa* data set to see if both methods are able to find these clusters correctly. The clusters found by k-MCMA and EMMA are shown in Figure 10.6a and b, respectively. The horizontal axis represents the time points in hours and the vertical axis represents the expression ratios. The k-MCMA and EMMA clusters in Figure 10.6a and b are vertically shifted by six points to distinguish them visually. The dashed black lines are the cluster centroids learned by EMMA (Figure 10.6b) and the centroids of the bacterium data computed using Equation 10.10.

The performance of the multiple-alignment method on the *Saccharomyces cerevisiae* data set is discussed in this section. The multiple-alignment method of Section 10.3 that we have used in k-MCMA and EMMA is based on NCS profiles. We extended the pairwise alignment formulas given by Rueda, Bari, and Ngom (2008) for PL profiles to multiple expression profile alignment. The procedure for extending pairwise profile alignment to multiple profile alignment was similar to the one described by Subhani et al. (2009) except that, in this case, we used PL profiles instead of NCS profiles. We combined k-means clustering

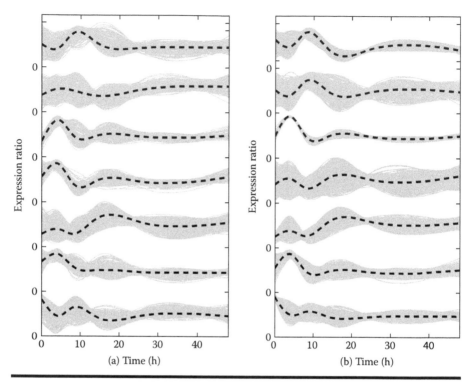

Figure 10.6 (a) *k*-means clustering with multiple-alignment (*k*-MCMA) clusters and (b) expectation-maximization with multiple-alignment (EMMA) clusters of the *Pseudomonas aeruginosa* data set, with centroids shown.

with our multiple-alignment approaches to cluster microarray time-series data. After setting $k = 5$, we applied k-MCMA on the yeast data set for both types of profiles (NCS and PL profiles) to find those phases as accurately as possible. Once the clusters were found, we compared the resulting clusters with those of the pre-clustered data set given by Cho et al. (1998). To achieve a better visual representation, we assigned each cluster obtained by k-MCMA to its corresponding phase mentioned in the work of Cho et al. (1998).

The five clusters found by k-MCMA using NCS profiles are shown in Figure 10.7b, whereas the corresponding phases mentioned by Cho et al. (1998) after the phase assignment are shown in Figure 10.7a. The horizontal axis represents the time points in minutes, and the vertical axis represents the expression values. The dashed black lines are the cluster centroids learned by k-MCMA (Figure 10.8b) and the known phase centroids of the yeast data (Figure 10.8a). Figure 10.7 shows the clustering on the same data set as Figure 10.8, in which k-MCMA used PL profiles. For the figures, each cluster and phase was vertically shifted up by three units,

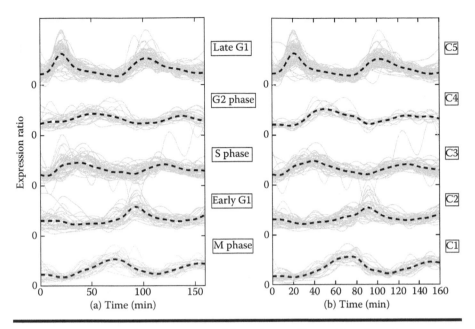

Figure 10.7 (a) *Saccharomyce cerevisiae* phases (From Cho, R., M. Campbell, E. Winzeler, L. Steinmetz, A. Conway, L. Wodicka, T. Wolfsberg, A. Gabriellan, D. Landsman, D. Lockhart, and R. Davis, *Mol Cell*, 2:65–73, 1998. With permission.) and (b) *k*-means clustering with multiple-alignment (*k*-MCMA) clusters using natural cubic spline profiles, with centroids shown.

in order to enhance visualization. After visually representing each cluster in both cases, we see that *k*-MCMA clusters with both representations are quite similar to exactly one of the yeast phases. We applied both *k*-MCMA and EMMA on the *M. luteus* data set for both types of profiles (NCS and PL profiles). The clusters found by *k*-MCMA and EMMA using NCS profiles are shown in Figure 10.9a and c, respectively. The clusters and phases in Figure 10.9 are vertically shifted up by four points to distinguish them visually. We clearly see that *k*-MCMA and EMMA clusters are similar to exactly one of the *M. luteus* phases.

10.6.2.8 Comparison with Previous Approaches

We have compared our approaches with the following two previously published approaches: (1) a clustering method that uses PL profiles (Rueda, Bari, and Ngom 2008), and (2) the VCD algorithm (Yin and Chiang 2008). We performed an objective measure for comparing the EMMA clusters with the yeast phases. The measurement was computed by taking the average classification accuracy as the number of genes that EMMA correctly assigned to one of the phases. Considering

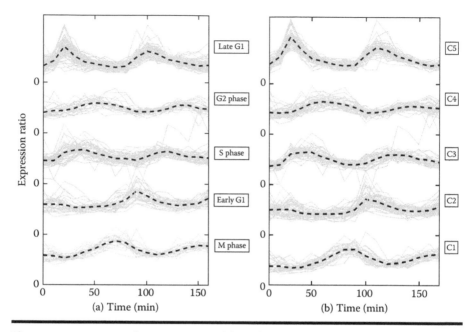

Figure 10.8 (a) *Saccharomyce cerevisiae* phases (From Cho, R., M. Campbell, E. Winzeler, L. Steinmetz, A. Conway, L. Wodicka, T. Wolfsberg, A. Gabriellan, D. Landsman, D. Lockhart, and R. Davis, *Mol Cell*, 2:65–73, 1998. With permission.) and (b) *k*-means clustering with multiple-alignment (*k*-MCMA) clusters using piecewise linear profiles, with centroids shown.

each EMMA cluster as a class, we trained a c-nearest neighbor (c-NN) classifier with clusters to classify the data with a 10-fold cross-validation procedure, in which c is the number of nearest profiles from the centroids. In our scenario, we found that $c = 5$ is the best number of clusters for the data set, and we used the distance function of Equation 10.16 to measure the distance between the centroids and their nearest profiles. We applied the same procedure for k-MCMA clusters. This criterion is reasonable as k-MCMA and EMMA are unsupervised learning approaches that do not know the phases beforehand and, hence, the aim is to "discover" these phases. In a study by Cho et al. (1998), the five phases were determined using biological information, including genomic and phenotypic features observed in yeast cell cycle experiments. The average classification accuracy of EMMA was 91.03%, whereas that of k-MCMA was 89.51%. We applied the same aforementioned objective measure to compare the EMMA clusters with *P. aeruginosa* clusters and obtained an average classification accuracy of 91.40%. In studies by Waite et al. (2006) and Rueda et al. (2008), the correlation coefficient was used as the distance measure between gene profiles, whereas we used the distance function defined in Equation 10.16.

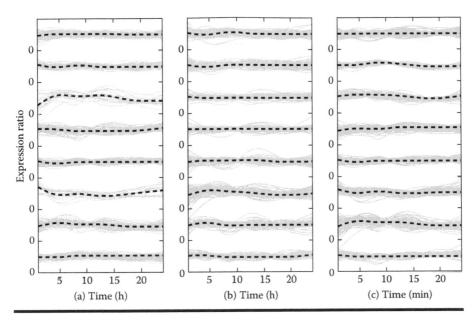

Figure 10.9 **(a)** **Expectation-maximization** **with** **multiple-alignment** **(EMMA)** clusters, **(b)** *Micrococcus luteus* phases (From Muller et al., *Proc Natl Acad Sci USA*, 94:11508–13, 1997. With permission.), and **(c)** *k*-means clustering with multiple-alignment (*k*-MCMA) clusters, with centroids shown.

Figure 10.9 shows that EMMA yields better results than *k*-MCMA and the methods used by Waite et al. (2006). We applied the same aforementioned objective measure to compare *k*-MCMA clusters using PL profiles (an approach suggested by Rueda, Bari, and Ngom [2008]) with the yeast phases, and obtained an average classification accuracy of 86.12%. For the bacterium data set, we obtained an average classification accuracy of 90.90%. Table 10.2 shows the average classification accuracies of our approaches with the approach of Rueda, Bari, and Ngom (2008). As shown in Figures 10.5 and 10.6 and Table 10.2, we observed that NCS profiles performed better than PL profiles. We also observed that *k*-MCMA and EMMA clusters using NCS profiles on both data sets obtained over 90% classification accuracy, which is very high considering that they are both unsupervised learning methods, whereas EMMA yields better performance results than *k*-MCMA. We also performed the same classification comparison for a method presented in the work of Yin and Chiang (2008), which is the VCD algorithm. In this approach, gene expressions are translated into gene variation vectors, and the cosine values of these vectors are used to evaluate their similarities over time. We compared the results of our EMMA approach and those of the VCD approach of Yin and Chiang (2008) on two data sets: (1) *Saccharomyces cerevisiae* and (2) *Schizosaccharomyces pombe*. The results are listed in Table 10.3.

Table 10.2   Experiment Results Overview of *k*-MCMA and EMMA with PL Profile

Profiles	Approaches	Saccharomyces cerevisiae	P. aeruginosa	Serum	M. luteus	E. Coli
NCS	*k*-MCMA	89.51%	91.4%	78.47%	85.24%	85.37%
PL	*k*-MCMA	86.12%	90.9%	77.21%	82.73%	81.33%
NCS	EMMA	91.03%	92.71%	85.83%	89.37%	88.36%
PL	EMMA	86.43%	89.37%	83.79%	87.76%	86.91%

Source: Adapted from Rueda, L., A. Bari, and A. Ngom. 2008. Clustering time-series gene expression data with unequal time intervals. *Springer Trans. on Computational Systems Biology* LNBI 5410:100–23.

Table 10.3   Experiment Results Overview of EMMA Approach and VCD Method

Approaches	Saccharomyces cerevisiae	Schizosaccharomyces pombe
k-MCMA	89.51%	87.63%
EMMA	91.03%	86.94%
VCD	80.68%	70.46%

Source: Adapted from Yin, Z. -X., and J. -H. Chiang. 2008. Novel algorithm for coexpression detection in time-varying microarray data sets. *IEEE/ACM Trans. on Computational Biology and Bioinformatics* 5:120–35.

In Figure 10.8, cluster number 5 of EMMA and *k*-MCMA is similar to the corresponding *S* phase, in which the VCD method assigns many differentially expressed genes. If we carefully examine the figure, we can see that the EMMA clusters are better than those of all other methods. The EMMA clusters are even better than precharacterized phases, at least visually. In Figure 10.11, we can see that VCD identifies three clusters that contain only two genes and that there are many genes misclassified to the clusters. In this data set, the *k*-MCMA clusters are better than the EMMA clusters.

For the *Saccharomyces cerevisiae* data set, both EMMA and VCD found 5 clusters, and the EMMA clusters obtained over 90% classification accuracy. On the *Schizosaccharomyces pombe* data set, we ran EMMA in conjunction with four validity indices. We found eight clusters (see Figure 10.11) in this data set.

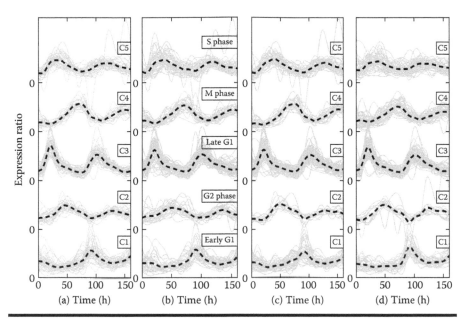

Figure 10.10 (a) Expectation-maximization with multiple-alignment (EMMA) clusters, (b) *Saccharomyce cerevisiae* phases (From Cho, R., M. Campbell, E. Winzeler, L. Steinmetz, A. Conway, L. Wodicka, T. Wolfsberg, A. Gabriellan, D. Landsman, D. Lockhart, and R. Davis, *Mol Cell*, 2:65–73, 1998. With permission.), (c) *k*-means clustering with multiple-alignment (*k*-MCMA) clusters, and (d) variation-based coexpression detection (VCD) clusters, with centroids shown.

The EMMA method, therefore, applied after setting $k = 8$ obtained 89.53% classification accuracy. After setting $\lambda = 0.59$ and $z_p = 7$, we applied the VCD algorithm of Yin and Chiang (2008) on the *Schizosaccharomyces pombe* data set as well to find the clusters. The VCD algorithm identified another 8 clusters and obtained a classification accuracy of 70.46%. This method identified 33 unique genes that did not belong to any cluster. According to the authors, they obtained 71 clusters in the *Schizosaccharomyces pombe* data set when they set the parameters $\lambda = 0.75$ and $z_p = 1.96$. In their method, λ covered the similarity between sets, and z_p determined the number of clusters. The 8 EMMA and *k*-MCMA clusters on the *Schizosaccharomyces pombe* data set obtained 86.94% and 87.63% classification accuracies, respectively, which shows that the *Schizosaccharomyces pombe* data set contains 8 clusters. In fact, EMMA and VCD are both unsupervised learning methods, although EMMA performs better than VCD (Yin and Chiang 2008).

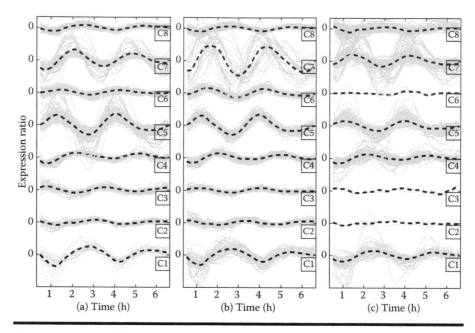

Figure 10.11 (a) *k*-means clustering with multiple-alignment (*k*-MCMA) clusters, (b) expectation-maximization with multiple-alignment (EMMA) clusters, and (c) variation-based coexpression detection (VCD) clusters on the *Schizosaccharomyces pombe* data set, with centroids shown.

10.7 Conclusion

The clustering of gene expression time-series data offers many computational challenges. Various studies have been done is this direction and some of the issues inherent to time-series data addressed; some other studies provide grounds for further studies and warrant the development of new computational tools. The current research on clustering time-series data and methods from a wide range of clustering approaches proposed in various studies are discussed and presented in this chapter. The methods differ substantially, each offering their own possibilities, and pros and cons. Most of the proposed methods for clustering time-series data are suitable for any data, but only a few incorporate profile alignment before applying a distance measure.

We proposed *k*-MCMA, a method that combines *k*-means with multiple profile alignment of gene expression profiles, to cluster microarray time-series data. The profiles are represented as NCS functions to compare profiles, in which the expression measurements are not necessarily taken at regular time intervals. A multiple-alignment is based on minimizing the sum of integrated squared errors over a time interval and is defined on a set of profiles. Another method, EMMA, is also proposed. The EMMA was performed by combining EM and the multiple-alignment of gene expression profiles to cluster microarray time-series data. We designed a distance function that suits the NCS profiles. Four cluster validity indices were used in conjunction with the aforementioned methods to determine the appropriate number of clusters and the goodness or the validity of the clusters.

An objective measure for comparing the k-MCMA and EMMA clusters using NCS profiles with the yeast phases (Cho et al. 1998) was computed by taking the average classification accuracy as the number of genes that k-MCMA and EMMA correctly assigned to a phase. We have used a supervised classification approach (c-NN) to consolidate the discrimination ability of the inferred classes and have obtained accuracies near 90%, which is very high considering that our clustering methods are unsupervised. The EMMA performed better than k-MCMA on all data sets. These results suggest that EMMA can also be used to correct manual-phase assignment errors.

Meanwhile, the results show that EMMA, using an NCS profiles approach for clustering microarray gene expression data presented in this work, finds clusters that are close to those of biologically characterized phases using NCS profiles. The EMMA using NCS profiles performed better than the PL profile method studied by Rueda, Bari, and Ngom (2008).

The EMMA also outperformed the VCD method of Yin and Chiang (2008), especially when the VCD did not assign some of the genes to any cluster. Both k-MCMA and EMMA outperformed published distance-based clustering algorithms (i.e., using PL profiles of Rueda, Bari, and Ngom [2008]) and the VCD method of Yin and Chiang (2008). Our experiments showed that the proposed EMMA algorithm finds better clusters than the algorithms of biologically characterized phases given in the work of Cho et al. (1998). The clustering algorithms proposed in this chapter can be modified to be used in well-known problems in bioinformatics and computational biology that are expressed as clustering visualization and supervised pattern recognition for microarray time-series gene expression analysis. In the future, we plan to study other distance-based clustering approaches using our multiple-alignment method. Investigations of support vector clustering (SVC) or spectral clustering with the multiple-alignment approach, other validity indices, and phase detection by aligning over a portion of the time-series expression could be interesting. The effect of a multiple-alignment method on visualizing clusters could be explored in the field of cluster visualization. It would be interesting to study the effectiveness of our clustering methods in dose-response microarray data sets. Cluster validity indices based on multiple-alignments could also be investigated. We argue that in real applications, data can be very noisy, and the use of cubic spline interpolation could lead to problems. The use of splines has the advantage of being tractable, however, although we also plan to study interpolation methods that incorporate noise. Although our focus is on clustering, the effect of using different imputation methods rather than natural cubic splines for representing the profiles should also be investigated.

References

Bar-Joseph, Z., G. Gerber, T. S. Jaakkola, D. K. Gifford, and I. Simon. 2003. *J Comput Biol* 10:341–56.

Brehelin, L. 2005. Clustering gene expression series with prior knowledge. *Lect Notes Comput Sci* 3692:27–8.

Cho, R., M. Campbell, E. Winzeler, L. Steinmetz, A. Conway, L. Wodicka, T. Wolfsberg, A. Gabriellan, D. Landsman, D. Lockhart, and R. Davis. 1998. A genome-wide transcriptional analysis of the mitotic cell cycle. *Mol Cell* 2:65–73.

Cho, K., C. Moller-Levet, F. Klawonn, and O. Wolkenhauer. 2005. Clustering of unevenly sampled gene expression time-series data. *Fuzzy sets syst* 152:49–66.

Conesa, A., M. J. Nueda, and A. Ferrer. 2006. maSigPro: A method to identify significantly differential expression profiles in time-series microarray experiments. *Bioinformatics* 22:1096–102.

Déjean, S., P. G. P. Martin, A. Baccini, and P. Besse. 2007. Clustering time-series gene expression data using smoothing spline derivatives. *EURASIP J Bioinform Syst Biol* 2007:1–11.

Ernst, J., G. J. Nau, and Z. Bar-Joseph. 2005. Clustering short time series gene expression data. *Bioinformatics* 21(Suppl 1):i159–68.

Hwang, J. T. G., and S. D. Peddada. 1994. Confidence interval estimation subject to order restrictions. *Ann Statist* 22:67–93.

Iyer, V., M. Eisen, D. Ross, G. Schuler, T. Moore, J. Lee, J. Trent, L. Staudt, J. Hudson, M. Boguski, D. Lashkari, et al. 1999. The transcriptional program in the response of human fibroblasts to serum. *Science* 283:83–7.

Kruglyak, S., L. Heyer, and S. Yooseph. 1999. Exploring expression data: Identification and analysis of coexpressed genes. *Genome Res* 9:1106–15.

Kuhn, H. W. 1955. The Hungarian method for the assignment problem. *Nav Res Logist Q* 2:83–97.

Laird, N., A. Dempster, and D. Rubin. 1977. Maximum likelihood from incomplete data via the EM algorithm. *J R Stat Soc* 39:1–38.

Maulik, U., and S. Bandyopadhyay. 2002. Performance evaluation of some clustering algorithms and validity indices. *IEEE Trans Pattern Anal Mach Intell* 24:1650–4.

Muller, H., G. Dimopoulos, A. Richman, and F. Kafatos. 1997. Molecular immune responses of the mosquito anopheles gambiae to bacteria and malaria parasites. *Proc Natl Acad Sci USA* 94:11508–13.

Peddada, S. D., L. Lobenhofer, L. Li, C. Afshari, C. Weinberg, and D. Umbach. 2003. Gene selection and clustering for time-course and dose-response microarray experiments using order-restricted inference. *Bioinformatics* 19:834–41.

Peddada, S. D., K. Prescott, and M. Conaway. 2001. Tests for order restrictions in binary data. *Biometrics* 57:1219–27.

Peng, X., R. K. M. Karuturi, L. D. Miller, L. Lin, Y. Jia, P. Kondu, L. Wang, L. -S. Wong, T. Liu, M. K. Balasubramanian, and J. Liu. 2005. Identification of cell cycle-regulated genes in fission yeast. *Mol Biol Cell* 16:1026–42.

Ramoni, M., P. Sebastiani, and I. Kohane. 2002. Cluster analysis of gene expression dynamics. *PNAS* 99:9121–6.

Ramsay, J., and B. Silverman. 2005. *Functional Data Analysis*. 2nd ed. New York: Springer.

Roth, V., J. Laub, M. Kawanabe, and J. Buhmann. 2003. Optimal cluster preserving embedding of non-metric proximity data. *IEEE Trans PAMI* 25:1540-51.

Rueda, L., and A. Bari. 2006. A new profile alignment method for clustering gene expression data. *LNCS* 4013:86–97.

Rueda, L., A. Bari, and A. Ngom. 2008. Clustering time-series gene expression data with unequal time intervals. *Springer Trans Comput Syst Biol* LNBI 5410:100–23.

Schliep, A., and D. C. Torney. 2003. Group testing with DNA chips: generating designs and decoding experiments. In *Proceedings of the 2nd IEEE Computer Society Bioinformatics Conference*, Vol. 99, 9121–6. Stanford, CA.

Subhani, N., A. Ngom, L. Rueda, and C. Burden. 2009. Microarray time-series data clustering via multiple alignment of gene expression profiles. *Pattern Recognit Bioinform*, Springer, LNBI 5780:377–90.

Tamayo, P., D. Slonim, J. Mesirov, Q. Zhu, S. Kitareewan, E. Dmitrovsky, E. Lander, and T. Golub. 1999. Interpreting patterns of gene expression with self-organizing maps; methods and application to hematopoietic differentiation. *Proc Natl Acad Sci USA* 96:2907–12.

Waite, R. D., A. Paccanaro, A. Papakonstantinopoulou, J. M. Hurst, M. Saqi, E. Littler, and M. A. Curtis. 2006. Clustering of *Pseudomonas aeruginosa* transcriptomes from planktonic cultures, developing and mature biofilms reveals distinct expression profiles. *BMC Genomics* 7:162.

Xu, R., and D. C. Wunsch II. 2009. *Clustering. IEEE Press Series on Computational Intelligence.* New York: Wiley.

Yin, Z. -X., and J. -H. Chiang. 2008. Novel algorithm for coexpression detection in time-varying microarray data sets. *IEEE/ACM Trans Comput Biol Bioinform* 5:120–35.

Chapter 11

Mining of Imaging Biomarkers for Quantitative Evaluation of Osteoarthritis

Xian Du

Contents

11.1 Introduction

The World Health Organization (WHO) ranked osteoarthritis (OA) as one of the top 10 global diseases (Andriacchi et al. 2004). The OA originates from the damage of joints, such as the loss of cartilage, wear and tear of joints, and overuse of joints. The disease develops progressively with age and normally goes unnoticed until it causes joint pain, disability, and health-care expenditure. Radiological technology, such as X-rays and magnetic resonance imaging (MRI), provides health-care facilitators with opportunities to make early diagnoses and decisions regarding therapy for sufferers of OA at low cost (Sinha and Tameem 2007; Conaghan and Wenham 2009). The most commonly employed imaging biomarkers of OA include the morphometric measurements of cartilage volume, thickness, and surface measurement, or a multifactor measurement comprising the aforementioned attributes.

Similar to humans, a variety of animals suffers from OA problems with increase in age. Overweight of bodies, overloads on knees, and irregular body postures can exacerbate the development of OA and the disease's incidence. Based on experiments, researchers report the progressions of OA in several animals, such as guinea pigs (Bendele and Hulman 1991), mice (Munasinghe et al. 1995; Pastoureau et al. 2003), and rats (Loeuille et al. 1997). They report the increase of cartilage thickness in the knees of guinea pigs (Bendele and Hulman 1991; Loeuille et al. 1997; Pastoureau et al. 2003) and demonstrate its correlation with OA progression. In this study, I use minimum thickness as a biomarker and focus on the segmentation of three-dimensional (3-D) MRI images of knee cartilage using a snake-based approach. The accurate segmentation of knee cartilage offers an opportunity to mine imaging biomarkers for the quantitative evaluation of OA.

An MRI is reproducible and can quantify and access the degenerations of cartilage given accurate segmentation. The 3-D cartilage segmentation of an MRI image on small animals is challenging due to poor resolution and the occurrence of noise. It is more complex than segmenting human cartilage, as the minimum cartilage thickness between femur and tibia of a small animal is around 300 μm, rather than a few millimeters, whereas the animal image maintains noise at the same level as a larger human image.

The snake method (hereafter referred to as snake) was introduced in image segmentation by Kass, Witkin, and Terzopoulos (1998). The method is called an "active contour model" because of its ability to minimize its energetic functional actively by finding the salient image contour. Normally, the active contour is driven toward the object contour by internal and external forces simultaneously. The classical active contour has two basic internal forces, extension and curvature of curves, which regularize the surface of a snake. External forces, which are derived from image information, for example, an edge, drive a snake to reach the boundary of the image. Incorporating both internal and external forces, researchers perform snakes that are successful in tasks related to boundary detection, such as still image segmentation and object tracking in videos.

In practice, a snake suffers from two problems: (1) contour initialization and (2) capture range of external forces. The first problem refers to difficulty in the initialization of contours. We have to guarantee that the edge forces are strong enough to attract the initial contours to the correct boundaries. The second problem refers to the difficulty of active contours to advance to the boundary concavities. Using balloon force (Cohen 1991), the initial contour is not strictly required to be very close to the detecting boundary. Meanwhile, balloon force prevents the active contour from shrinking and pushes the contour into the concavities. However, the setup of balloon forces poses another problem: The strength of the forces must stay in an acceptable range such that the balloon forces will not overwhelm the weak edge.

Researchers have introduced a number of solutions to the aforementioned problems over the past decades. The following are the most commonly employed techniques in knee cartilage segmentation: Gradient vector flow (GVF) uses the diffusion of the gradient vector of an edge map to capture a large range and concavity; but it is sensitive to noise and parameter (Prince and Xu 1997). Vector field convolution (VFC) convolves a vector-field kernel function with boundaries to improve the robustness to noise and initialization; but VFC does not work well when the detecting boundaries stay in homogeneous areas (Li 2006). When applied to 3-D MRI image segmentation (especially on small animals), these methods were largely constrained by MRI image resolution, noise, manual interaction, and algorithms.

Prior knowledge has also been introduced to recover the missing cartilage boundaries. The active shape model (ASM) enables cartilage segmentation to be accurate and reproducible (Solloway et al. 1997; Fripp, Crozier, and Warfield 2005). However, such techniques require a tedious training process, and they are only consistent with training data. In addition, these technologies are based on a 2-D extension. The automatic segmentation of articular cartilage was obtained from 3-D MRI volume based on the 2-D statistical shape model (Pirnog 2005; Kapur 1999). The initialization step includes mainly patellar bone segmentation, which provides the bone-cartilage interface (BCI) of the patellar cartilage. A cartilage model was obtained by moving the BCI vertices in the normal direction with average thickness. The method provided superior reproducibility and speed when compared with the results of manual and semiautomatic segmentation methods. However, the 3-D mesh is not accessible for the user to adjust directly, which makes it difficult to correct inaccuracies. The robustness of the automatic segmentation algorithm remains an unresolved issue.

Ghosh et al. (2000) developed a reliable watershed algorithm for cartilage segmentation. They performed immersion thresholding in the watershed algorithm, derived the mean and standard deviation of each region of interest per cartilage slice, and performed feature extraction through regional intensity perturbation.

The immersion-based watershed algorithm uses the volume metric (VM) to standardize and compare OA in cross-sectional studies. This standardization and comparison eliminates the effect of variations in knee sizes among the segmented image populations. The researchers demonstrated the accuracy of the proposed algorithm on three OA groups labeled as (1) normal, (2) mild, and (3) severe. However, the threshold algorithm suffers from high computation cost and from strong dependence on human operation and a priori knowledge, as well as experience. The local feature extraction also implicates possible oversegmentation.

Lynch et al. (2000) introduced a hybrid segmentation method to reconstruct 3-D cartilage volumes. The segmentation included human intervention in building up cubic splines activated by Canny filters and other active contour functions. They demonstrated that the proposed framework obtained higher reproducibility while maintaining lesser expert intervention than conventional region-growing techniques. It is more reliable than conventional snakes for which manual corrections are needed.

Camio et al. (2005) developed a semiautomatic segmentation algorithm, which builds up Bezier splines by interactively detecting the boundaries. The semiautomatic segmentation method applies a minimum 3-D Euclidean distance to measure the cartilage thickness. Moreover, Lösch et al. (1997) evaluated four thickness measurement methods for cartilages. They suggested that among these four metrics, the minimum distance between articular surface and BCI is most suitable for investigating the thickness of articular cartilages.

As shown in the literature review, no methods have been developed to solve cartilage segmentation automatically, robustly, accurately, and efficiently, especially for 3-D image mining. Meanwhile, basis spline (B-spline) has been implemented, and it has accurately measured biomarkers in segmentation algorithm implementation. The B-spline has the potential to be applied in 3-D knee cartilage segmentation in a more robust way.

We present a 3-D segmentation algorithm using the 3-D smoothing B-spline active surface (SBAS; Du et al. 2008). This algorithm implements external and internal forces in an integrated framework that can integrate other types of forces or prior knowledge in various applications. We also introduce an adaptive external force into this framework in order to improve the accuracy and robustness of the segmentation algorithm. We applied and evaluated the proposed algorithm on real MRI image data of guinea pigs. This algorithm can obtain BCI accurately and quickly. Note that minimal cartilage thickness is an example of biomarkers that will be found by determining the BCI of a femur and a tibia and by finding in three dimensions the minimum distance between these two interfaces as detailed in the study by Bolbos et al. (2007).

The BCI segmentation process is divided into several parts. First, for all MRI images in a data set, the BCI snake is initialized according to the shape and position of the segmentation target. Such initialization is performed because small animals like guinea pigs have similar features in their knees and cartilage. Second, we propose an adaptive and smoothing BCI segmentation method: The SBAS is introduced and an external force, which incorporates the relationship between evolutionary process and external force range, governs snake evolution. In addition, the adaptive scaling of forces improves the accuracy of the surface-evolving process. Last, accuracy and reproducibility are introduced to validate the segmentation result. The 3-D segmentation result of guinea pig cartilage is compared with the result given in the study by Bolbos et al. (2007).

11.2 Knee Bone–Cartilage Structure

Figure 11.1 shows the eight main structures of knee bone cartilage: (1) femur, (2) femur BCI (FBCI), (3) femur cartilage interface (FCI), (4) femur cartilage-fluid layer, (5) tibia, (6) tibia BCI (TBCI), (7) tibia cartilage interface (TCI), and (8) tibia cartilage-fluid layer. The femur, a large bone in the thigh, is attached to the knee by ligaments and to the tibia by a capsule. Cartilage is also called the "meniscus" or the "meniscal cartilage." Knee cartilage covers the end of bones, tibia, and femur and fits into the joint between them. Hence, knee cartilage appears as a C-shaped piece of tissue.

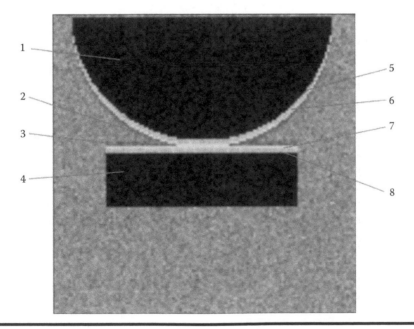

Figure 11.1 Knee bone–cartilage structure: (1) femur, (2) femur cartilage-fluid layer, (3) tibia cartilage-fluid layer, (4) tibia, (5) femur bone-cartilage interface, (6) femur cartilage interface, (7) tibia cartilage interface, and (8) tibia bone-cartilage interface.

11.3 Active Surface Framework

In Section 11.3, we introduce the fundamental of active surface framework, including B-spline function and snake.

11.3.1 Cubic Spline

Mathematically, a snake is an active contour function. Snake algorithms search for the minimum energies accumulated along and around the active contour. A spline can mathematically approximate the contour. The spline evolves through external forces, such as image forces caused by intensity gradients on boundaries and homogeneous features across regions. A spline is a well-known piecewise polynomial function. The function approximates curves and surfaces by optimally connecting polynomial pieces at knots (Boor 2001). Theoretically, polynomial pieces can present a spline in any desired order. In practical applications, the most commonly employed are cubic polynomials. The following is an example of a 2-D cubic polynomial for a group of data points $\{(x_i, y_i)\}$, $i = 0,1,\ldots,n$. The ith piece of the spline is computed by the following cubic polynomial functions:

$$x_i(s) = a_{x_i} + b_{x_i}s + c_{x_i}s^2 + d_{x_i}s^3$$
$$y_i(s) = a_{y_i} + b_{y_i}s + c_{y_i}s^2 + d_{y_i}s^3 \tag{11.1}$$

where s is a variable such that $s \in [0,1]$; and $a_{x_i}, b_{x_i}, c_{x_i}, d_{x_i}, a_{y_i}, b_{y_i}, c_{y_i}$, and d_{y_i} are coefficients for different orders to determine the shape of the spline.

A complex spline requires high-order polynomial pieces to describe the local details, such as curvature; but higher orders can cause some unexpected nonlocal behaviors in polynomials. For example, the polynomial value at one point has significant effects on other faraway points. To control the nonlocal effects, it is critical to place an appropriate number of control knots or control points at the effective positions. For example, a cubic polynomial (as shown in Equation 11.1) is the lowest-order function that describes the inflection points in curves. It constrains the maximum number of inflection points to two. In order to describe more inflection points and more elaborate shapes for a spline, the finer decomposition of the spline makes sense, although this decomposition introduces the challenges of controlling more knots and more coefficients. We can derive the polynomial coefficients from the critical prosperities or constraints of the cubic spline, for example, the continuity of derivatives of various orders of the spline at boundary corners.

Let us define an evolving 2-D snake $\vec{v}(s)$ using its Cartesian coordinates (x, y) on an image, which we can parameterize by variable s as follows:

$$\vec{v}(s) = [x(s), y(s)] \tag{11.2}$$

where $s \in [0,1]$ is a linear parameter. We will define the energy of a spline function or snake using polynomials. Then, using mathematical optimization techniques or other programming tools, we can derive the polynomial coefficients to minimize the defined energy. In this research, we use a cubic spline curve to represent the snake; $x(s)$ and $y(s)$ are cubic piecewise polynomials.

11.3.2 Basic Energies in the Snake

Originally, Kass, Witkin, and Terzopoulos (1998) defined the energy of the snake as follows:

$$E_{\text{snake}} = \frac{1}{2} \int [E_{\text{internal}}(\vec{v}(s)) + E_{\text{image}}(\vec{v}(s))] \, ds \qquad (11.3)$$

In Equation 11.3, E_{internal} denotes the internal energy of the snake, which highly depends on the geometric features along the contour. The term E_{image} denotes image energy, which is derived from the image intensity distribution along the intersectional regions of the contour. In addition, other energy terms can be defined according to specific applications, for example, constraint energy.

Kass, Witkin, and Terzopoulos (1998) defined internal energy as follows:

$$E_{\text{internal}} = \alpha(s) \left| \frac{\partial v}{\partial s} \right|^2 + \beta(s) \left| \frac{\partial v^2}{\partial s^2} \right|^2 \qquad (11.4)$$

In Equation 11.4, the coefficient $\alpha(s)$ controls the stretching rate, which is the longitudinal contraction of the curve that the snake undergoes. The coefficient works as an elastic coefficient, whereas the snake works like an elastic string. Large values of $\alpha(s)$ will make a large contraction of the snake in the direction of the stretch, whereas small values of $\alpha(s)$ will preserve the robustness of the energy function in case of local stretch variations. Parameter $\beta(s)$ controls the rate of convexity or the curvature, which regulates curve change in a direction normal to the curve. It works as a rigid coefficient, whereas the snake works like a rigid string. Large values of $\beta(s)$ make the snake hard and resistant to bending, which preserves the curve's smooth shape, whereas small values of β preserve the robustness of the energy function in case of reflection variations in the snake and allow the snake to develop corners at the correct points. Both the $\alpha(s)$ and $\beta(s)$ parameters can release geometric constraints on the snake in case of lesser options. Thus, the combinational tuning of both coefficients can lead to an appropriate elasticity so that the snake embraces the object desirably.

Kass, Witkin, and Terzopoulos (1998) defined the image energy of the snake as follows:

$$E_{\text{image}} = \theta_1 I(x, y) + \theta_1 \left| \nabla I(x, y) \right|^2 + \dots \qquad (11.5)$$

In Equation 11.5, $I(x, y)$ denotes the grayness level of the image, θ_1 constrains line energy, and θ_2 constrains edge energy.

The signs and values of parameter θ_1 activate the snake cooperatively according to image intensity. For example, large positive (negative) values of θ_1 can attract the snake along the line features of high-intensity (low-intensity) areas in the given image $I(x,y)$. Similarly, the combination of signs and values of the parameter θ_1 controls the snake according to intensity variations in the image. For example, large positive (negative) values of θ_2 can attract (retard) the snake along the inhomogeneous features, for example, boundaries, in the image.

11.4 Smoothing Basis-Spline Active Surface

In Section 11.4, we present the fundamental of using Basis-spline in active contour.

11.4.1 Cubic Basis Spline

A B-spline function is defined as a polynomial function that linearly connects all pieces of basic splines to describe a contour with sufficient smoothness, continuity, and partitioning. A B-spline function is among the most frequently employed curve representations in computer vision and computer graphics. The B-spline functions allow C^2 regularity at any point along the curve. Meanwhile, B-spline functions perform efficiently for a large number of points.

Let us start with the definition a spline function. For two points, c_1 and c_2, a linear spline curve between them can be represented by the following equation:

$$p_{i,0}(s) = \frac{s_{i+2} - s}{s_{i+2} - s_{i+1}} c_i + \frac{s - s_{i+1}}{s_{i+2} - s_{i+1}} c_{i+1}, \quad s \in [s_{i+1}, s_{i+2}] \tag{11.6}$$

The points $(c_i)_{i=1,\dots,n}$ denote the control points of the spline curve, and the parameters $s = (s_i)_{i=2,\dots,n+1}$ denote the knots of the spline curve, which assign the values of s at the control points. A piecewise linear curve based on some given points $(c_i)_{i=1,\dots,n}$ is constructed by connecting all neighboring points with a straight line. The linear curve $f(s)$ is defined with parameters $(s_i)_{i=1,\dots,n+1}, s_i < s_{i+1}$ for $i = 2,3,\dots,n$ by

$$f(s) = \begin{cases} p_{2,1}(s), & s \in [s_2, s_3], \\ p_{3,1}(s), & s \in [s_3, s_4], \\ \vdots & \vdots \\ p_{n,1}(s) & s \in [s_n, s_{n+1} \end{cases} \tag{11.7}$$

To make the expression succinct, the piecewise constant functions $B_{i,0}(s)$ are defined by

$$B_{i,0}(s) = \begin{cases} 1, & s_i \le s < s_{i+1} \\ 0, & \text{otherwise} \end{cases} \tag{11.8}$$

Using the defined function in Equation 11.8, Equation 11.7 can be simplified as follows:

$$f(s) = \sum_{i=2}^{n} p_{i,1}(s) B_{i,0}(s) \tag{11.9}$$

To fit the practical curve accurately, a higher-order spline function is introduced. A spline curve of order d with n control points is defined as

$$
f(s) = \begin{cases}
p_{d+1,d}(s), & s \in [s_{d+1}, s_{d+2}], \\
p_{d+2,d}(s), & s \in [s_{d+2}, s_{d+3}], \\
\vdots & \vdots \\
p_{d+n,d}(s), & s \in [s_n, s_{n+1}]
\end{cases}
\tag{11.10}
$$

Similar to Equation 11.8, a high-order curve can be expressed as

$$
f(s) = \sum_{i=d+1}^{n} p_{i,d}(s) B_{i,0}(s)
\tag{11.11}
$$

The point $p_{i,d}(s)$ on the spline curve $p_{i,d}$ of order d is obtained by the following computation using convex combinations:

$$
p_{j,r}(s) = \frac{s_{j+d-r+1} - s}{s_{j+d-r+1} - s_j} p_{j-1,r-1}(s) + \frac{s - s_j}{s_{j+d-r+1} - s_j} p_{j,r-1}(s)
\tag{11.12}
$$

for $j = i - d + r$ and $r = 1,2,\ldots,d$. The points are initialized by the given control points $p_{j,0}(s) = c_j$. It turns out that spline curves of order d have continuous derivatives up to degree $d - 1$.

We formulate a cubic B-spline curve using a piecewise polynomial function (order $k = 4$) as follows:

$$
p_i(s) = V_{i-1} B_{i-1}(s) + V_i B_i(s) + V_{i+1} B_{i+1}(s) + V_{i+2} B_{i+2}(s)
\tag{11.13}
$$

where s denotes curvilinear abscissa, V_i denotes the control points along the polynomial curve, B_i denotes the piecewise B-spline functions, and i denotes the spacing between knots.

The B-splines can approximate a desired contour with local regulation and continuity. First, the decomposition of B-splines into pieces makes it possible to modify the position of control points and to adjust a piece of the whole spline curve locally. Second, B-splines maintain continuity at each control point. According to the definition in Equation 11.13, we know that B-splines of order k can achieve C^{k-2} continuity at control points. Moreover, by varying the number and positions of control points, we can fine-tune the smoothness and continuity at control points. For μ, the multiplicity order of a control point, we can achieve the continuity $C^{k-1-\mu}$ at this point. We define the continuity at control points as C^0 if $\mu = k - 1$, and the curve interpolates these control points. This property facilitates the calculation of the polygon by fitting the given sample data points to B-spline curves and minimizing least-square error. We can also select the number of control points automatically by adjusting the tolerance of the least-square fitting errors.

11.4.2 The Basis Spline Snake

We can obtain the B-snake energy in Equation 11.3 by substituting V with $p_i(s)$ in the equation. Then we can minimize this energy by fine-tuning the control points. First, we start by solving the problem of minimization of snake energy in Equation 11.3.

Minimization of the energy E_{snake} in Equation 11.3 is performed in a study by Kass, Witkin, and Terzopoulos (1998) using its discretized version as follows:

$$E_{snake} = \sum_{i=1}^{n} (E_{int}(i) + E_{ext}(i)) \tag{11.14}$$

where i is the discretized version of the parameter s.

Let $f_x(i) = \partial E_{ext}/\partial x_i$ and $f_y(i) = \partial E_{ext}/\partial y_i$, where we can derive the derivatives by analytical solutions or approximate them by finite differences. Then, we can use Euler–Lagrange equations to solve the minimization in Equation 11.14. We obtain a force balance equation as follows:

$$F_{int} + F_{ext} = 0 \tag{11.15}$$

$$F_{int} = (Ax, Ay) \tag{11.16}$$

$$F_{ext} = (f_x(x,y), f_y(x,y)) \tag{11.17}$$

where A is called a "pentadiagonal banded matrix" (Kass, Witkin, and Terzopoulos 1998). In this model, F_{int} generally regularizes the extension and curvature of curves. The value F_{ext} is the sum of traditional potential forces, such as image forces, that attract the snake to the contour of image. To increase the capture range of snakes, researchers introduced multiple forces to Equation 11.17, including balloon force, GVF, VFC, etc.

To solve Equations 11.16 and 11.17, Kass, Witkin, and Terzopoulos (1998) derived negative time directives from the left-hand sides of the two equations. Then they obtained products of the directives and defined step size, and set the right-hand sides of the two equations equal to these products. The authors (Kass, Witkin, and Terzopoulos 1998) derived the following evolving equations from Equations 11.16 and 11.17 at the time when derivatives vanish:

$$Ax_t + f_x(x_{t-1}, y_{t-1}) = -\gamma(x_t - x_{t-1}) \tag{11.18}$$

$$Ay_t + f_y(x_{t-1}, y_{t-1}) = -\gamma(y_t - y_{t-1}) \tag{11.19}$$

where γ denotes step size in evolution.

Equations 11.18 and 11.19 can be solved by the following equations:

$$x_t = (A + \gamma I)^{-1}(x_{t-1} - f_x(x_{t-1}, y_{t-1})) \tag{11.20}$$

$$y_t = (A + \gamma I)^{-1}(y_{t-1} - f_y(x_{t-1}, y_{t-1})) \tag{11.21}$$

Using B-spline functions, Equations 11.20 and 11.21 can be equivalently represented by the following equations:

$$c_{x,t} = (A^b + \gamma I)^{-1}(\gamma \cdot c_{x,i-1} - f_x^b(x_{t-1}, y_{t-1})) \tag{11.22}$$

$$c_{y,t} = (A^b + \gamma I)^{-1}(\gamma \cdot c_{y,i-1} - f_y^b(x_{t-1}, y_{t-1})) \tag{11.23}$$

where $c_{x,i}$ and $c_{y,i}$ refer to the coefficient vectors at iteration t. Matrix A^b refers to the integration between the two coefficients of matrix A, and the first-order and second-order derivatives of the B-spline function. Using the respective equations of external force and the B-spline function at snake point k, an evolving force $f^b(k)$ can be obtained. Hence, the coefficient vectors in Equations 11.22 and 11.23 can drive the evolution of such a snake. Meanwhile, the internal energy in Equation 11.14 can be removed because we can regularize the smoothness of cubic B-splines by controlling the coefficient vectors. Controlling the difference between neighboring coefficients can regularize local smoothness along the B-spline curve. A big difference between two neighboring coefficient values generally results in a smooth part of the B-spline curve. By setting the two coefficients of matrix A to 0, the snake evolution process can be described as follows:

$$c_{x,t}(k) = c_{x,t-1}(k) - \gamma^{-1} \cdot f_x^b(x_{t-1}, y_{t-1}, k) \tag{11.24}$$

$$c_{y,t}(k) = c_{y,t-1}(k) - \gamma^{-1} \cdot f_y^b(x_{t-1}, y_{t-1}, k) \tag{11.25}$$

Furthermore, the snake model can be regularized by controlling the B-spline curve points. The following equation transforms the B-spline coefficients $c(k)$ to the B-spline curve points $g(k)$ as follows:

$$g(k) = b(k) \times c(k) \xrightarrow{z} G(z) = B(z) \cdot C(z) \tag{11.26}$$

where $G(z)$, $B(z)$, and $C(z)$ denote the z-transforms of $g(k)$, $b(k)$, and $c(k)$, respectively. The transform $B(z)$ can be obtained as follows:

$$B(z) = \frac{z + 4 + z^{-1}}{6} \tag{11.27}$$

By assigning the inverse result of Equation 11.27 to Equation 11.26, the B-spline coefficients $c(k)$ can be obtained (Velut, Benoit-Cattin, and Odet 2006). The aforementioned transformation from $g(k)$ to $c(k)$ is called "B-spline filtering."

Now, the snake curve can be deformed through snake points directly without resorting to the regularization of B-spline coefficients. Let $g(k) = (x(k), y(k))$ denote the kth snake curve point. Equations 11.24 and 11.25 can be presented as follows:

$$x_t(k) = x_{t-1}(k) - \gamma^{-1} \cdot f_x(x_{t-1}, y_{t-1}, k) \tag{11.28}$$

$$y_t(k) = y_{t-1}(k) - \gamma^{-1} \cdot f_y(x_{t-1}, y_{t-1}, k) \tag{11.29}$$

where $x_t(k)$ and $y_t(k)$ denote the 2-D cardinal coordinates of the kth snake point at iteration t.

11.4.3 Smoothing Basis-Spline Active Surface

A B-spline function maintains its continuity and smoothness through interpolating a set of coefficients. A B-spline snake can be constructed by the interpolation of a set of snake points $g(k)$. Reinsch (1967) illustrated that an exact interpolation is not reliable. He proposed a smoothing spline function to solve the problem. The function approximates the given data set by minimizing the following equation:

$$\varepsilon_s^2 = \sum_{k=-\infty}^{+\infty} (g(k) - \hat{g}(k))^2 + \lambda \int_{-\infty}^{+\infty} \left(\frac{\partial^2 \hat{g}(s)}{\partial^2(s)} \right)^2 \mathrm{d}s \tag{11.30}$$

In Equation 11.30, $g(k)$ denotes the kth desired snake point, $\hat{g}(k)$ denotes the approximate value of $g(k)$, $\hat{g}(s)$ denotes the interpolation function of $\hat{g}(k)$, and λ denotes the regularization parameter. Equation 11.30 combines fitness errors in the curve approximation and cost of nonsmoothness along the curve.

In Equation 11.30, the first term represents generalized approximation error between the approximating data set and the original data set. The second term represents accumulated curvature values of the spline. Parameter λ panelizes the big curvature, or favors the smooth spline. In a study by Velut, Benoit-Cattin, and Odet (2006), it was proved that a cubic B-spline can solve the minimization in Equation 11.30 by applying B-spline filtering to the smoothing B-spline snake. It was shown that an infinite impulse response (IIR) filter S_λ could facilitate the computation of coefficients in the approximation of the smoothing B-spline snake. Moreover, any B-spline snake point $\hat{g}(k)$ could be approximated by transforming the coefficients $\hat{c}(k)$ through the B-spline filter.

In a study by Unser, Aldroubi, and Eden (1993), the smoothing B-spline snake points $\hat{g}(k)$ were obtained by convolving the input signal $g(k)$ and a smoothing B-spline filter as follows:

$$\hat{g}(k) = g(k) \times b(k) \times s_\lambda(k) = g(k) \times sb_\lambda(k) \tag{11.31}$$

where $sb_\lambda(k)$ denotes the impulse response (IR) of the B-spline filter SB_λ.

In the study by Velut, Benoit-Cattin, and Odet (2006), an improved B-spline filter SB_λ was introduced, which controls the curvature of a contour or regularizes the smoothness of a snake according to the regulaization parameter λ. Furthermore, the smoothing B-spline snake was extended to a smoothing SBAS. A 3-D segmentation method was obtained based on the SBAS given in the work by Du et al. (2008).

According to Du et al. (2008), a quadrangular mesh represents the initial surface of segmentation. Then, the smoothing snake algorithm deforms the mesh points through an external force filter, as shown in Figure 11.2. In this framework, any snake point (k, l) can be obtained through an iterative deformation:

$$\hat{g}_{t+1}(k, l) = \hat{g}_t(k, l) - \gamma^{-1} \cdot (sb_{\lambda,\mu} \times f_t(k, l)) \tag{11.32}$$

Here, (k, l) refers to the indices of the initial snake point along x and y coordinates: kth in x coordinate and lth in y coordinate. The equation $\lambda = \mu$ denotes the regularization parameter along x and y coordinates, respectively. The iteration index is denoted by t. The terms $\hat{g}_{t+1}(k, l)$ and $\hat{g}_t(k, l)$ denote the respective final coordinates of snake point (k, l) at iteration $t + 1$ and t. The terms $F_t(k, l)$ denote the deformation vector generated by the sum of external forces at each snake point (k, l) and at each iteration t. The step size is denoted by γ. The impulse response of $SB_{\lambda,\mu}$ is denoted by $sb_{\lambda,\mu(k)}$.

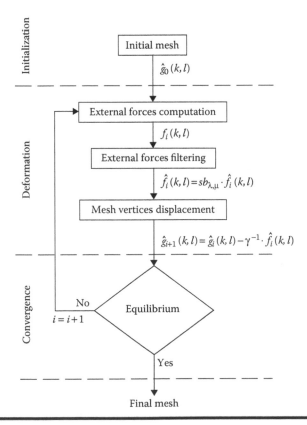

Figure 11.2 Smoothing basis-spline snake algorithm. (From Du, X., J. Velut, R. Bolbos, O. Beuf, C. Odet, and B. C. Hugues, 3-D knee cartilage segmentation using a smoothing B-spline active surface, *IEEE International Conference on Image Processing*, 2924–7. © 2008 IEEE. With Permission.)

The parameter λ controls the regularization process in snake evolution by fine-tuning the smoothing B-spline filter and smoothing the deformation force in iterations. Such an algorithm provides a framework to integrate prior knowledge or local image information in the B-snake.

As shown in Figure 11.2, the procedure of SBAS is described as follows: First, the active surface is manually initialized, and each snake point (k, l) has coordinates $g_{t=0}(k, l)$ in the initial mesh. The smoothing B-spline coefficients can be obtained by the following equation:

$$\hat{c}_{t,0}(k, l) = s_\lambda(k, l) \times g_{t,0}(k, l) \tag{11.33}$$

Here $s_\lambda(k, l)$ denotes the IR of filter $S_\lambda(Z)$. The smoothing B-spline snake algorithm can be performed as follows:

Step 1: Calculate the coordinates of snake points as follows:

$$\hat{g}_t(k, l) = b(k, l) \times \hat{c}_t(k, l) \tag{11.34}$$

Step 2: Derive the evolution forces $f_t(k, l)$ as follows:

$$f_t(k, l) = v \cdot N(k, l) \tag{11.35}$$

where v denotes the evolving speed of snake points and N denotes the unit vector normal to the local region at snake point (k, l).

Step 3: Obtain the next snake coordinates $\hat{g}_t(k, l)$ by moving $\hat{g}_t(k, l)$ by $f_t(k, l)$ as follows:

$$g_{t+1}(k, l) = \hat{g}_t(k, l) + f_t(k, l) \tag{11.36}$$

Step 4: Derive the B-spline coefficients as follows:

$$\hat{c}_{t+1}(k, l) = s_\lambda(k, l) \times g_{t+1}(k, l) \tag{11.37}$$

Step 5: Run steps 1 through 4 until convergence.

In the aforementioned process, the snake converges and ceases iterations when it reaches an equilibrium condition or sufficient convergence. The convergence criteria can be determined by a set of features on the smoothing snake surface, such as smoothness.

In the aforementioned steps, each iteration implements a regularization performance in steps 1 and 4. In step 1, $b(k, l)$ denotes the IR of the B-spline filter $B(z)$ at the snake point (k, l). From these steps, we can obtain an evolving snake with the extraction of forces $f_t(k, l)$. In Section 11.5, we define the forces for snake evolution.

11.5 Deformation Forces Definition and Active Control

For BCI segmentation, the SBAS algorithm is applied with external forces defined by a combination of image force vectors (denoted as F_{image}) and balloon force (denoted as $F_{balloon}$):

$$\vec{F}_{ext} = \alpha \cdot \vec{F}_{image} + \beta \cdot \vec{F}_{balloon} \tag{11.38}$$

where α and β are weighting parameters that control the contribution of the two forces.

In this study, image force is defined by the Laplacian applied to the image that is first smoothed with a Gaussian smoothing filter in order to reduce its sensitivity to noise. Hence, the image force and the corresponding external energy are described as follows:

$$F_{image} = -\nabla E_{ext}(x, y, z) \tag{11.39}$$

$$E_{ext}(x, y, z) = -\left|\nabla(G_\sigma(x, y, z) \times I(x, y, z))\right|^2 \tag{11.40}$$

where $G_\sigma(x, y, z)$ is a 3-D Gaussian function with standard deviation σ, ∇ is the gradient operator, and $I(x, y, z)$ is the image gray level located at coordinate (x, y, z).

Balloon force defined by Cohen (1991) acts in a direction normal to the curve. In such a scheme, it is tedious to define by trial and error an optimal-scale combination of image and balloon forces. A simple factor selection can be as follows:

$$\alpha = c \tag{11.41}$$

$$\beta = 1 - \alpha \tag{11.42}$$

where c is the constant of Equation 11.32 located between 0 and 1.

The settings of Equations 11.41 and 11.42 cannot guarantee that the snake slows down its evolving step, as it is near or on the image boundary. We propose in Equations 11.43 and 11.44 an adaptive scaling of these weighting parameters that improves the robustness of snake evolution and avoids the trial-and-error tuning of these parameters:

$$\alpha = c.IterationDone/NumberOfIterations \tag{11.43}$$

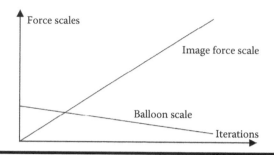

Figure 11.3   Scale factors of forces. (From Du, X., J. Velut, R. Bolbos, O. Beuf, C. Odet, and B. C. Hugues, 3-D knee cartilage segmentation using a smoothing B-spline active surface, *IEEE International Conference on Image Processing*, 2924–7. © 2008 IEEE. With Permission.)

$$\beta = (1 - IterationDone/NumberOfIterations).(1 - c) \qquad (11.44)$$

where *NumberOfIterations* is the total iteration number to be performed, and *IterationDone* is the finished iteration number.

As shown in Figure 11.3, scale factors α (for image force) and β (for balloon force) change with iteration numbers, which is different from the study by Wang and Li (1999) where the balloon force adapts only spatially. Our scale factors follow the phenomenon of snake evolution. When the initial snake starts running, the edge information is not enough and the balloon force should be stronger; when the snake approaches the image edge, the balloon force should fade as edge-based forces become stronger.

11.6   Results

In Section 11.6, we present the experimental results using the proposed 3D smoothing Basis-spline active surface algorithm.

11.6.1   Magnetic Resonance Imaging Data Set

We worked on the same MRI data set as Du et al. (2008). The MRI set has been obtained with a 7 T Biospec system (Brüker, Germany). A fat-suppressed 3-D gradient-echo fast imaging (GEFI) sequence has been used with a flip angle α of 25°, echo time = 3.6 milliseconds, repetition time = 50 milliseconds, and 42-kHz receiver bandwidth. A total of 64 slices (312-μm thick) were acquired in the sagittal plane with a field of view of 30 mm and an imaging matrix size of 512 × 384 pixels, corresponding to a reconstructed spatial resolution of 59 × 59 μm^2.

Finally, we applied the segmentation algorithm on a resliced volume of interest of 64^3 voxels with an isotropic resolution of 59 μm. We worked on two groups of guinea pigs: (1) Sham controlled guinea pigs (SHAM) with 9 animals used as control and (2) meniscectomized (MNX) guinea pigs with 10 animals that had undergone meniscectomy.

11.6.2 Segmentation Experiments

The SBAS algorithm associated with the adaptive combination of deformation forces was implemented using the visualization toolkit (VTK) library and Python language on the MRI images data set. The process aims at the segmentation of FBCI and TBCI. The initial surfaces for FBCI and TBCI are quadrangular patches extracted from a sphere and a plane, respectively (Figure 11.4). We chose the smoothing regulator $\lambda = 500$ and iteration number $N_1 = 600$ for FBCI and $N_2 = 300$ for TBCI, respectively.

As illustrated in Figure 11.4, the proposed algorithm is able to segment the whole TBCI and FBCI and can be used to compute a thickness map of the cartilage (Figure 11.5). In order to conduct a quantitative assessment of the proposed algorithm similar to the one mentioned in the study by Bolbos et al. (2007), we consider only the central part of the cartilage (Figure 11.6).

11.6.3 Evaluation of Segmentation Result

The quantitative segmentation assessment is based on the minimum cartilage thickness (Bolbos et al. 2007) that equals the minimum distance between FBCI and TBCI. This parameter enables us to compare the segmentation accuracy quan-

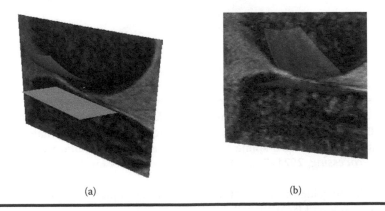

(a) (b)

Figure 11.4 Local segmentation of femur bone-cartilage interface (the grey plain-shaped surface) and tibia cartilage interface (the grey sphere-shaped surface): (a) initialization with two quadrangular mesh patches and (b) final segmentation results. (From Du, X., J. Velut, R. Bolbos, O. Beuf, C. Odet, and B. C. Hugues, 3-D knee cartilage segmentation using a smoothing B-spline active surface, *IEEE International Conference on Image Processing*, 2924–7. © 2008 IEEE. With permission.)

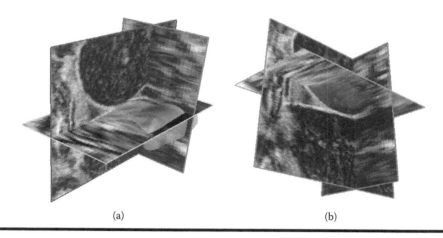

(a)   (b)

Figure 11.5 Whole segmentation of tibia and femur bone-cartilage interfaces: (a) up-down view; (b) the down-up view. (From Du, X., J. Velut, R. Bolbos, O. Beuf, C. Odet, and B. C. Hugues, 3-D knee cartilage segmentation using a smoothing B-spline active surface, *IEEE International Conference on Image Processing*, 2924–7. © 2008 IEEE. With permission.)

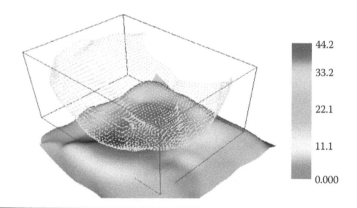

Figure 11.6 Three-dimensional map of cartilage thickness. (From Du, X., J. Velut, R. Bolbos, O. Beuf, C. Odet, and B. C. Hugues, 3-D knee cartilage segmentation using a smoothing B-spline active surface, *IEEE International Conference on Image Processing*, 2924–7. © 2008 IEEE. With Permission.)

titatively with histological results (HISTO) working as previous 3-D snake-based segmentation method (SOVITA; Bolbos et al. 2007).

The segmentation mean and standard error of mean (SEM) are analyzed for the two animal groups (SHAM and MNX). The SEM is defined by

$$SEM = \frac{\sigma}{\sqrt{n}} \qquad (11.45)$$

where σ is the standard deviation of the minimum thickness and n is the number of images by group.

11.6.4 Results and Discussion

The segmentation results of SBAS that are shown in Tables 11.1 and 11.2 were obtained for the SHAM and MNX groups, respectively, and they represent the published approach (SOVITA) and histological data (HISTO) segmentation results for each group. The comparison shows that the proposed SBAS method obtains a more accurate measurement of minimum thickness than the SOVITA method compared with HISTO.

The SEM behavior is similar in the two data groups. For both these groups, the SEM values are similar to those of SOVITA. The results indicate that the accuracy is improved without losing the reproducibility of cartilage thickness measurement.

Table 11.1   Segmentation Results Given as the Minimum Cartilage Thickness on SHAM

SHAM Group	SBAS	HISTO	SOVITA
Number of samples	9	9	9
Minimum thickness average (µm)	324	329	314
SEM	8.2	4.7	8.5

Source: Du, X., J. Velut, R. Bolbos, O. Beuf, C. Odet, and B. C. Hugues, 3-D knee cartilage segmentation using a smoothing B-spline active surface, *IEEE International Conference on Image Processing*, 2924–7. © 2008 IEEE. With permission.

Table 11.2   Segmentation Results Given as the Minimum Cartilage Thickness on MNX

MNX Group	SBAS	HISTO	SOVITA
Number of samples	10	10	10
Minimum thickness average (µm)	310	317	294
SEM	13.3	4.5	12.4

Source: Du, X., J. Velut, R. Bolbos, O. Beuf, C. Odet, and B. C. Hugues, 3-D knee cartilage segmentation using a smoothing B-spline active surface, *IEEE International Conference on Image Processing*, 2924–7. © 2008 IEEE. With permission.

The SBAS method simplifies the manual tuning of image force scale and balloon force scale with the integration of the two scales by using an adaptive parameter. In addition, with such a deformation force combination and the proper selection of iteration number, initial surfaces are not limited by external force range due to different data sets, which improves the robustness of the segmentation.

The current research results focus on mining the minimum thickness of an imaging biomarker of OA. Moreover, the SBAS algorithm generates 3-D FBCI and TBCI surfaces that can be used for mining other imaging biomarkers for the quantitative evaluation of OA. For example, software can be developed to integrate bone-cartilage surface mining or bone-cartilage volume.

11.7 Conclusions

This chapter presents a 3-D bone-cartilage segmentation approach based on SBAS coupled with adaptive deformation forces. This SBAS algorithm and the adaptive scheme facilitate the tuning of the segmentation algorithm and enhance the robustness of segmentation to initialization.

A quantitative assessment based on histological reference shows that the proposed approach improves the accuracy of cartilage thickness measurements.

As illustrated in Figure 11.6, the proposed method is not limited to the measurement and analysis of cartilage thickness. Other criteria, such as cartilage volume, topography, articular contact areas, and surface feature, can also be extracted from the quantitative analysis of segmentation results.

Acknowledgments

The MRI acquisition has been realized on ANIMAGE, the small animal imaging platform of Lyon, France. This project has been funded by the Institut de Recherches Servier, Croissy-sur-Seine, France. I thank the National Center for Scientific Research, France, for their postdoctoral financial support.

References

Andriacchi, T. P., A. Mündermann, R. L. Smith, E. J. Alexander, C. O. Dyrby, and S. Koo. 2004. A framework for the in vivo pathomechnics of osteoarthritis at the knee. *Ann Biomed Eng* 32:447–57.

Bendele, A. M., and J. F. Hulman. 1991. Effects of body weight restriction on the development and progression of spontaneous osteoarthritis in guinea pigs. *Arthritis Rheum* 34:1180–4.

Bolbos, R., O. Beuf, H. Benoit-Cattin, A. Chomel, E. Chereul, C. Odet, P. Pastoureau, M. Janier, and O. Beuf. 2007. Knee cartilage thickness measurements using MRI: A 4 1/2-month longitudinal study in the meniscectomized guinea pig model of OA. *Osteoarthritis Cartilage* 15:656–65.

Boor, C. D. 2001. *A Practical Guide to Splines.* Revised Edition. New York: Springer-Verlag.

Camio, J. C., J. S. Bauer, K. Y. Lee, S. Krause, and S. Majumdar. 2005. Combined image processing techniques for characterization of MRI cartilage of the knee. In Xu, D. and D. Bhaskar: *The 27th IEEE Annual International Conference of Engineering in Medicine and Biology Society (EMBS)*, 3043–6, Shanghai.

Cohen, L. D. 1991. On active contour models and balloons. *CVGIP Image Underst* 53:211–21.

Conaghan, C. Y. J., and P. G. Wenham. 2009. Imaging the painful osteoarthritic knee joint: What have we learned? *Nat Clin Pract Rheumatol* 5:149–58.

Du, X., J. Velut, R. Bolbos, O. Beuf, C. Odet, and B. C. Hugues. 2008. 3-D knee cartilage segmentation using a smoothing B-spline active surface. *IEEE Int Conf Image Proc* 25:2924–7. San Diego, Ca.

Fripp, J., S. Crozier, and S. Warfield. 2005. Automatic initialization of 3D deformable models for cartilage segmentation. In Lovell B. C., A. J. Maeder, T. Caelli, and S. Ourselin. *Proceedings of Digital Imaging Computing: Techniques and Applications (DICTA)*, 513–8. IEEE. Cairns, Australia.

Ghosh, S., O. Beuf, M. Ries, N. E. Lane, L. S. Steinbach, T. M. Link, and S. Majumda. 2000. Watershed segmentation of high resolution magnetic resonance images of articular cartilage of the knee. In *Proceedings of the 22th Annual EMBS International Conference*, 4:3174–6. Chicago.

Kapur, T. 1999. Model based three dimensional medical image segmentation. Ph.D. Thesis. MIT, Cambridge, MA.

Kass, M., A. Witkin, and D. Terzopoulos. 1998. Snakes: Active contour models. *Int J Comput Vis* 1:321–31.

Li, B. 2006. Vector field convolution for image segmentation using snakes. *IEEE Int Conf Image Proc* 1:1637–40.

Loeuille, D., P. Gonord, C. Guingamp, P. Gillet, A. Blum, M. Sauzade, et al. 1997. In vitro magnetic resonance microimaging of experimental osteoarthritis in the rat knee joint. *J Rheumatol* 24:133–9.

Lösch, A., F. Eckstein, M. Haubner, and K. H. Englmeier. 1997. A non-invasive technique for 3-dimensional assessment of articular cartilage thickness based on MRI part 1: Development of a computational method. *Magn Reson Imaging* 15:795–804.

Lynch, J. A., S. Zaim, J. Zhao, A. Stork, C. G. Peterfy, and H. K. Genant. 2000. Cartilage segmentation of 3D MRI scans of the osteoarthritic knee combining user knowledge and active contours. *Proc SPIE* 3979:925–35.

Munasinghe, J. P., J. A. Tyler, T. A. Carpenter, and L. D. Hall. 1995. High resolution MR imaging of joint degeneration in the knee of the STR/ORT mouse. *Magn Reson Imaging* 13:421–8.

Pastoureau, P., S. Leduc, A. Chomel, and F. De Ceuninck. 2003. Quantitative assessment of articular cartilage and subchondral bone histology in the meniscectomized guinea pig model of osteoarthritis. *Osteoarthritis Cartilage* 11:412–23.

Pirnog, C. D. 2005. Articular cartilage segmentation and tracking in sequential MR Images of the knee. Ph.D. Thesis. ETH. Zurich, Swiss.

Prince, C., and J. L. Xu. 1997. Gradient vector flow: A new external force for snakes. *IEEE Proc Conf Comput Vis Pattern Recognit (CVPR)*, 11:1189–1211.

Reinsch, C. H. 1967. Smoothing by spline functions. *Numer Math* 10:177–83.

Sinha, H. Z., and U. S. Tameem. 2007. Automated image processing and analysis of cartilage MRI: Enabling technology for data mining applied to osteoarthritis. *AIP Conf Proc* 262–76.

Solloway, S., C. E. Hutchinson, J. C. Waterton, and C. J. Taylor. 1997. The use of active shape models for making thickness measurements of articular cartilage from MR images. *Magn Reson Med* 37:943–52.

Unser, M., A. Aldroubi, and M. Eden. 1993. B-spline signal processing. Part I. Theory. *IEEE Trans Signal Process* 41:821–33.

Velut, J., H. Benoit-Cattin, and C. Odet. 2006. Segmentation by smoothing B-spline active surface. In *IEEE International Conference on Image Processing*, 209–12. Atlanta, USA.

Wang, X. B., and J. K. Li. 1999. Adaptive balloon models. *IEEE Comput Soc Conf Comput Vis Pattern Recognit* 2:24–34.

Chapter 12

Supervised Classification of Digital Mammograms

Harpreet Singh and Sumeet Dua

Contents

12.1 Introduction

The latest report published by the American Cancer Society (ACS; 2009–2010) estimated that 192,370 cases of invasive breast cancer and 62,280 cases of in situ breast cancer would be diagnosed among American women in 2009 (Breast Cancer Fact 2009–2010). The ACS report also estimated that 1910 men would be diagnosed with the disease. The report further predicted that 40,170 women and 440 men would die from breast cancer in 2009. Based on these rates, ACS estimated that 13.2% of women born in 2009 would be diagnosed with breast cancer at some time in their life (Cancer.gov 2009).

Age is considered the biggest risk factor for breast cancer; as a woman ages, her risk of developing breast cancer increases (Cancer.gov 2009). According to the latest report of the National Cancer Institute (NCI), the chances of a woman being diagnosed with breast cancer vary according to age (Cancer.gov 2006). According to the NCI, the chances of occurrence of breast cancer in women are as follows:

- 1 in 233, if the woman is between 30 and 39 years of age
- 1 in 69, if the woman is between 40 and 49 years of age
- 1 in 38, if the woman is between 50 and 59 years of age
- 1 in 27, if the woman is at least 60 years old

12.2 What Is Breast Cancer?

Cancer is the name given to a group of diseases that result in mutation and uncontrollable growth of cells in the body. These cells eventually result in the formation of a lump, called a "tumor" (Breast Cancer Fact 2009–2010). Cancer is usually named after the body part where the tumor originates. Breast cancer begins in the breast tissue, which consists of lobules (glands for milk production) and ducts that connect the lobules to the nipple (Breast Cancer Fact 2009–2010). Breast cancer can be divided into two types:

1. In situ breast cancer is usually confined to the ducts or the lobules. Oncologists commonly believe that this type of cancer is not a true cancer but an indicator of an increased risk of developing invasive breast cancer.
2. Invasive cancer begins at the ducts or lobules of the breast but eventually spreads to different parts of the breast.

The American Joint Committee on Cancer classifies invasive cancer into four progressive stages based on how much the cancer spreads. The stages are I, II, III, and IV. Stage I is the earliest stage. In this stage, the cancerous tumor is relatively small (approximately an inch in diameter) and is isolated within the breast. Stage IV is the most advanced stage. In this stage, the cancer has already spread throughout the body. Although age is commonly considered the most important risk factor for the development of breast cancer, other factors like breast density, body weight, family medical history, hormone therapy, race/ethnicity, and radiation exposure are also indicators that an individual may be at risk of developing cancer. Since there is no known way to prevent breast cancer, early diagnosis is considered the best option to decrease the mortality rate of breast cancer. Mammograms are the most commonly used tool for early diagnosis of breast cancer.

12.3 Mammograms

A mammogram, considered the first line of defense for breast cancer, is the most important tool for diagnosing and evaluating the disease and for providing follow-up information for women who have had breast cancer (Breastcancer.org 2009). Since breast cancer cannot be prevented, the best treatment option is early diagnosis. Regular mammogram screenings can help decrease the mortality rate for breast cancer. In severe cases, the breast is removed to stop the spreading of cancer; early diagnosis of breast cancer using a mammogram followed by proper medical procedure may prevent this extreme measure.

12.3.1 What Is a Mammogram and What Is Mammography?

A mammogram is a radiographic (low-dose X-ray) image of the breast. For a mammogram, two images of each breast are taken: (1) the top-down view, called the "craniocaudal" (CC) view; and (2) the side view, called the "mediolateral oblique" (MLO) view. In most cases, images of both breasts are taken, although the cancer might be present in only one breast. The process of acquiring a mammogram is known as mammography. During mammography, a technician compresses the breast using a dedicated mammography machine between two plates. Then, MLO and CC views of the breast are taken using a specialized camera (Breastcancer.org 2009). Two types of mammograms are typically used: (1) "screening" mammogram and (2) "diagnostic" mammogram. Screening mammograms are regular mammograms that are performed every year to detect early signs of cancer. Diagnostic mammograms are focused on the areas of risk found by screening mammograms. Although not perfect, mammograms can detect suspicious regions of a tumor, which can go undetected during self-examination or physical checkup by a doctor. Radiologists visually examine mammograms for specific abnormalities depicting possible cancerous cells; they also look for the following masses or structures for further investigation (Breastcancer.org 2010):

- ■ "Calcifications" are tiny flecks of calcium that are similar to grains of salt and can sometimes indicate the presence of early breast cancer. Mammography is the only way to find calcifications, and a doctor may recommend further tests.

Macrocalcifications, or big calcifications, are not associated with cancer; but the shape, size, and number of microcalcifications, or small calcifications, in a group may be early signs of cancer.

■ "Cysts" are common fluid-filled masses in the breast that are rarely associated with cancer. An ultrasound is the best way find out if a cyst is cancerous, since sound waves will pass through a liquid-filled cyst but will bounce back to the film if the cyst is a solid cancerous tumor.

■ "Fibroadenomas" are movable, solid, round lumps of normal breast cells that may grow. Although fibroadenomas are not usually cancerous, they need to be removed so that they do not develop into cancer.

Although mammograms are not perfect (15%–20% of breast cancers are not visible using this technique), they are the most powerful and most widely used tool for the early diagnosis of breast cancer (Breastcancer.org). Two types of mammograms are currently in use: (1) film mammograms and (2) digital mammograms. The only difference between film and digital mammograms lies in the storage of mammographic data. In film mammograms, the image is recorded in black and white on a large X-ray film; in digital mammograms, the image is recorded directly on the computer. Digital mammographic images can be enlarged, and suspicious areas can be highlighted for further analysis. Although X-ray film mammograms are more commonly used by technicians today, digital mammograms are slowly replacing the use of film mammograms. Figure 12.1 shows digital mammographic images that belong to normal, benign, and malignant classes.

Since digital images are directly stored on a computer, a technician has greater control over digital mammograms. These images can be easily stored in and retrieved from a database, transmitted electronically over long physical distances for quick reviews, and enhanced to concentrate on specific areas.

Whereas film-based mammograms have been used for breast cancer detection since the 1970s, digital mammography has started gaining in importance only recently. The U.S. Food and Drug Administration (FDA) approved the use of digital mammography in January 2000 (Cancer.gov 2009). A large clinical trial was performed in 2005 to compare the results obtained using digital mammography with those obtained using film mammography. Although the results did not indicate much difference between the methods in detecting breast cancer in women over the age of 50 years, the researchers concluded that premenopausal or perimenopausal women who are under the age of 50 and have dense breast tissue might benefit from digital mammograms (Pisano et al. 2005). Although there are many clinical advantages to using digital mammograms, the technique is still not widely used. Taking and reading digital mammograms requires advanced technical expertise and equipment that is too expensive for many clinics. Further, in order to perform digital mammography, a clinic must first be able to perform conventional mammography and be certified by the FDA (Cancer.gov 2009).

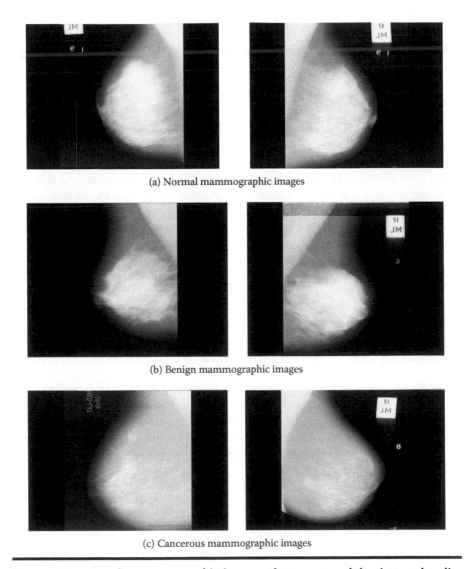

(a) Normal mammographic images

(b) Benign mammographic images

(c) Cancerous mammographic images

Figure 12.1 Sample mammographic images of types normal, benign, and malignant, from the mammographic image analysis data set.

12.3.2 How Useful Are Mammograms for Breast Cancer?

Mammography, like most medical tests, is not perfect; but it can detect 80%–90% of breast cancers (Breast Cancer Fact 2009–2010). Early diagnosis of breast cancer can reduce the risk of a person dying from this disease by 25%–30% or more (Breastcancer.org). Therefore, early detection of breast cancer improves a patient's chances of survival. According to studies by the NCI, mammograms lower the risk of death associated with breast cancer by 35% in women over 50 years and by 25%–35% in women between the ages of 40 to 50 years (Breastcancer.org). Experts from the NCI, ACI, and the American College of Radiology (ACR) suggest annual mammograms for women who are over 40 years old.

12.3.3 Computer-Aided or Automated Mammogram Classification

Automated mammogram classification systems can aid radiologists in classifying mammograms as benign or cancerous. These systems can highlight areas of risk that might be missed by a radiologist. Further, computer-aided systems can work as a second radiologist by giving a second opinion on diagnosis. Studies indicate that dual reading of mammograms, that is, mammogram reading and diagnosis by two radiologists, reduces the chance of erroneous diagnosis by 10%–15% (Breastcancer.org 2009). However, obtaining this dual reading can be difficult because of the following three reasons:

1. There are too few radiologists who are skilled enough to perform quick and efficient diagnoses. Analyzing a mammogram is a skill that radiologists learn over time.
2. Not all medical facilities employ multiple radiologists because of a lack of available radiologists and a shortage of funds.
3. Insurance companies do not always pay for dual readings of mammograms.

Therefore, a plausible solution is an automated system that can aid doctors in making better diagnostic decisions. Such a system does not exist. Automated mammogram classification systems can fill this void. These systems flag the suspicious regions on mammograms for further analysis by expert radiologists. Such a system can save lives by reducing the amount of time a patient has to wait for a radiologist to read the mammogram.

12.3.4 Computational Challenges for Image Mining

The recent advances in image acquisition and database technology have resulted in the availability of large and detailed image data sets. This huge amount of data, if analyzed properly, can provide significant insight to users and aid in the discovery of useful underlying patterns. Due to the current lack of automated tools for this analysis, image mining has gained significant popularity. However, image mining is a challenging task since image data sets are multidimensional databases that are difficult to handle. There is high variance among images that belong to the same class as well as some semantic similarities among images that belong to different classes, which share some semantic regions. For example, images of beach and mountains classes both contain the region sky.

Image mining is different from computer vision. The motivation for using image mining is not to extract better features from individual images, rather it is to uncover underlying patterns among a collection of images (Zhang, Hsu, and Lee 2001).

These implicit underlying patterns can then be used for image classification and retrieval purposes. The biggest challenge for image mining is mapping low-level pixel representation into high-level semantic representation. A common misconception is that image mining is simply the application of data mining tools for image data sets. However, image mining is significantly different from mining relational databases. Zhang, Hsu, and Lee (2001) identified the following three differences between image mining and relational database mining:

1. Individual pixel values in images may not be significantly useful unless the context supports them. A pixel value (for instance, x) could appear darker or brighter than the actual pixel value (for instance, y) depending on the surrounding pixel values in its neighborhood.
2. Since individual pixel values do not provide much information about an image, spatial information is critical for image interpretation.
3. Image pattern representation is one of the most important issues in image mining. A user has to decide whether to use region-based or global features and how to incorporate contextual and spatial information into the pattern representation. All the issues adversely affect the type of classification or retrieval models that will be used for image analysis.

Another potential problem affecting image mining is the heterogeneous nature of image data. There could be a high variation in size of the images for one class and among images of different classes. There could also be a high degree of imbalance for data for each class, that is, one class could contain a large amount of data and another could contain only a small amount of data.

In addition to the problems that exist for image mining in general, medical images like mammograms pose additional problems. First, although huge quantities of medical data are available, there is no set standard for data storage and maintenance. This lack of a set baseline standard for capturing the data results in high variation among these images. Second, digitized medical images usually contain intentionally added labels, which need to be removed before processing. Most mammograms are usually low-contrast images, which are captured to be analyzed by humans and not automated machines. This design further complicates the image mining task. These images need to be preprocessed to remove any noise and unwanted information, for example, a black noisy background, before technicians can perform additional data mining tasks.

12.4 Mammogram Data Sets

Due to privacy concerns, most medical data sets are not freely available. As a result, there are not many medical image data sets available online for researchers to compare their results. Currently, there are only two publicly available mammogram data sets: (1) mammographic image analysis (MIAS) database and (2) digital database for screening mammography (DDSM). Most studies in mammographic image analysis use one or both of these data sets to evaluate their techniques.

12.4.1 Mammographic Image Analysis Data Set (The Mini-Mammographic Image Analysis of Mammograms)

The MIAS data set, one of the most commonly used databases for mammography, was created by research groups based in the United Kingdom. The database consists of left and right breast images for 161 women, taken from the UK National Breast Screening Program. All the images in the data set were digitized at a resolution of 1024 × 1024 with an 8-bit gray level. The data set contains 322 mammograms. Of the 322 mammograms, 208 belong to class normal, 63 belong to class benign, and 51 belong to class malignant. For each abnormal case (benign and malignant), the location information of the abnormality is also provided as an approximate radius with center positions x and y. Further, the type of suspected abnormality detected on the mammogram is provided for each abnormal case. The different classes of abnormalities are calcifications (well-defined/circumscribed masses), speculated masses (ill-defined masses), architectural distortion, asymmetry, and normal tissue. Each mammogram is provided with information regarding the breast position (left or right) and the density class: fatty, glandular, or dense.

12.4.2 Digital Database for Screening and Mammography Data Set

The DDSM is another publicly available data set (Heath et al. 2001). It is a result of the collaborative effort of researchers at Massachusetts General Hospital (D. Kopans and R. Moore); Sandia National Laboratories, California (P. Kegelmeyer); and the University of South Florida, Department of Computer Science and Engineering (K. Bowyer). The data set consists of 2620 cases. Each case further consists of two images, each of which contains a visual representation of both the left and right breasts. Additionally, information, such as age of the patient at the time of the mammogram, the ACR breast density rating, subtlety rating for abnormality, the ACR keyword description of abnormality, and image information like scanner and spatial resolution, is also provided for each case. Files for all suspicious cases also contain added information regarding the location of the suspected abnormality. The DDSM database is maintained by the University of South Florida and is arranged into cases and volumes.

Each volume is a collection of several cases. A single case represents a mammography exam of a single patient. As such, each case consists of the mammograms of a single patient. The 2620 cases are arranged into volumes. Normal cases consist of

Table 12.1   Contents of the DDSM Database

Institution	Digitizer	Number of Cases by Most Severe Finding				
		Normal	Benign without Callback	Benign	Malignant	Total
MGH	DBA M2100 ImageClear	430	0	0	97	527
	Howtek 960	78	0	446	323	847
WFU	Lumisys 200 Laser	82	93	126	159	460
SH		0	48	202	234	484
WU	Howtek MultiRad850	105	0	96	101	302
Total		695	141	914	914	2620

Source: Heath, M. D., K. Bowyer, D. Kopans, R. Moore, and W. P. Kegelmeyer, In *Proceedings of the Fifth International Workshop on Digital Mammography*, ed. M. J. Yaffe, 212–8, Medical Physics Publishing, Madison, WI, 2001. With permission.

mammograms that were previously diagnosed as normal at a screening and at an additional screening 4 years later. Benign cases consist of mammograms that were previously flagged by radiologists as containing something suspicious but were later diagnosed as not malignant. The cancer volumes consist of cases in which malignant cancer was found. Table 12.1 shows the composite information from the 2620 cases. The normal, benign, and malignant classes have already been discussed. The class benign without callback includes those cases in which a suspicious abnormality was found but was later diagnosed as not malignant and no additional recalls from the patient were required.

12.5 Mammogram Classification Techniques

Most mammogram classification techniques can be sorted as density-based or abnormality-based classification, as shown in Figure 12.2. Density-based classification techniques categorize mammograms into tissue density classes like fatty, glandular, and dense or into the breast imaging reporting and data system (BIRADS) I–IV categories. Abnormality-based classification techniques perform categorization based on whether and how a tissue is abnormal. Such classification includes normal and abnormal; or normal, benign cancerous, and malignant cancerous classes. In this chapter, we first explain density-based classification procedures and, then, we explain abnormality-based classification.

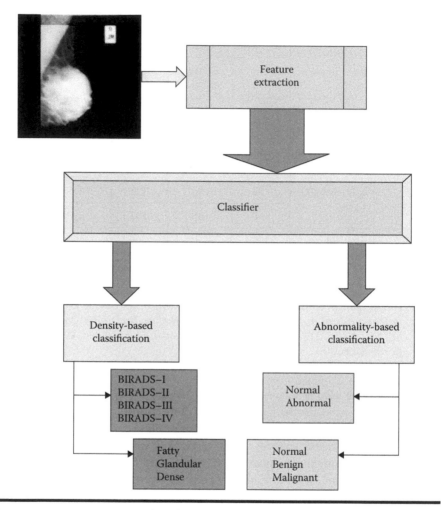

Figure 12.2 Mammogram classification techniques.

12.6 Tissue Density–Based Mammogram Classification

Tissue density is an important indication of breast cancer. The first researcher to show a connection between tissue density and the risk of breast cancer is Wolfe (1976). Since Wolfe's discovery, tissue density–based mammographic classification has been rigorously studied.

Oliver et al. (2005) used morphological and texture features to classify mammograms into the four BIRADS categories. The underlying hypothesis of their technique is that mammograms belonging to different categories are represented by tissues with different texture features. Initially, they used a fuzzy c-means algorithm to segment the breast part from the mammographic image. The four steps of their segmentation procedure are explained as follows:

1. Smooth the breast region with a median filter of size 5×5 to normalize the effects of microtexture.
2. Calculate the histogram of the smoothed image.
3. Pick two seeds for clustering: (1) one with gray-level values representing 15% and (2) the other with gray-level values representing 85% of the histogram.
4. Perform fuzzy *c*-means clustering using the seeds found in the last step as initial cluster centers to minimize the following function:

$$e^2(P,M) = \sum_{i=1}^{N} \sum_{j=1}^{K} m_{ij} \left\| x_i - c_j \right\|^2 \tag{12.1}$$

where P = partition of the mammographic image, N = total number of pixels in the image, K = number of clusters (two in this case), and M = membership matrix of size $N \times K$ such that a cell m_{ij} represents the membership of pixel x_i for cluster j. In the study by Oliver et al. (2005), the center of the cluster was represented by $c_j = \sum_{i=1}^{N} m_{ij} x_i$.

Once the breast tissue was segmented from the image, 39 features, that is, 3 morphological features, relative area, center of masses, and medium; and 9 co-occurrence matrix–based texture features, contrast, energy, entropy, correlation, sum average, sum entropy, difference average, difference entropy, and homogeneity for 4 co-occurrence matrix directions, were extracted from each segmented region. The four directions used for the co-occurrence matrix were 0°, 45°, 90°, and 135°. A pixel distance of 1 pixel was used to generate the co-occurrence matrix in each direction. Two classifiers, *k*-nearest neighbors and decision tree, were used to evaluate the extracted features. All features were normalized to unit variance and zero mean to remove any bias during weighting in *k*-NN. Additionally, a third classifier that combines the fuzzy memberships of both the classifiers for every class, takes the average of each of the four class memberships, and then classifies the image into the class with the highest average was developed. Oliver et al. (2005) used 300 images from the DDSM database to evaluate their proposed algorithm.

Out of the 300 mammograms contained in the DDSM database, 50 belong to each of BIRADS I and IV categories, and 100 belong to each of BIRADS II and III categories. Tables 12.2 and 12.3 represent the confusion matrices for *k*-NN and ID3-based classification. The average classification accuracies using the leave-one-out method for *k*-NN and ID3 were 40.3% and 43.3%, respectively. From the tables, it is evident that ID3 performs better than *k*-NN. This behavior is attributed to the fact that ID3 uses a feature-selection discrimination process that ensures non-discriminate features are not used in weighting for classification. Table 12.4 shows the confusion matrix for the combined fuzzy classifier (third technique). As can be

Table 12.2 Confusion Matrix for *k*-NN

		Automatic Classification			
		BIRADS I	*BIRADS II*	*BIRADS III*	*BIRADS IV*
Truth	BIRADS I	17	24	5	4
	BIRADS II	27	35	30	8
	BIRADS III	3	25	54	18
	BIRADS IV	3	9	23	15

Source: Oliver, A., J. Freixenet, and J. Zwiggelaar, Automatic classification of breast density, *IEEE ICIP*, vol. 2, 1258–61. © 2005 IEEE. With permission.

Table 12.3 Confusion Matrix for ID3

		Automatic Classification			
		BIRADS I	*BIRADS II*	*BIRADS III*	*BIRADS IV*
Truth	BIRADS I	24	20	3	3
	BIRADS II	34	49	10	7
	BIRADS III	18	17	40	25
	BIRADS IV	9	5	19	17

Source: Oliver, A., J. Freixenet, and J. Zwiggelaar, Automatic classification of breast density, *IEEE ICIP*, vol. 2, 1258–61. © 2005 IEEE. With permission.

Table 12.4 Confusion Matrix for the Fuzzy Combination of *k*-NN and ID3

		Automatic Classification			
		BIRADS I	*BIRADS II*	*BIRADS III*	*BIRADS IV*
Truth	BIRADS I	23	21	3	3
	BIRADS II	32	48	12	8
	BIRADS III	3	22	52	23
	BIRADS IV	5	7	20	18

Source: Oliver, A., J. Freixenet, and J. Zwiggelaar, Automatic classification of breast density, *IEEE ICIP*, vol. 2, 1258–61. © 2005 IEEE. With permission.

seen from the table, the overall performance is increased for this case. The average classification accuracy for this case using the leave-one-out method was 47%.

Bosch et al. (2006) introduced a generative classifier called "probabilistic latent semantic analysis" (pLSA) based on local descriptors for classifying mammograms. Two local descriptors, textons and scale-invariant feature transform (SIFT), were used as representative features for the MIAS and DDSM data set mammograms. Their technique can be divided into three major steps:

1. Segmentation of breast region from the mammogram is a vital preprocessing step that has been shown to affect the accuracy of classifiers. In the study by Bosch et al. (2006), the authors introduced a two-phase method for segmentation. The segmentation procedure is explained as follows:
   - Compute a global gray-level histogram and represent this computation using eight bins.
   - Set the threshold to be the minimum value over the eight histograms.
   - Use the threshold to divide the image into regions, and label each region using the connected component-labeling algorithm.
   - Keep the largest connected region, which is always the combination of breast part and the pectoral muscle, and delete all the other connected components.
   - Use the polynomial modeling–based pectoral muscle identification method proposed by Ferrari and Rangayyan (2004) and Bosch et al. (2006) to extract and delete the pectoral muscle tissue.
2. Bag of words–based tissue representation is the second step of the method detailed by Bosch et al. (2006). Two types of local features, textons (Muhimmah and Zwiggelaar 2006) and SIFT (Hadjidemetriou, Grossberg, and Nayar 2004), descriptor were extracted from each segmented breast part of the training image. In order to extract textons, a pixel was selected and then an $N \times N$ square was formed around this pixel (the square is called a patch). All the pixels in this patch were rearranged according to the row-first format. These rows were concatenated one after the other to form a new N^2-dimensional vector. The sizes of N used for this method were 3, 5, 7, 11, 15, and 21. Multiples of such patches with a spacing of M pixels on a regular grid were placed over the breast tissue. Two sizes of M (2 and 7) were used for placing the patches. There was an overlap of information for $M = 2$ and $N = 3$, 5, and 7, and for $M = 7$ and $N = 11$, 15, and 21. Assuming P total patches, one would have P vectors of dimension N^2. For extracting SIFT features, the authors initially selected pixels separated by M pixels on a regular grid. However, unlike textons, a circular patch of radius r was formed over these pixels. Then, 128-dimensional SIFT descriptors were extracted from these patches. The number of descriptors varied according to the size of the tissue. The descriptors extracted from training images were quantized using a k-means algorithm to make a visual vocabulary of V words. Then, each image

was represented by a $1 \times V$ vector, in which each row represented the count of V words in a particular image. Assuming I total training images, the training data set was represented as an $I \times V$ co-occurrence matrix in which a cell (i,j) in the matrix $1 \le i \le |V|, 1 \le j \le |I|$ represented how often the visual word i repeated in the image j.

3. The pLSA-based feature representation for classification was used to find tissue distribution for each image using the latent variable concept. This distribution was provided as the input to a support vector machine (SVM) and k-NN for classifying unseen testing instances. The training procedure was defined as follows: The authors

   – Constructed a joint probability model $P(i,j) = P(j)P(i|j)$ over the co-occurrence matrix $I \times V$, where $P(i|j) = \sum_{c \in C} P(i|c)P(c|j)$, $c \in C = \{c_1, \ldots, c_k\}$, represented the class label and $P(i|c)$ was the tissue-specific distribution
   – Modeled each image as a mixture of $P(c|j)$
   – Used the set of training images to learn the tissue-specific distribution $P(i|c)$ and represented each training image by a c-vector $P(c|j_{train})$

Once both $P(i|c)$ and $P(c|j_{train})$ were determined, unseen test images were classified. For each test image $P(c|j_{test})$, a c-dimensional vector was computed, and then k-NN or SVM was used to classify this test image.

Bosch et al. (2006) used both the MIAS and the DDSM data sets to evaluate their technique. Whereas images in the DDSM data set have BIRADS classification, images in the MIAS data set have the MIAS annotation of fatty (106), glandular (104), and dense (112) tissue. Therefore, two expert mammogram readers were consulted to classify the MIAS image data set into the BIRADS I–IV categories. Of the 322 images in the MIAS data set, the experts performed the following classification: BIRADS I (128), BIRADS II (80), BIRADS III (70), and BIRADS IV (44). For the 500 images present in the DDSM data set, 125 belonged to each of the categories. Three sets of experiments were performed: (1) with the MIAS data set, using the MIAS annotation of fatty, glandular, and dense tissues; (2) with the BIRADS-classified data for both the MIAS and DDSM data sets; and (3) with both the MIAS and DDSM data sets compared with the results of published studies.

Figure 12.3 represents the results of the first set of experiments. It can be seen from the experiments that SVM outperforms k-NN in most of the cases. The best results were reported with $V = 1600$, $C = 20$, and $= 6$ (the k value is applicable for k-NN only). Further, it was found that texton features perform better than SIFT features. The best classification accuracy, 80% for k-NN and 91.39% for SVM, using texton was reported with $N = 7$ and $M = 2$. Figure 12.4 shows a representation of the results for the second set of experiments. As with the first set, we found that SVM outperforms k-NN in most of the cases. The best accuracy was reported for

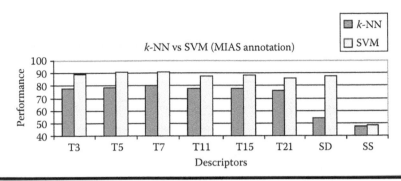

Figure 12.3 Performance according to mammographic image analysis annotation. (From Bosch, A., X. Munoz, A. Oliver, and J. Marti, Modeling and classifying breast tissue density in mammograms, *IEEE CVPR*, vol. 2, 1552–8. © 2006 IEEE. With permission.)

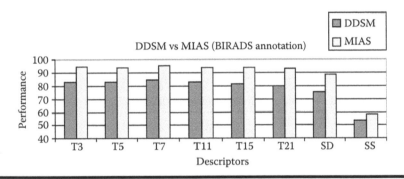

Figure 12.4 Performance according to breast imaging reporting and data system (BIRADS) annotation. (From Bosch, A., X. Munoz, A. Oliver, and J. Marti, Modeling and classifying breast tissue density in mammograms, *IEEE CVPR*, vol. 2, 1552–8. © 2006 IEEE. With permission.)

the combination $V = 1600$, $C = 20$, and $k = 7$. The best accuracy reported for the MIAS data set was 95.42% using texton features with $N = 7$ and $M = 2$. For the DDSM data set, the best accuracy reported was 84.75% with the same combination of textons and N and M values as the MIAS data set. Finally, in order to make a comparative evaluation the authors compared their results with those of the existing literature. Table 12.5 shows these results. The annotations of both the MIAS data sets (BIRADS and pLSA) are used for comparison. It can be seen from the results that the pLSA-based classifier provides better accuracy than the classifiers of existing studies.

Muhimmah and Zwiggelaar (2006) used a "multiresolution histogram"–based approach to classify a mammogram into one of three density classes, that is, fatty, glandular, or dense. The underlying hypothesis of their study is that mammographic densities correspond to image intensities, and this information can be captured

Table 12.5 Comparative Evaluation with Other Works

References	Database	Annotation	Author (%)	Our (%)
Blot and Zwiggelaar 2001	MIAS	MIAS	50	91.39
Oliver et al. 2005; automatic classification of breast tissue	MIAS	MIAS	73	91.39
Oliver et al. 2005; automatic classification of breast density according to BIRADS categories	MIAS	BIRADS	50	95.42
Oliver et al. 2005; automatic classification of breast density	DDSM	BIRADS	47	84.75
Bovis and Singh 2002	DDSM	BIRADS	71	84.75
Petroudi, Kadir, and Brady 2003	OXFORD	BIRADS	76	—

Source: Bosch, A., X. Munoz, A. Oliver, and J. Marti, Modeling and classifying breast tissue density in mammograms, *IEEE CVPR*, vol. 2, 1552–8. © 2006 IEEE. With permission.

using intensity histograms. The steps that Hadjidemetriou, Grossberg, and Nayar (2004) followed are outlined as follows:

■ Extract the breast region from the mammogram and generate a five-level Gaussian pyramid (Burt and Adelson 1983) for this region (labeled I_0, I_1, I_2, I_3, and I_4 for levels 0–4, respectively). In the case study conducted by Hadjidemetriou, Grossberg, and Nayar (2004), the original image (level 0) was represented by I_0.

■ For every $I_i, 0 \leq i \leq 4$, compute the histogram h_i, where h_i is a row vector. The combined length of the histogram is 1280.

■ Normalize each histogram with respect to breast area using the L_1 norm.

■ Smooth each histogram with a Gaussian filter of window size 5, and compute the cumulative histograms. Five histograms are formed. Find the difference in cumulative histograms of two consecutive levels, and label them as difference histograms. Four histograms are formed for five levels.

■ Use a factor of $2^{2/3}$ to subsample each difference histogram.

■ Normalize each subsampled histogram again.

■ Combine all normalized, subsampled histograms, one after the other, to form a feature vector for a mammogram. This feature vector is called the multi-resolution histogram feature.

The authors used the MIAS data set to evaluate their multiresolution feature vector. A directed acyclic graph (DAG)–SVM (DAG-SVM) with the leave-one-out method was used as the classifier to train and test the feature vector. The mammograms were classified into the three categories, fatty, glandular, and dense, provided by the MIAS data set. The images were first divided into left or right mammogram categories depending on the position of the breast. Then, classification into the three density classes was performed individually for these categories. After classification, the results were combined to report the classification accuracy for each of the three classes. Classification was performed for both the individual histogram-based features and the multiresolution histogram-based features. The confusion matrices for the h_0 and h_4 histogram-based features can be seen in Tables 12.6 and 12.7, respectively. The average classification accuracy was reported as 67.60% for h_0 and 71.34% for h_4. The confusion matrix for multiresolution histogram-based features is presented in Table 12.8. As expected, the classification accuracy using multiresolution features was higher, reported as 77.57%. Whereas there was significant improvement in individual class-level accuracies for both fatty and dense classes, the accuracy for the glandular class was lower than h_4.

Table 12.6 MIAS Classification for h_0

		MIAS Classification		
		F	G	D
	F	88	22	7
Automatic Classification	G	14	65	40
	D	4	17	64

Source: Muhimmah, I., and R. Zwiggelaar, *ITAB* '06, 2006. With permission.

Table 12.7 MIAS Classification for h_4

		MIAS Classification		
		F	G	D
	F	91	19	9
Automatic Classification	G	11	73	37
	D	4	12	65

Source: Muhimmah, I., and R. Zwiggelaar, *ITAB* '06, 2006. With permission.

Table 12.8 MIAS Classification for Multiresolution-Based Features

		MIAS Classification		
		F	G	D
Automatic Classification	F	97	19	5
	G	3	65	19
	D	6	20	87

Source: Muhimmah, I., and R. Zwiggelaar, 2006. *ITAB '06,* 2006. With permission.

From the confusion matrices, it can be seen that there was a major misclassification between glandular and dense classes. The average multihistogram features for all classes were plotted, and the graphs for glandular and dense classes showed similar patterns. This similarity was further strengthened by the χ^2 statistic value, which was low, at 0.046. This similarity among the information from glandular and dense classes caused the misclassification of images from these two classes.

Oliver et al. (2005) improved their technique by incorporating more features and a better ensemble classification mechanism based on the Bayesian classifier defined in the works by Chiracharit et al. (2004) and Oliver et al. (2008). This new technique was evaluated over both the MIAS and the DDSM data sets. The entire procedure can be divided into three steps: (1) segmentation of breast region, (2) feature extraction, and (3) classifier training and classification. The three steps are explained as follows:

1. The segmentation procedure reported by Chiracharit et al. (2004) and Oliver et al. (2008) is similar to the one used by Oliver et al. (2005); therefore, we do not elaborate on it here. Please refer to the second paragraph of Section 12.6 for a detailed explanation of this process. In some mammograms, the distinction between dense and fatty components was not clear. The result of segmentation for these cases was two clusters: (1) one grouping the breast tissue; and (2) one grouping the less-compressed tissue, like the region of pixels along the skin line.

2. The segmentation procedure always results in two clusters. As in their previous work, the authors extracted morphological and texture features, but used more features than in their previous work. Whereas Oliver et al. (2005) in their previous study extracted only three morphological features and nine texture features (four directions and only one pixel distance), they (Oliver et al. 2008) extracted five morphological features and nine texture features (four directions and three pixel distances) in this study, similar to the work of Chiracharit et al. (2004). The five morphological features extracted by

Chiracharit et al. (2004) and Oliver et al. (2008) are relative area of the cluster; and the four moments of the cluster histogram, "mean intensity," "standard deviation," "skewness," and "kurtosis." For texture feature extraction, the co-occurrence matrices for the pixels in each cluster were formed for four directions, 0°, 45°, 90°, and 135°, and three pixel distances, 1, 5, and 9. Then, nine texture features, contrast, energy, correlation, sum average, sum entropy, difference average, difference entropy, and homogeneity, were extracted from each of these co-occurrence matrices. This extraction provided 108 texture features and 5 morphological features for each class, resulting in a 113-dimensional feature vector for both clusters. Every mammogram in the data set was represented by two 113-dimensional row vectors.

3. Classification was performed using three classifiers: (1) k-NN, (2) decision tree, and (3) ensemble classifiers. For k-NN classification, two preprocessing steps, (1) feature normalization to unit variance and zero mean and (2) sequential forward selection–based feature selection, were performed over the features. For decision tree–based classification, the ID3 classifier with a boosting procedure was used. The ensemble classifier used in the studies of Chiracharit et al. (2004) and Oliver et al. (2008) is a Bayesian logic-based combination of the k-NN and ID3 classifiers. Binary classification values for both classifiers were transformed into class-level memberships, which were combined using the classic Bayes probabilistic equation.

The classification results were presented in the form of confusion matrices and a coefficient, kappa (κ), was provided as a means of estimating agreement. The following formula was used to calculate the coefficient: $\kappa = \dfrac{P(D) - P(E)}{1 - P(E)}$, where $P(D)$ equals the proportion of times the model value is equal to actual value and $P(E)$ is the expected proportion by chance. The relationship between κ and 1 determines the value of the classifier: A value of κ that is near to 1 indicates a good classifier, whereas a value of κ that is away from 1 indicates a poorly performing classifier. Table 12.9 shows some commonly used interpretations of the κ value.

Two data sets, MIAS and DDSM, were used to evaluate the proposed technique. Whereas the DDSM provides a BIRADS categorization label for each image, no such information is present for images from the MIAS data set. Hence, three expert radiologists were consulted to classify the MIAS data set images into BIRADS I, II, III, and IV categories. Each mammogram was classified as a BIRADS category based on the consensus of all the radiologists. A consensus was reached by assigning a category label to a mammogram if at least two radiologists agreed on the label. The median value was selected as the consensus label in cases where radiologists assigned different category labels to an image. Table 12.10 shows the confusion matrices for the three expert radiologists. The difficulty in categorizing the mammograms according to density is replicated in the manual classification by radiologists. As is clear from the table, there is a great deal of divergence among the readings of the radiologists.

Table 12.9   Common Interpretations of κ Values

κ	*Agreement*
<0	Poor
[0, 0.20]	Slight
[0.41, 0.60]	Fair
[0.41, 0.60]	Moderate
[0.61, 0.80]	Substantial
[0.81, 1.00]	Almost perfect

Source: Oliver, A., J. Freixenet, and J. Zwiggelaar, Automatic classification of breast density, *IEEE ICIP*, vol. 2, 1258–61. © 2005 IEEE. With permission.

Table 12.10   Confusion Matrices for Three Experts

Consensus	*B-I*	*B-II*	*B-III*	*B-IV*
Expert A (κ = 0.70)				
B-I	85	2	0	0
B-II	43	60	0	0
B-III	1	17	70	7
B-IV	0	0	0	37
Expert B (κ = 0.85)				
B-I	85	2	0	0
B-II	1	93	9	0
B-III	0	17	72	6
B-IV	0	0	0	37
Expert C (κ = 0.61)				
B-I	59	28	0	0
B-II	0	58	45	0
B-III	0	0	88	7
B-IV	0	0	10	27

Source: Oliver, A., J. Freixenet, and J. Zwiggelaar, Automatic classification of breast density, *IEEE ICIP*, vol. 2, 1258–61. © 2005 IEEE. With permission.

12.6.1 Mammographic Image Analysis Data Set Classification

Once the BIRADS classification of all 322 MIAS data set mammograms was gathered using the consensus information from the radiologists, two sets of experiments were performed. One set of experiments was performed on the mammograms categorized individually by a radiologist, and the other set of experiments was performed on the consensus information–based BIRADS-categorized mammograms. Table 12.11 shows the results for the first set of experiments. From the results, we can see that for expert A, the sequential forward selection (SFS) + k-NN ($k = 7$) combination classifier performed better (accuracy of 78%) than the C4.5 classifier (accuracy of 74%). However, the Bayesian classifier provided the best classification accuracy (83%). The same trend can be seen in the classification accuracy for classification based on mammograms categorized by both the experts B and C.

The results for classification based on the consensus data can be seen in the last row of Table 12.11. The classification accuracy for SFS + k-NN ($k = 7$) is 77%, and the accuracy for C4.5 is 72%. Once again, the best classification accuracy was achieved by using the Bayesian classifier at 86%. The kappa value of 0.81 reconfirmed these results, as it belonged to the top category of "almost perfect." The authors combined the data from BIRADS I and BIRADS II (low density) into one class and from BIRADS III and BIRADS IV (high density) into another class to perform binary classification on the data. An overall classification of 91% was achieved with individual class-level accuracies of 89% and 94% for low-density and high-density classes, respectively.

12.6.2 Digital Database for Screening Mammography Data Set Classification

The proposed technique was also evaluated on 831 mammograms from the DDSM data set. The BIRADS labels were attached to these mammograms, so the expert information was not needed for them. Of the 831 mammograms, 106 belong to category I, 336 belong to category II, 255 belong to category III, and 134 belong to category IV. There are four mammograms in the DDSM data set for each patient. These four mammograms include MLO views and CC views of both breasts. To remove any bias, only the right breast MLO views were used for experiments. The confusion matrices for classification using SFS + k-NN, C4.5, and Bayesian classifier I are presented in Table 12.12. Once again, the best classification accuracy (77%) was obtained using the Bayesian classifier. The value of $\kappa = 0.67$ belonged to the "substantial" category. The individual class-level accuracies for the DDSM data set were lower than the individual class-level accuracies of the MIAS data set. The possible explanation for this behavior is the relative tissue similarity of mammograms from categories I and II in DDSM. The authors combined the data into low-density and high-density classes for binary classification. The overall

Table 12.11 Confusion Matrices for MIAS Classification Based on Individual Expert Categorization

		SFS + k-NN				C4.5				Bayesian			
		B-I	B-II	B-III	B-IV	B-I	B-II	B-III	B-IV	B-I	B-II	B-III	B-IV
Expert A	B-I	113	10	5	1	114	12	2	1	118	6	5	0
	B-II	8	59	9	3	18	47	12	2	7	60	10	2
	B-III	4	13	46	7	2	11	48	9	0	6	53	11
	B-IV	1	3	6	34	0	1	13	30	0	2	7	35
Expert B	B-I	75	8	2	1	69	15	2	0	78	6	2	0
	B-II	7	85	16	4	13	73	22	4	10	93	8	1
	B-III	1	20	55	5	1	27	46	7	0	16	55	10
	B-IV	2	7	11	23	0	1	13	29	0	1	10	32
Expert C	B-I	50	5	1	3	43	14	0	2	51	5	1	2
	B-II	13	53	19	1	15	49	22	0	9	64	12	1
	B-III	0	21	115	7	2	15	119	7	1	16	122	4
	B-IV	3	3	7	21	1	0	13	20	0	2	6	26
Consensus	B-I	70	13	1	3	72	13	1	1	79	1	3	4
	B-II	9	80	13	1	13	68	20	2	3	86	6	8
	B-III	1	17	73	4	0	21	68	6	0	2	85	8
	B-IV	3	2	8	24	0	2	11	24	0	6	4	27

Source: Oliver, A., J. Freixenet, R. Marti, J. Pont, E. Perez, E. R. E. Denton, and R. Zwiggelaar, A novel breast tissue density classification methodology, *IEEE Trans Inf Technol Biomed*, 55–65. © 2008 IEEE. With permission.

Table 12.12   DDSM Data Set Classification Results

		sfs + kNN				C4.5				Bayesian			
		B-I	*B-II*	*B-III*	*B-IV*	*B-I*	*B-II*	*B-III*	*B-IV*	*B-I*	*B-II*	*B-III*	*B-IV*
	B-I	54	40	12	0	51	30	25	0	58	25	23	0
Truth	B-II	44	266	25	1	22	279	35	0	15	295	26	0
	B-III	9	60	177	9	16	59	178	2	12	46	196	1
	B-IV	0	21	30	83	8	14	25	87	5	18	18	93

Source: Oliver, A., J. Freixenet, R. Marti, J. Pont, E. Perez, E. R. E. Denton, and R. Zwiggelaar, A novel breast tissue density classification methodology, *IEEE Trans Inf Technol Biomed*, 55–65. © 2008 IEEE. With permission.

classification accuracy was 84%, with individual class accuracies of 89% and 79% for low-density and high-density mammograms, respectively.

12.6.3 Abnormality-Based Mammogram Classification

Chiracharit et al. (2004) used an SVM classifier based on transformed feature spaces to classify abnormal and normal mammograms. The underlying hypothesis of their method is that features with crossed distributions make classification harder for any classifier; therefore, a new transformed feature space with better knowledge representation is needed to perform accurate classification. Their technique is summarized as follows:

- First, extract 86 features for each mammogram in the training database. (The features extracted were 18 curvilinear features, 16 gray-level co-occurrence matrix (GLCM)-based texture features, 32 Gabor features, and 20 multiresolution statistical features.)
- Given these 86 features for each mammogram, 3655 pairs of two features are possible ($N(N-1)/2$). For each pair of two features, say f_m and f_n, find the correlation coefficient $R(f_m, f_n)$ for all normal and abnormal mammogram training data.
- Mark the pair of features f_m and f_n as a crossed distribution pair if $\underset{\text{normal}}{R}(f_m, f_n) \times \underset{\text{abnormal}}{R}(f_m, f_n) \leq -\epsilon$ where ϵ is the threshold, such that $0 \leq \epsilon \leq 1$. The greater the value of the threshold ϵ, the more distinct the crossed pattern will be.
- For a crossed feature pair, use two standard kernel functions and one specialized function to uncross them. The standard kernel functions used are the polynomial function $K(f_m, f_n) = (f_m^T f_n + c)^d$, where $d > 0$ is the constant that defines the kernel width, and the Gaussian radial basis function (RBF)

$K(f_m, f_n) = \exp\left(-\dfrac{\|f_m - f_n\|^2}{2\sigma^2}\right)$. The specialized transformation function is

$\Phi(f_m, f_n) = -1 + 2f_m + 2f_n - 4f_m f_n$.

■ Train the SVM classifier on these transformed features and classify the test instances.

The proposed technique was evaluated on the mammograms from the DDSM data set. A total of 50 normal and 50 abnormal mammograms were used for training, whereas 20 normal and 20 abnormal mammograms were used for testing. A threshold of $\epsilon = 0.1$ was used to find the cross-distributed feature pairs from 3655 total pairs. These pairs resulted in 90 crossed feature pairs, consisting of 54 features. Of the 54 features, 12 were curvilinear, 11 were GLCM-based texture features, 20 were Gabor features, and 11 were multiresolution statistical features. The first set of experiments was performed separately on the four types of features with transformation based on polynomial and Gaussian kernels only. The results are presented in Table 12.13.

Three of the four types of features showed high overall classification accuracy: (1) curvilinear features combined with Gaussian RBF kernel at 75%, (2) texture features combined with Gaussian RBF kernel at 72.5%, and (3) Gabor features combined with polynomial kernel at 92.5%. Only these features were used in the second set of experiments, in which the test mammogram was classified based on a majority vote of the three features. In this set of experiments, only the crossed

Table 12.13   Classification Results for Individual Feature Types

Feature Type	Kernel Functions	TN	FP	TP	FN	%
Curvilinear	Polynomial	2	18	20	0	55
	Gaussian	17	3	13	7	75
GLCM	Polynomial	0	20	20	0	50
	Gaussian	19	1	10	10	72.5
Gabor	Polynomial	19	1	18	2	92.5
	Gaussian	2	18	18	2	50
Multiresolution	Polynomial	2	18	20	0	55
	Gaussian	2	18	19	1	52.5

Source: Chiracharit, W., Y. Sun, P. Kumhom, K. Chamnogthai, C. Babbs, and E. J. Delp, Normal mammogram classification based on a support vector machine utilizing crossed distribution features, *Twenty-Sixth International Conference of the IEEE EMBS*, San Francisco, CA, 1581–3. © 2004 IEEE. With permission.

Table 12.14   Classification Results Based on Only Crossed Distribution–Based Features

	True Negative	False Positive	True Positive	False Negative	%
	19	1	16	4	87.5
%	95		80		

Source: Chiracharit, W., Y. Sun, P. Kumhom, K. Chamnogthai, C. Babbs, and E. J. Delp, Normal mammogram classification based on a support vector machine utilizing crossed distribution features, *Twenty-Sixth International Conference of the IEEE EMBS*, San Francisco, CA, 1581–3. © 2004 IEEE. With permission.

distribution–based transformed features were used for classification. Table 12.14 gives the results for this set of experiments. An overall classification accuracy of 85% was achieved with specific accuracies of normal and abnormal classes being 95% and 80%, respectively.

Shah, Bruce, and Younan (2004) used bagging and boosting ensemble–based approaches (called modular classification [MoC]) to classify mammograms into benign and malignant classes. The primary aim of this research was to compare and contrast the classification of mammograms using traditional classifiers (TrCs) and using MoC. They used 200 mammograms from the DDSM data set to evaluate their technique. Each mammogram was manually evaluated for breast segmentation by an expert radiologist. Then, seven features, tumor circularity, mean, standard deviation, normalized entropy, zero crossings, area ratio, and roughness index, were extracted from the segmented portion of the mammogram. Patient's age formed the eighth feature for the segmented mass. Once the features were extracted from all the mammograms in the training set, Fisher's linear discriminant analysis (LDA) was performed to find a reduced set of optimal features for classification. Two TrCs, nearest mean (NM) and maximum likelihood (ML), and two MoCs using bagging and boosting were used to evaluate the features. Three approaches were used for each MoC, resulting in six MoC designs:

1. Classification using NM classifiers only
2. Classification using ML classifiers only
3. Classification based on the combination of both NM and ML classifiers employing the minimum training error mechanism for class label prediction

The number of component classifiers for each MoC was varied from 3 to 15, and the training data was varied from 50% to 80% with an increment of 10%. Then, the leave-one-out method was used for training and testing. In these experiments, the malignant class was considered the positive class and the benign class was considered the negative class. The results (without feature selection) are presented in terms of sensitivity, specificity, positive predictive value, and negative predictive value in Table 12.15. As can be seen from table, the sensitivity for both TrCs was close: 74% with NM and 75% with ML. Whereas there is not much difference in sensitivity for bagging and boosting, the highest sensitivity was noted for the combination of NM and ML classification schemes at 83%. In terms of positive predictive value and specificity, ML-based methods outperform all the other methods.

Table 12.16 shows the results found using an optimized set of features using Fisher's LDA. An interesting observation from these results is that there is a small improvement in sensitivity for this case, that is, about 4%–5%. Another interesting

Table 12.15   Results for Classification without Feature Selection

Classification Method		Sensitivity	(PV+)	Specificity	(PV−)
TrSCS–NM		0.740	0.771	0.780	0.750
TrSCS–ML		0.750	0.843	0.860	0.775
MoCs–bagging	NM	0.780	0.821	0.830	0.791
	ML	0.820	0.837	0.840	0.824
	NM/ML	0.830	0.822	0.820	0.828
MoCs–AdaBoost	NM	0.760	0.809	0.820	0.774
	ML	0.800	0.842	0.850	0.810
	NM/ML	0.830	0.814	0.810	0.827

Source: Shah, V. P., L. M. Bruce, and N. H. Younan, Applying modular classifiers to mammographic mass classification, *Twenty-Sixth International Conference of the IEEE EMBS*, San Francisco, CA, 1585–8. © 2004 IEEE. With permission.

Table 12.16   Results for Classification with LDA-Based Selected Features

Classification Method		Sensitivity	(PV+)	Specificity	(PV−)
TrSCS–NM		0.830	0.822	0.820	0.828
TrSCS–ML		0.830	0.822	0.820	0.828
MoCs–bagging	NM	0.880	0.822	0.810	0.837
	ML	0.860	0.819	0.810	0.828
	NM/ML	0.880	0.822	0.810	0.871
MoCs–AdaBoost	NM	0.870	0.845	0.840	0.853
	ML	0.870	0.829	0.820	0.871
	NM/ML	0.870	0.813	0.8	0.866

Source: Shah, V. P., L. M. Bruce, and N. H. Younan, Applying modular classifiers to mammographic mass classification, *Twenty-Sixth International Conference of the IEEE EMBS*, San Francisco, CA, 1585–8. © 2004 IEEE. With permission.

observation is that there is a small variation in sensitivity for all three approaches, NM, ML, and the combination of NM and ML. This variation shows that the feature set obtained using LDA was stable for all three approaches. Further, the positive predictive value, specificity, and negative predictive values were also stable, as there was little change in these values.

Vibha et al. (2006) proposed random forest-based decision tree classifiers for mammogram classification. Their technique involved classification of mammograms from the MIAS data set into three classes: (1) normal, (2) benign, and (3) malignant. It involved four primary steps: (1) data preprocessing, (2) feature extraction, (3) data postprocessing, and (4) classifier training and classification. We describe the four steps as follows:

1. For data preprocessing, the authors used two steps: (1) cropping and (2) "histogram equalization." Initially, every mammogram was cropped to remove existing labels and the black background that were not useful for classification. These cropped mammograms were normalized using histogram equalization to reduce over- or underexposure.
2. For feature extraction, the preprocessed image was divided into four quadrants of equal size and then into four statistical features, which were extracted: (1) mean, (2) variance, (3) skewness, and (4) kurtosis.
3. For data postprocessing, a supervised resample filter was used over the data set to generate random subsamples. The bootstrap procedure was used for sampling. The individual class distributions were maintained in the subsamples.
4. For classifier training and classification, a random forest-based classifier was used for training on the data set. The random tree classifier was a decision tree–based ensemble method in which a tree was generated for each subsampled data set. A test instance was then categorized to a class based on the majority vote of all the trees.

The proposed method was tested on 200 images from the MIAS data set. In the method, 10-fold repetition was used to minimize bias in classification. Of the 200 images, 151 were normal, 27 were malignant cancerous, and 22 were benign cancerous. A freely available software application called WEKA was used to construct the random forest classifier. The confusion matrix for the classification is shown in Table 12.17. As can be seen from the confusion matrix, no images from the normal class were misclassified to other classes. However, some mammograms that belonged to the cancerous and benign classes were misclassified to the normal class. Table 12.18 shows the overall classification accuracy for the 10 runs of the algorithm. It is evident from the results that the algorithm proposed by Vibha et al. (2006) is more stable than the previously considered algorithms.

Zaiane et al. (2002) proposed an association rule–based classification mechanism, called ARC-BC, for mammograms. They used the mammogram data from the MIAS database to evaluate ARC-BC. The mammograms were classified into

Table 12.17 Confusion Matrix of Random Tree

	N	C	B
N	151 (C_{11})	000 (C_{12})	000 (C_{13})
C	007 (C_{21})	020 (C_{22})	000 (C_{23})
B	007 (C_{31})	000 (C_{32})	015 (C_{33})

Source: Vibha, L., G. M. Harshvardhan, K. Pranaw, P. D. Shenoy, K. R. Venugopal, and L. M. Patnaik, Classification of mammograms using decision trees, *IEEE IDEAS* '06, © 2006 IEEE. With permission.

Table 12.18 Classification Accuracy for 10 Runs

	Results Using the Algorithm Proposed by Vibha et al. 2006		Previous Results	
Sets	Random Tree	RFDC	BP NN	ARC-BC
1	87.0	92.0	96.9	80.0
2	84.0	90.0	90.6	93.3
3	90.5	93.0	90.6	86.7
4	82.5	90.0	78.1	76.7
5	83.0	89.0	81.3	70.0
6	85.5	90.5	84.4	76.7
7	86.0	89.5	65.6	83.3
8	86.5	89.0	75.0	76.7
9	86.5	91.0	56.3	76.7
10	89.5	93.0	93.8	83.3
Average	86.1	90.7	81.26	80.34

Source: Vibha, L., G. M. Harshvardhan, K. Pranaw, P. D. Shenoy, K. R. Venugopal, and L. M. Patnaik, Classification of mammograms using decision trees, *IEEE IDEAS* '06, © 2006 IEEE. With permission.

the three base classes of normal, benign, and malignant. Their method can be divided into four major parts: (1) data preprocessing, (2) feature extraction and data representation, (3) association rule mining, and (4) classifier building and classification. We explain the four parts as follows:

1. As is common with most digital mammograms, the images used by Zaine et al. (2002) contained a black background and digital artifacts like names and logos and required data preprocessing. During preprocessing, the extra information was deleted so that it would not affect feature extraction. Initially, a mammographic image was cropped by sweeping through it horizontally to delete most of the black background and artifacts. Since sizes of the resulting images differ (because of the different breast sizes and, hence, different background information), each image was rescaled to size 0–255 pixels. This normalization was followed by another preprocessing method called histogram equalization to balance the over- or underexposure of images. The same procedure was repeated for each mammogram in the database.

2. After the mammogram was preprocessed, it was divided into four quadrants of equal size for feature extraction and data representation. Six features were extracted from each of these four quadrants. The features extracted were four moments of data, which were mean, variance, skewness, and kurtosis; mean over the histogram; and peak of the histogram. Each of these features was discretized over its interval. Once the features had been extracted from each quadrant, they were arranged in a transaction database format. If the training image belonged to a normal class, then features from all the quadrants were combined to form a transaction. However, if the image belonged to an abnormal class (benign or malignant) only features belonging to those quadrants that contained the abnormality were used. This transaction format (in which each row represents the features extracted from an image) was then used for association rule generation in the next step. A separate transaction database was generated from the images of each class. Therefore, in the MIAS database, where there were three classes, there would be three separate transactional databases. Each object/image in the database was represented as a transaction of the following form: $I_j = \{c_i, f_1, f_2, \ldots, f_n\}$, where I_j was an image, c_i was the class which this image belonged to, and $\{f_1, f_2, \ldots, f_n\}$ were the features for this image.

3. Given a transaction database for a class, the Apriori association rule–mining algorithm was used to generate association rules for this class. A constraint was pushed into the rule-making process, so that only class-constrained association rules (CARs), where the class label was on the right-hand side of the rule, resulted. This constraint resulted in predictive association rules of the form $f_1 \cap f_2 \cap f_n \Rightarrow c_i$ for each class.

4. Classifier training and classification are important since association rule mining generates a huge number of rules. The large number of rules makes it necessary to keep only the most useful rules for building the classifier. The first step performed by Zaine et al. (2002) in rule pruning was to keep only general rules with high confidence. For two rules R_1 and R_2, they determined that R_2 is a general rule if $R_2 \subseteq R_1$. Once all the general rules were found, they were sorted, first according to the highest confidence and then according to the highest support for each confidence. If two rules R_i and R_j had the same

support and confidence, then a rule that had fewer attributes on the left-hand side of the rule was given a higher weight. Hence, if R_i had three attributes on the left-hand side and R_j had four, then R_i was ranked higher than R_j. After rule ranking, each rule was used to classify instances from the training database. Only those rules that correctly classified instances in their own classes and did not misclassify an instance to another class were kept. Once the final set of rules was extracted, four types of classifiers were developed. The first type of classifier that was developed classified a testing image only to the class of the highest ranked rule. Therefore, classification depended only on the single highest ranked, matched rule. The second classifier found all the possible matched rules and grouped them according to their class labels. Then, the class label of the group with maximum aggregate confidence was assigned as the class label for the testing image. The third classifier was similar to the second classifier, and the maximum aggregate confidence was used to provide the class label. However, in this classifier, some rules that pointed to more than one class for the same combination of feature values (conflicting rules) were deleted after rule ranking. The fourth classifier was also based on the aggregate confidence like the second classifier, except that some low-confidence rules were deleted after rule ranking.

To evaluate the efficacy of the proposed method, Zaine et al. (2002) used the MIAS database for training and testing. They used 90% data for training from each class and the rest for testing. For association rule generation, the minimum support and confidence were set at 25% and 50%, respectively. The actual support with which a class was mined for rules was adapted so that the support and confidence values were changed until a classification accuracy of 95% was reached over the training data set. As a result, the support and confidence values were different for different classes. The support value was low (8%) for abnormal class images due to class imbalance. (There were fewer abnormal cases than normal cases.) Classification accuracies of the 4 classifiers computed over 10 random repetitions of the data were 71.02%, 80.33%, 77.33%, and 62.67%, respectively. From the results, it can be seen that the best accuracy was achieved using the second classifier. These results show that rule pruning does not always guarantee the best results.

Zaine et al. (2002) also generated the precision and recall graphs for normal versus abnormal cases to judge the accuracy of the classifier. Assuming that TP represents normal images that were classified as normal, FP represents abnormal images that were falsely classified as normal, TN represents abnormal images that were classified as abnormal, and FN represents images that were normal but were falsely classified as abnormal, the following formulas were used to calculate precision and recall:

$$\text{Precision} = \frac{\text{TP}}{\text{TP} + \text{FP}}$$

and

$$Recall = \frac{TP}{TP + FN}$$

The average precision and recall values for the normal cases over 10-fold repetition were 93% and 98%, respectively. These high values show that the proposed classifier is stable and has few false positives and false negatives. Both of these attributes are desirable in the medical image classification domain.

Jaffar et al. (2009) proposed a multidomain features-based classification technique for mammograms. They employed SVM and multilayer perceptron (MLP) for classifying the mammograms from the MIAS data set to benign and malignant classes. Their technique can be summarized as follows:

■ First, they used a detail-preserving fuzzy filter (Hussain et al. 2009) to remove any noise present in the mammograms. Since medical images require noise-removal algorithms that can preserve details while removing any unwanted noise, Hussain et al. (2009) considered the fuzzy filter–based approach a good choice.
■ After the noise was removed from the mammogram, Jaffar et al. (2009) extracted the region of interest, that is, the breast segment from the image. Many images in the MIAS data set have the nipple position on the left. In order to preserve symmetry, these images were turned over so that the nipple was on the right side of the breast.
■ This transformed image was then changed to its binary counterpart.
■ After binary transformation, a stack-based region-growing method was used to segment the breast part of the image from the mammogram. The binary image was read one line at a time until a foreground pixel was found. This foreground pixel was labeled as the seed for region growing. If this foreground pixel was unlabeled, it was marked with a new label, and the position of this pixel was pushed onto a stack. The neighboring foreground pixels of the seed pixel were pushed onto the stack and were marked with the label of the seed. As long as the stack was not empty the pixels on the stack continued to be labeled, and when the stack became empty the procedure continued in order to find the next seed for float fill. In this way, all foreground pixels were labeled with the seed label.
■ Next, cropping was performed to delete those regions that did not belong to the breast part. These regions usually correspond to intentionally added artifacts or labels for manual mammogram analysis, which are not needed for automated classification.
■ After the images were cropped, a commonly used image enhancement procedure called histogram equalization was used to compensate for over-brightness or overdarkness.

■ Image enhancement was followed by pectoral muscle separation. A simple region-growing method (Mat-Isa, Mashor, and Othman 2005) was used to separate the pectoral muscle from the breast image. Initially, a seed pixel was chosen, and its neighborhood was examined to find similar pixels. All the pixels that were similar to this pixel were joined to grow in a region. When the region growth stopped, a new seed pixel was taken. The new seed pixel did not belong to any region, and the region-growing procedure was repeated. Once every pixel was assigned to a region, the region with pectoral muscle was deleted.

■ Once the breast region was segmented from the mammogram, eight features were extracted from this region: (1) mean, (2) variance, (3) skewness, (4) kurtosis, (5) entropy, (6) energy, (7) contrast, and (8) homogeneity.

In order to evaluate their technique, Jaffar et al. (2009) used two classifiers, SVM and the multilayer backpropagation neural network (MLP), to classify the images from the MIAS data set into the benign and malignant classes. Three measures of evaluation, accuracy, sensitivity, and specificity, were used to measure the goodness of the classifier. The following formulas were used to calculate these measures:

$$\text{Accuracy} = \frac{TP + TN}{TP + TN + FP + FN}, \ \text{Sensitivity} = \frac{TP}{TP + FN}, \ \text{Specificity} = \frac{TN}{TN + FP}$$

where TP represents the number of malignant cases that were accurately classified as malignant, TN represents the number of benign cases that were accurately classified as benign, FP represents the number of benign cases that were inaccurately classified as malignant, and FN represents the number of malignant cases that were inaccurately classified as benign. Table 12.19 shows the results for both the SVM classifier and the MLP classifier.

Table 12.19   Classification Results for Both SVM and MLP Classifiers

Performance Measures	SVM with Eight Features	SVM with Four Features (Suckling et al. 1994)	MLP with Eight Features	MLP with Four Features (Suckling et al. 1994)
Accuracy (%)	96.781	85.60	94.119	84.80
Sensitivity (%)	98.662	95.45	96.112	94.55
Specificity (%)	98.213	77.86	95.413	77.14

Source: Jaffar, M. A., B. Ahmed, A. Hussain, N. Naveed, F. Jabeen, and A. M. Mirza, Multidomain features based classification of mammogram images using SVM and MLP, *IEEE Fourth International Conference on Innovative Computing, Information and Control*, 1301–4. © 2009 IEEE. With permission.

It can be seen that the overall accuracy of the SVM classifier (96.78%) is better than that of the MLP classifier (84.80%). Further, sensitivity, which is critical for medical images, is high (98.66%) for the SVM classifier. These results show that the SVM classifier is a better fit for the classification problem that uses the multi-domain features used in this proposed technique.

References

Blot, L., and R. Zwiggelaar. 2001. Background texture extraction for the classification of mammographic parenchymal patterns. In *MIUA*, Birmingham, UK: 145–8.

Bosch, A., X. Munoz, A. Oliver, and J. Marti. 2006. Modeling and classifying breast tissue density in mammograms. In *IEEE CVPR*, vol. 2, 1552–8.

Bovis, K., and S. Singh. 2002. Classification of mammographic breast density using a combined classifier paradigm. In *IWDM*, 177–80.

Breast Cancer Fact 2009–2010. http://www.cancer.org/downloads/STT/BCFF-Final.pdf.

Breastcancer.org. Symptoms and testing with mammograms. http://www.breastcancer.org/symptoms/testing/mammograms.

Breastcancer.org. 2009. Mammography technique and types. http://www.breastcancer.org/symptoms/testing/types/mammograms/types.jsp.

Breastcancer.org. 2010. What mammograms show. http://www.breastcancer.org/symptoms/testing/types/mammograms/mamm_show.jsp.

Burt, P. J., and E. H. Adelson. 1983. The Laplacian pyramid as a compact image code. *IEEE Trans Commun* 31:532–40.

Cancer.gov. 2006. Probability of breast cancer in American women. http://www.cancer.gov/cancertopics/factsheet/Detection/probability-breast-cancer.

Cancer.gov. 2009. Mammograms. http://www.cancer.gov/cancertopics/factsheet/detection/screening-mammograms.

Chiracharit, W., Y. Sun, P. Kumhom, K. Chamnogthai, C. Babbs, and E. J. Delp. 2004. Normal mammogram classification based on a support vector machine utilizing crossed distribution features. In *Twenty-Sixth International Conference of the IEEE EMBS*, San Francisco, CA, 1581–3.

Ferrari, R., and R. Rangayyan. 2004. Automatic identification of pectoral muscle in mammograms. *IEEE Trans Med Imaging* 23:232–45.

Hadjidemetriou, E., M. D. Grossberg, and S. K. Nayar. 2004. Multiresolution histograms and their use for recognition. *IEEE Trans Pattern Anal Mach Int* 26:831–47.

Heath, M. D., K. Bowyer, D. Kopans, R. Moore, and W. P. Kegelmeyer. 2001. The digital database for screening mammography. In *Proceedings of the Fifth International Workshop on Digital Mammography*, ed. M. J. Yaffe, 212–8. Madison, WI: Medical Physics Publishing.

Hussain, A., M. A. Jaffar, A. M. Mirza, and A. Chaudhary. 2009. Detail preserving fuzzy filter for impulse noise removal. *Int J Innovative Comput Inf Control* 5:3583–91.

Jaffar, M. A., B. Ahmed, A. Hussain, N. Naveed, F. Jabeen, and A. M. Mirza. 2009. Multi domain features based classification of mammogram images using SVM and MLP. In *IEEE Fourth International Conference on Innovative Computing, Information and Control*, 1301–4.

Mat-Isa, N. A., M. Y. Mashor, and N. H. Othman. 2005. Seeded region growing features extraction algorithm: Its potential use in improving screening for cervical cancer. *Int J Comput Internet Manage* 31:61–70.

Muhimmah, I., and R. Zwiggelaar. 2006. Mammogram density classification using multi-resolution histogram information. In *ITAB* '06, Ioannina, Greece.

Oliver, A., J. Freixenet, A. Bosch, D. Raba, and R. Zwiggelaar. 2005. Automatic classification of breast tissue. In *IbPRIA*, 431–8.

Oliver, A., J. Freixenet, R. Marti, J. Pont, E. Perez, E. R. E. Denton, and R. Zwiggelaar. 2008. A novel breast tissue density classification methodology. *IEEE Trans Inf Technol Biomed*, 55–65.

Oliver, A., J. Freixenet, and J. Zwiggelaar. 2005. Automatic classification of breast density. In *IEEE ICIP*, vol. 2, 1258–61.

Oliver, A., J. Marti, J. Freixenet, J. Pont, and R. Zwiggelaar. 2005. Automatic classification of breast density according to BIRADS categories using a clustering approach. *International Computer Assisted Radiology and Surgery Congress*, Berlin, Germany.

Petroudi, S., T. Kadir, and M. Brady. 2003. Automatic classification of mammographic parenchymal patterns: A statistical approach. In *IEEE EMBS*, vol. 2, 798–802.

Pisano, E. D., C. Gatsonis, E. Hendrick, M. Yaffe, J. K. Baum, S. Acharya, E. F. Conant et al. 2005. Diagnostic performance of digital versus film mammography for breast cancer screening. *N Engl J Med* 353:1773–83.

Shah, V. P., L. M. Bruce, and N. H. Younan. 2004. Applying modular classifiers to mammographic mass classification. In *Twenty-Sixth International Conference of the IEEE EMBS*, San Francisco, CA, 1585–8.

Suckling, J., J. Parker, D. Dance, S. Astley, I. Hutt, and C. Boggis. 1994. The mammographic image analysis society digital mammogram database. *Exerpta Med Int Congr Ser* 1069:375–8.

The Mini-MIAS Database of Mammograms. http://peipa.essex.ac.uk/info/mias.html.

Vibha, L., G. M. Harshvardhan, P. Pranaw, P. D. Shenoy, K. R. Venugopal, and L. M. Patnaik. 2006. Classification of mammograms using decision trees. In *IEEE IDEAS*, Delhi, India.

Wolfe, J. N. 1976. Risk for breast cancer development determined by mammographic parenchymal pattern. *Cancer* 37:2486–92.

Zaiane, O. R., M.-L. Antonie, and A. Coman. 2002. Mammography classification by an association rule-based classifier. In *ACM SIGKDD*.

Zhang, J., W. Hsu, and M. L. Lee. 2001. Image mining: Issues, frameworks and techniques. In *Proceedings of Second ACM SIGKDD International Workshop on Multimedia Data Mining (MDM/KDD'01)*, San Francisco, CA.

Chapter 13

Biofilm Image Analysis: Automatic Segmentation Methods and Applications

Dario Rojas, Luis Rueda, Homero Urrutia,
Gerardo Carcamo, and Alioune Ngom

Contents

13.1 Introduction

It has been known for several years that bacteria can form societies by means of emergent behavior, which they use to complete complex tasks that would be impossible to carry out individually (Johnson 2008). One of the most remarkable emergent behaviors of bacteria is the formation of biofilms. The behavior of bacteria in a biofilm state is different from their behavior in the planktonic state (free-living/floating bacteria; Stewart and Franklin 2008), and bacterial behavior in a biofilm state can produce negative consequences in clinical, agricultural, and industrial environments (Sunner, Beech, and Hiraoka 2005; Hall-Stoodley, Costerton, and Stoodley 2004). The biological success of these environments is not trivial, as biofilms involve several factors such as the use of genetic information, food, and energy supply, which have profound consequences on bacterial physiology and survival. A biofilm is a complex aggregate of bacteria stuck to each other, which may or may not be attached to a surface.

Bacteria in a biofilm are embedded within a protective self-generated matrix of extracellular polymeric substances (EPSs; McBain 2009). Biofilms can form and grow in many environments, including living or nonliving surfaces, and represent a prevalent mode of microbial life. Bacteria aggregate in a biofilm in response to certain stresses in their environment. Biofilms are ubiquitous and are a result of a complex biological process (Hall-Stoodley, Costerton, and Stoodley 2004; Kaplan 2010) known as "quorum-sensing" (QS), which is a cell-to-cell communication circuit used by most bacteria that enables them to keep track of their numbers and, hence, to grow. In the bacterial QS process, bacteria sense and respond to their population density via chemical signals called "autoinducers" (AIs) that accumulate as density increases. As a consequence, the concentrations of some AIs increase, and on reaching critical intra- and extracellular threshold concentrations these AIs bind to certain proteins called "receptors" that then activate (or repress) a particular gene expression pattern. Depending on the bacterial species or its current environment, the genes that are activated can include those that encode phenotypes such as bioluminescence, virulence factors, antibiotic production, and biofilm growth capability.

Bacteria in biofilm states are extremely resistant to most forms of environmental or antimicrobial stresses, and they express a high resistance to antimicrobial compounds (antibiotics and biocides) when compared with their planktonic counterparts (free-living/floating bacteria; Stewart and Costerton 2001; Stewart and Franklin 2008). Hence, they can produce negative consequences in their environments (Sunner, Beech, and Hiraoka 2005). At least 60% of all microbial infections are now believed to involve biofilms. In this state, bacteria can tolerate the highest deliverable doses of antibiotics, which make them impossible to eradicate. The most important property of bacteria in a biofilm is their resistance to antimicrobial compounds (Stewart and Costerton 2001). One reason for this phenomenon, among others, is that a biofilm assumes a certain complex structure (including three-dimensional [3-D] structure, temporal evolution, physiological makeup of bacteria, and a number of distinct bacterial species in the biofilm), which makes it hard for antibacterial agents to penetrate.

The study of the structural parameters in a biofilm is a novel and important research field in which the aims are to understand how biofilms develop and grow and how their formation and evolution can be inhibited. In order to understand biofilm structures, different techniques are used. Flow-cell reactors (i.e., small chambers with transparent surfaces, where bacteria are submerged in an environment continually refreshed with nutrients; Branda et al. 2005) are one of the traditional tools used in monitoring biofilm formation in bench environments. These have allowed researchers to capture biofilm images in real time and provide images of submerged biofilms, which consist of mushroom-like structures separated by water-filled channels in structural biofilms (Branda et al. 2005). However, biofilm formation is a complex process that involves several stages: The first step involves the absorption of inorganic or organic molecules by the surface, creating a conditioning layer that stimulates bacterial attachment (Gotz 2002). The next step in biofilm formation is the adhesion of organisms to that layer, which is mediated by the use of fimbriae, pili, flagella, and EPSs that form a bridge between bacteria and the conditioning film. As bacteria grow, they excrete larger volumes of EPSs that provide a protective barrier around the cells (Hall-Stoodley and Stoodley 2009).

Depending on the environmental and physiological characteristics of biofilms, the bacteria can grow quickly and become "mature" within 24–48 hours. In a matter of days, they grow to millimetric proportions (it may take months in case of anaerobic biofilms). In the last step, large chunks of the biofilm will periodically detach from the biofilm structure due to flow rate dynamics, fluid shearing effects, chemicals within the fluid, or even changing properties of the bacteria present in the biofilm. The released bacteria restart the biofilm formation process once they are transported to a new surface. In this step, the bacteria in the biofilm are in a steady state showing huge morphological and structural diversity. Some biofilms form a thin uniform layer, whereas others are thick and uneven with mushroom-like structures that extend to the bulk media. Some mixed-species biofilms exhibit a high degree of patchiness, whereas others have a clear stratification of different species.

Special microscopes are used to obtain digital images of live bacterial biofilm structures grown in laboratories, including confocal laser scanning microscopy

(CLSM) and optical microscopy (OM; Gorur et al. 2009). The appreciation of different types of biofilm structures in digital images can be subjective and depends on the observer (Beyenal, Donovan et al. 2004) hence, it is necessary to quantify the complex structure of a biofilm in useful parameters through image analysis for the microbiologist to use them. In order to quantify these images, preprocessing is necessary to distinguish relevant elements and characteristics from the structures. Preprocessing involves an image segmentation process, which if done correctly will not propagate the errors of appreciation in image quantification.

13.2 Biofilm Image Analysis

Most modern confocal systems are based on the ideas developed by Marvin Minsky in the mid 1950s (patented in 1961) when he was a postdoctoral student at Harvard University. However, the technology that existed at that time did not allow him to demonstrate the full potential of this new microscopy technique. The scanning process with confocal microscopy allows one to obtain transversal sections of a biofilm. Each transversal section corresponds to an individual digital image, composing a stack of images that represent the biofilm in three dimensions. The resolution of the images is dependent on hardware, and the most common resolutions are 512×512 and 1024×1024 pixels with 16-bit color depth. These images are preferably used in raw format for the analysis and morphological transformations for image enhancement; the use of filters is deliberately avoided, and they are used for visualization purposes only and not for quantification. These methods remove noise by operating on local neighborhoods in the images. However, an important side effect of this method is the removal of small details in the images, which can be important for the calculation of the Minkowski sausage fractal dimension and the surface area.

Image segmentation is one of the most important steps in processing images for analysis, and it is one of the first stages on which other high-level processes such as image quantifications are based (Sing-Tze 2002). For general image segmentation, several approaches exist, including the following:

Clustering-based methods: Clustering has been used in image analysis, allowing segmentation in an iterative form of classification without supervision and assigning the pixels of an image to one of k clusters, where k is indicated by an expert. These algorithms do not guarantee an optimal solution, because they require the specification of good initial cluster centers for correct convergence and the specification of the number of clusters (Theodoridis and Koutroumbas 2006).

Histogram-based methods: These methods are efficient and fast and are based on thresholding a histogram, following this procedure: Obtain a histogram that represents the intensities of the color or grayscale channels of an image, and find the thresholds that decide which values of the pixel channels belong to an individual class. The disadvantage of these algorithms is their little tolerance to noise, which can be avoided by using image filters prior to performing the segmentation process (Russ 2007).

Region-growing methods: These methods take a specified number of seeds (set of pixels or initial clusters), which are grown based on the vicinity of each seed. The nearest neighbors in terms of color/intensity/texture are grouped in such clusters. The method ends when there are no more pixels left to be assigned. Although this method is more tolerant to noise, it requires the specification of an appropriate initial set of seeds to produce accurate results (Adams and Bischof 1994).

Classifier-based methods: These methods include pattern recognition techniques that partition the feature space obtained from an image using labels or a priori knowledge. These techniques are called "supervised learning" methods, because they require manually segmented images in the training preprocessing stage (Xu et al. 1998).

Image segmentation approaches based on supervised classification, in many cases, are not viable since the analysis of the biofilm is dependent on the setup of the experiments, which makes it difficult to obtain a training set. The application of region-growing methods has its own intricacies as well, because biofilm images can have disconnected regions between layers and region-growing methods tend to eliminate the noise from the images. In biofilm image analysis, this noise represents structural parameters that are important to quantify. In the same context, clustering-based methods require, in general, determination of the number of clusters, good initialization, and appropriate convergence conditions. On the other hand, thresholding methods are widely used for image segmentation. However, when dealing with multiple thresholds, finding optimal thresholds with traditional methods is not efficient and, hence, the applicability of such methods in finding five or more thresholds tends to be prohibitive. A few suboptimal schemes have been proposed in this regard, which are not capable of finding the optimal thresholds. Fortunately, this problem has been solved by a dynamic programming–based multilevel thresholding algorithm with polynomial-time complexity $O(kn^2)$ (Rueda 2008), in which the number of thresholds is k and the number of gray levels in the image is n. Furthermore, this algorithm, as in Section 13.4.3 is able to use various thresholding criteria to find the optimal thresholds on irregularly sampled histograms.

13.3 Related Works on Biofilm Image Segmentation

As discussed in Section 13.1, biofilm structures contain information about phenotypic features of bacteria. The studies of Costerton et al. (1995), Johnson (2008), Jorgensen et al. (2003), and Klapper (2006) indicate that the structural heterogeneity of a biofilm can affect its dynamic activities and functional properties. In the work by

Yang et al. (2000), a biofilm image–processing approach was proposed to obtain the structural characteristics of the biofilm. In the same work, an extension to the studies of Beyenal, Tanyolac, and Lewandowski (1998); Yang and Lewandowsky (1995); and Lewandowski et al. (1992) was proposed, incorporating the computation of parameters like porosity, fractal dimension, diffusion capacity, and entropy. In order to study the structural features of biofilms in a study by Heydorn et al. (2000), a novel software called COMSTAT was proposed to quantify the characteristics of biofilms obtained through CLSM image stacks; the authors further used manual and Otsu's (1979) thresholding criterion for image segmentation. Another important aspect of this work is the use of an image-filtering stage that eliminates noise, which concluded that the use of some of these filters is not recommended due to the elimination of small image details that are important for the quantification of some parameters such as fractal dimensions. Another software package for general image quantification is MAPPER, which has also been used to quantify biofilm images (Tolle, McJunkin, and Stoner 2003). However, the segmentation methods in this software package are manual. The strength of this software is the parallelization of image processing in several nodes of a network through secure communication protocols.

Another work related to quantifying the parameters of biofilm structures is the one by Beyenal et al. (2004); a detailed explanation of quantification methods can be found in this study. The algorithms for segmentation used by these approaches are the traditional Otsu's criterion (1979) and an iterative method for finding thresholds that was proposed by Beyenal et al. (2004). In addition, the advantage of automatic thresholding methods is discussed in the context of not inducing subjectivity.

An increasing number of studies use CLSM, which can automatically determine the thresholds for image segmentation and for biofilm visualization. In the study by Merod et al. (2007), it was found that certain images of CLSM stacks contain errors including atypical pixels, which can lead to erroneous thresholds in the segmentation process. In order to resolve this problem, a software package based on the PHOBIA LSM Image Processor (PHLIP) was proposed (Mueller et al. 2006). The PHLIP-based software makes an automatic image exclusion of the stack by calculating the biomass covering each image stack. If an image deviates considerably from the biomass cover obtained by the other images, it is excluded from the process of computing the thresholds. However, automatic thresholding is carried out only with one threshold for image binarization by means of Otsu's thresholding criterion (1979).

In the same context, in the work by Yang et al. (2001), a survey of several automatic thresholding algorithms was presented, including local entropy, joint entropy, relative entropy, Renyi's entropy, and iterative selection. For comparison, 10 investigators at Montana State University, Bozeman, Montana, performed manual thresholding. From the 10 thresholding results, the average and standard deviation of each process was obtained, and the result was compared with those of each of the aforementioned algorithms. In order to evaluate the differences between manual and automatic segmentation, the mean of sum of squares of relative residuals (MSSRR) was calculated, concluding that only the iterative selection method is consistent with the manual thresholding done by the expert on different types of images. Validation of segmentation methods is important in biofilm analysis because the quantification of structural characteristics should be close to real structural characteristics in order to be useful. In this context, in the study by Zhang (1996), a survey of different methods proposed for segmentation evaluation is presented. In this work, the evaluation methods are classified into three groups: (1) analytical, (2) empirical goodness, and (3) empirical discrepancy groups. Moreover, an approach for objective comparison of segmentation methods was proposed in which a direct comparison of images through similarity indices such as Rand index (RI) was performed (Unnikrishnan, Pantofaru, and Hebert 2007).

Not all approaches for automatic image segmentation based on thresholding proposed so far allow the segmentation of different kinds of biofilm images in an optimal way without the intervention of an expert. Since they need to set the parameters manually, they produce suboptimal solutions and subjective results, and are prone to inefficient use of computational resources. Moreover, in general, a nonobjective evaluation of the process and results usually takes place in the segmentation of biofilm images. In Section 13.4, a combination of multilevel thresholding criteria and a set of clustering validity indices are discussed. The methods aim to perform completely automatic, efficient, and optimal multilevel thresholding segmentation of different kinds of biofilm images. A quantification process that was performed in a wet laboratory and was used to compare the results with those of applying biofilm image–processing techniques is also discussed.

13.4 Segmentation of Biofilm Images

An efficient method for the segmentation of biofilm images, described in Section 13.4.1 is implemented using a multilevel thresholding algorithm to perform automatic optimal segmentation in polynomial time. Different criteria and clustering validity indices are used to measure the performance of the segmentation methods and to determine the best number of thresholds, respectively. Then, the resulting segmented images are compared against a set of images segmented manually by an expert and through a quantification process. Prior to discussing the combination of methods, a few definitions are presented and then the algorithms and the indices are reviewed in detail.

13.4.1 The Thresholding Problem

An image can be considered a two-dimensional (2-D) discrete function of two possibly independent spatial variables. A grayscale image, which is composed only of pixel intensities, can be separated as a 2-D function $f : N^2 \rightarrow N$, where $0 \leq f(x;y) \leq 2^n - 1$. In this equation, n is the number of bits needed to store the intensity values (x, y);

$1 \le x \le N$ and $1 \le y \le M$ are independent spatial variables indicating the position of the pixel in the 2-D image; and N and M are width and height of the image, respectively. A frequency histogram H_f of an image with discrete intensity values in $[(0, 2^n - 1)]$ is an ordered set $H_f = \{h_f(0), h_f(1), h_f(2), \ldots, h_f(2^n - 2), h_f(2^n - 1)\}$, where $h_f(i)$ is a discrete intensity function that represents the frequencies of pixels with intensity i in the image. Thus, frequency $h_f(i)$ of the ith intensity is $h_f(i) = \sum_{x=1}^{N} \sum_{y=1}^{M} (f(x, y) = i)$.

Also, the histogram can be defined in terms of probabilities $\{h_p(0), h_p(1), h_p(2), \ldots, h_p(2^n - 2), h_p(2^n - 1)\}$ as a function based on the histogram of frequencies

$$h_p(i) = \frac{1}{\sum_{f=0}^{2^n-1} h_f(j)} h_f(i).$$

The multilevel thresholding problem consists of obtaining an ordered set $T = \{t_0, t_1, \ldots, t_k, t_{k+1}\}$ of k thresholds, where $0 \le t_i \le 2^n - 1$ and t_0 and t_{k+1} are the boundaries of the histogram, so that an arbitrary thresholding criterion is optimized. In other words, thresholding consists of partitioning a histogram into $k + 1$ classes or groups of pixels with consecutive intensity values. Thus, finding an optimal set T of thresholds is equivalent to maximizing or minimizing a function $\psi' = H_p^k \times [0,1]^{2^n-1} \rightarrow R^+$, which defines a thresholding criterion over a set of thresholds, where T partitions the set H_p into $k + 1$ classes as follows:

$$\varsigma_1 = \left\{ f(x, y) / t_0 = 0 \le f(x, y) \le t_1 \right\}$$

$$\varsigma_2 = \left\{ f(x, y) / t_1 + 1 \le f(x, y) \le t_2 \right\}$$

$$\varsigma_3 = \left\{ f(x, y) / t_2 + 2 \le f(x, y) \le t_3 \right\}$$

$$\vdots$$

$$\vdots$$

$$\vdots$$

$$\varsigma_k = \{ f(x, y) / t_{k-1} + 1 \le f(x, y) \le t_k \}$$

and

$$\varsigma_{k+1} = \{ f(x, y) / t_{k+1} + 1 \le f(x, y) \le t_{k+1} \}$$

Section 13.4.2, below, describes an algorithm based on dynamic programming, which obtains the optimal set T based on various thresholding criteria.

13.4.2 Local and Global Thresholding

Traditionally, software packages for biofilm analysis such as COMSTAT proceed by determining only one threshold to segment the entire stack of images that represent the biofilm. For this, a histogram is generated through the aggregation of frequencies of all intensity levels of all images in the stack and, hence, manual or automatic segmentation is carried out in one histogram only. In the manual approach, the user selects one layer of the stack and determines the threshold in a visual form only in the image that is being visualized. Then, the thresholding process is performed on all layers using the unique thresholds that are obtained. In automated segmentation, the methodology is the same, except that the threshold is selected by Otsu's thresholding criterion (1979). Another approach is to determine the thresholds for each layer in the stack independently, which is a difficult process when the threshold is selected manually but viable when the thresholding criterion is automated. In order to determine the best choice between local and global thresholds, the microbiologist manually segments each layer in a stack independently. In Figure 13.1, each layer of a biofilm and the threshold selected for these layers are shown. In this figure, the expert selects different thresholds for different levels of the stack in order to visualize the image correctly. It is not clear whether the use of only one global threshold is the

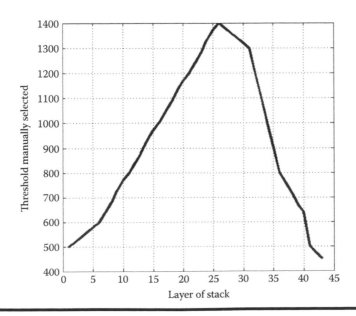

Figure 13.1   Local thresholds of a biofilm selected individually by an expert.

best approach to obtain an ideal visualization of the segmentation results according to the expert; therefore, it is necessary to compare the two approaches.

13.4.3 Polynomial-Time Optimal Multilevel Thresholding

In the work by Rueda (2008), a polynomial-time algorithm for multilevel thresholding is proposed. This algorithm is polynomial not only on the number of bins of the histogram but also on the number of thresholds. Moreover, it runs in polynomial time independently of the thresholding criterion, with Ψ defined as follows:

$$\Psi(T_{0,m}) = \Psi(\{t_0, t_1, \ldots, t_m\}) = \sum_{j=1}^{m} \psi_{t_{j-1}+1, t_j} \tag{13.1}$$

where $\Psi : H_p^m \times [0,1]^n \rightarrow R^+ \cup \{0\}, 1 \le m \le k+1, t_0$ is the first element of $T, \psi_{l,r} :$
$T, \psi_{l,r} : H_p^2 \times [0,1]^{l-r+1} \rightarrow \mathfrak{R}^+ \cup \{0\}$, and for $\psi(1, 1+1, \ldots, r, h_p(1), h_p(l+1), \ldots, h_p(r))$,
$\psi(l, r, h_p(l+1), \ldots, h_p(r))$ or, for short, $\psi_{l,r}$ $l < r$ must satisfy the following conditions:

- Condition 1: For any histogram H_p and any threshold set T, $\Psi > 0$ and $\psi \ge 0$.
- Condition 2: For any $m, \Psi(T_0, m)$ can be expressed as $\Psi(T_0, m) = \Psi(\{t_0, t_1, \ldots, t_{m-1}\}) + \psi_{t_{m-1}+1, t_m}$.
- Condition 3: If $\Psi_{t_{m-1}+1, t_j}$ is known, then $\Psi_{t_{m-1}+2, t_j}$ can be computed in O(1) time.

The dynamic programming algorithm can be characterized in terms of solutions to subproblems. Solving smaller problems in the order $\Psi(T_{0,1}), \Psi(T_{0,2}), \Psi(T_{0,3})$, and so on will avoid the re-solving of any subproblem two or more times. These smaller subproblems are incrementally used to solve larger subproblems until the whole problem, $\Psi(T_{0,k+1})$, is solved. The dynamic programming algorithm proposed by Rueda (2008) is depicted in Figure 13.2, where the function $\Psi_{l,r}$ corresponds to any thresholding criterion that satisfies conditions 1–3. The three main criteria are defined as follows (a complete description of the implementation of these three criteria can be found in the study by Rueda (2008):

- Otsu's thresholding criterion (1979; OTSU):

$$\psi_{t_{j-1}+1, t_j} = \omega_j \mu_j^2 \tag{13.2}$$

- Minimum-error criterion (MINERROR):

$$\Psi_{t_{j-1}+1, t_j} = 2\omega_j \{\log \sigma_j + \log \omega_j\} \tag{13.3}$$

and

Multilevel Thresholding Algorithm

Input: Probabilities, $H_p = \{h_p(0), h_p(1), \ldots, h_p(2^n - 1)\}$. Number of thresholds, k.

Output: A threshold set, $T = \{t_0, t_1, t_2, \ldots, t_k, t_{k+1}\}$

$\min T_j, \max T_j \leftarrow \text{findThresholdRanges}(k)$

$C(0,0) \leftarrow 0; D(0,0) \leftarrow 0$

for $j \leftarrow 1$ **to** $k+1$ **do**

    for t_j do $\leftarrow \min T_j(j)$ to $\max T_j(j)$ **do**

        $C(t_j, j) \leftarrow 0; \text{ psi} \leftarrow \psi_{j,t_j}$

        for $i \leftarrow \min T_j(j-1)$ to $\min\{\max T_j(j-1), t_j - 1\}$ **do**

            if $C(i, j-1) + \text{psi} > C(t_j, j)$ **then**

                $C(t_j, j) \leftarrow C(i, j-1) + \text{psi}$

                $D(t_j, j) \leftarrow i$

            end if

            $\text{psi} \leftarrow \text{Compute } \psi_{i+2,t_j} \text{ from psi}, i+1 \text{ and}/p_{i+1}$

        end for

    end for

end for

return $\text{findThresholds}(D)$

procedure $\text{findThresholdRanges}(k: \text{integer})$

for $j \leftarrow 0$ to $k+1$ **do**

    if $j = k+1$ **then** $\min T_j(j) \leftarrow n$

        else $\min T_j(j) \leftarrow j$

    end if

    if $j = 0$ **then** $\max T_j(j) \leftarrow 0$

        else $\max T_j(j) \leftarrow n - k + j - 1$

    end if

end for

return $\min T_j, \max T_j$

end procedure

procedure $\text{findThresholds}(D: \text{table})$

$T(k+1) \leftarrow n$

for $j \leftarrow k$ **downto** 0 **do**

    $T(j) \leftarrow D(T(j+1), j+1)$

end for

return T

end procedure

Figure 13.2 **General algorithm for multilevel thresholding based on dynamic programming.**

■ Entropy-based criterion (ENTROPY):

$$\Psi_{t_{j-1}+1,t_j} = -\sum_{i=t_{j-1}+1}^{t_j} \frac{h_p(i)}{\omega_j} \log \frac{h_p(i)}{\omega_j} \tag{13.4}$$

where $\omega_j = \sum_{i=t_{j-1}+1}^{t_j} h_p(i), \mu_j = \frac{1}{\omega_j}\sum_{t=t_{j-1}+1}^{t_j} i h_p(i), \sigma_j = \frac{i}{\omega_j}\sum_{t=t_{j-1}+1}^{t_j} h_p(i)(i-\mu_j)^2,$
and t_j is the *j*th threshold of *T*.

It is important to highlight that biofilm images lead to "sparse" histograms (many bins have zero probabilities) and, thus, for the sake of efficiency the algorithm for irregularly sampled histograms is described and used in the discussions and experiments presented in this chapter.

Despite it being possible to obtain various optimal thresholding sets *T*, it is not possible to establish the value of *k* using the thresholding criterion by itself. Section 13.4.4 describes the use of clustering validity indices in order to obtain the best value of *k* based on the selected thresholding criterion.

13.4.4 Optimal Number of Thresholds

Thresholding algorithms by themselves are not capable of determining the number of thresholds *k* in which an image can be segmented correctly. However, *k* has a direct relationship with the number of classes, *k* + 1, in which a histogram is partitioned by means of a multilevel thresholding algorithm. By viewing thresholding as a problem of clustering pixel intensities, clustering validity indices can be used to obtain the best number of classes *k* + 1 in which the histogram can be clustered and, hence, the number of thresholds. In this section, we discuss four clustering validity indices that are used to determine the best number of thresholds (Maulik and Bandyhopadhyay 2002):

1. Davies–Bouldin (DB) index: This index is defined as the ratio between the within-cluster scatter and the between-cluster scatter. The motivation for using this index is minimization of the value of the DB function, which is defined as follows:

$$DB = \frac{1}{k+1}\sum_{i=1}^{k+1} R_i \tag{13.5}$$

where *k* + 1 is the number of clusters and $R_i = \max_{j,j\neq i}\left\{\dfrac{S_i + S_j}{d_{ij}}\right\}$, where $1 \le i \le k+1, 1 \le j \le k+1$, and $S_j = \dfrac{1}{|\zeta_j|}\sum_{t=t_{j-1}+1}^{t_j} h_p(i)\left\| i - \mu_j \right\|$ is the within-cluster scatter of clusters ζ_i and ζ_j.

2. Dunn's index (DN): This index is based on the relationship between cluster size and distances between clusters. The aim is to maximize the function DN defined as follows:

$$DN = \min_{i \leq j \leq k+1}$$ (13.6)

$$\min_{i \leq j \leq k+1, j \neq i} \left\{ \frac{\delta(\zeta_i, \zeta_j)}{\max_{1 \leq r \leq k+1}(\Delta(\varsigma_r))} \right\}$$

where $\Delta(\zeta_i) = \max_{x,y \in \varsigma_i}\{d(xy)\}$ is the diameter of cluster ζ_i, $\delta(\zeta_i, \zeta_j) = \min_{x \in \varsigma_i, y \in \varsigma_j}\{d(xy)\}$ is the distance between clusters ζ_i and ζ_j, and $k + 1$ is the number of clusters.

3. Index I (IndexI): This index is composed of three terms, taking into account the number of clusters, scatter of the clusters, and distance between the clusters. The goal is to maximize the function IndexI defined as follows:

$$\text{IndexI} = \left(\frac{1}{k+1} \frac{E_1}{E_k} D_k \right)^p$$ (13.7)

where $E_k = \sum_{j=1}^{k+1} \sum_{t=t_{j-1}+1}^{t_j} \mu_{ij} h_p(i)\|i - \mu_j\|$ is the scatter of the clusters, μ_{ji} is the membership of intensity i to cluster E_1 and is a constant for the data set, $D_k = \max_{i,j=1}^{k+1} \|\mu_i - \mu_j\|$ is the distance between the farthest clusters, the power p is used to control the contrast between different cluster configurations, and $k + 1$ is the number of clusters.

4. Calinski–Harabasz (CH) index: This index is defined as the ratio between within-cluster scatter and between-cluster scatter. The motivation for using this index is maximization of the function CH defined as follows:

$$\text{CH} = \frac{traceB}{k} \bigg/ \frac{traceW}{n_t - k + 1}$$ (13.8)

where $traceB = \sum_{i+1}^{k+1} |\varsigma_i| \|\mu - \mu_i\|^2$ is the sum of between-cluster scatters, μ is the centroid of the entire data set H_p, $traceW = \sum_{j=1}^{k+1} \sum_{i=t_{j-1}+1}^{t_j} \|\varsigma_j\| h_p(i) \|i - \mu_j\|^2$ is the sum of within-cluster scatters, n_t is the number of elements in all clusters, and k is the number of thresholds.

Once various thresholding criteria and clustering validity indices are defined, a way to measure the performance of the combination of the two techniques can be defined. Section 13.4.5 describes a similarity index that allows us to measure the similarity of two images that have been segmented in two different ways.

13.4.5 Manual Segmentation versus Automatic Segmentation

In order to determine the best combination of thresholding criterion and clustering validity index, different indices can be used. In this section, two of them are discussed: (1) probabilistic RI and (2) sum of squares of relative residuals.

Probabilistic RI is the percentage of pairs for which there is an agreement. For example, let $L = \{l_1, l_2, \ldots, l_{NM}\}$ and $L_r = \{l_1^r, l_2^r, \ldots, l_{NM}^r\}$ be the ordered sets of labels l_i for each element of an image defined as a set of ordered pixels without spatial relationship $X = \{x_1, x_2, \ldots, x_{NM}\}$, where NM is the number of pixels in an image, R is the number of reference segmentations L_r to be compared against a specific segmentation L, and $1 \leq r \leq R$. The RI is defined as a function over segmented images $RI(L, L_{\{1,\ldots,R\}}) : L \times L^R \rightarrow [0,1]$, which is defined as follows:

$$RI(L, L_{(1,\ldots,R)}) = \frac{1}{\binom{R}{2}} \sum_{i,j}^{NM} {}_{i \neq j} [I(l_i = l_j) P(l_i = l_j) + I(l_i \neq l_j) P(l_i \neq l_j)] \quad (13.9)$$

where $l_i, l_j \in L$ are the labels of pixels x_i and x_j, respectively; I is the identity function; and $P(l_i = l_j) = \frac{1}{R} \sum_{r=1}^{R} I(l_i^r = l_j^r)$ and $P(l_i \neq l_j) = \frac{1}{R} \sum_{r=1}^{R} I(l_i^r \neq l_j^r)$ are the probability that $l_i = l_j$ and the probability that a label $l_i \neq l_j$, respectively. This index takes a value of 1 when L and $L_r = \{L_1, \ldots, L_R\}$ are equal, and 0 if they do not agree on anything at all.

The MSSRR measure is used to evaluate the differences between the threshold levels selected by automatic thresholding and the manual method. The MSSRR is defined as follows (Yang et al. 2001):

$$MSSRR = \frac{1}{M} \sum_{i=1}^{M} \left(\frac{t_i - t_i'}{t_i} \right)^2 \quad (13.10)$$

where t_i and t_i' are the thresholds of the ith image found manually and automatically, respectively, and M is the total number of images.

13.4.6 Determining the Best Combination of Techniques

Based on empirical studies and analyses, we have determined the best combination of techniques for automatic segmentation by means of multilevel thresholding of biofilm images combined with other methods for selecting the best number of thresholds and indices of validity. A method that combines automatic multilevel thresholding and clustering validity indices is discussed here. The general scheme for these indices is depicted in Figure 13.3. The first step in the process is image

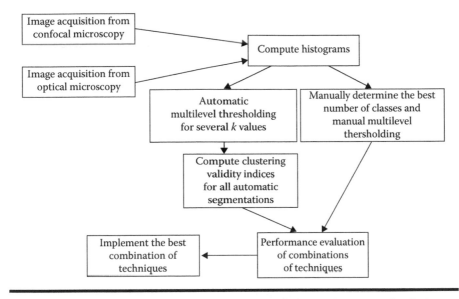

Figure 13.3   General scheme for determining the best combination of techniques.

acquisition from optical or confocal microscopy and calculation of the histogram for each image. Next, manual multilevel segmentation is performed by means of trial and error, in order to determine k and T for each original image. This process is performed by an expert, obtaining the segmented images and the best number of thresholds for each image in the stack. Additionally, each original image is automatically segmented by means of the optimal multilevel thresholding algorithm for three thresholding criteria and for several values of k (number of thresholds). Clustering validity indices are calculated for each image that is segmented automatically. Finally, RI is calculated for each segmented image by means of manual and automatic thresholding, in order to determine the best combination of thresholding criteria and clustering validity indices.

13.5 Experimental Analysis

This section presents a comprehensive experimental study on the performance of different thresholding algorithms and clustering validity indices on various biofilm images of different nature. In this regard, a data set of 649 images was used to evaluate the proposed biofilm segmentation approach. These images were obtained as follows: Mature biofilms of *Pseudomonas syringae* strains were developed within the Biofilm and Environmental Microbiology Laboratory (Biotechnology Centre, University of Concepción, Chile).* The biofilms were then scanned using CLSM

---

*http://www.udec.cl/~bem-lab

Table 13.1   Description of the Data Set of Biofilm Images for Determining the Best Combination of Techniques

k	Number of Images	Microscopy	Resolution
1	616	Confocal	512×512
2	10	Optical	1040×1392
3	10	Optical	1040×1392
4	6	Optical	1040×1392
5	6	Optical	1040×1392
6	1	Optical	1040×1392

and OM, generating stacks of images that represent the 3-D structure of biofilms. The images were segmented individually. Table 13.1 shows the features of all the images, which are 12-bit grayscale images. In addition to the number of thresholds, an expert finds k manually.

Since many biofilm images in the data set are appropriately segmented using a single threshold and a few images are best segmented with more than one threshold, the data set is divided into two subsets of images to perform the experiments. This division helps to avoid any bias introduced by the difference in the number of images. We should note that for this data set, the images that are best segmented with one threshold are those obtained by using CLSM, whereas those with more than one threshold correspond to the images obtained by using OM. However, the relationship between the procedure utilized for the acquisition of images and the appropriate number of thresholds may not necessarily be the case for other images.

13.5.1 Performance of Thresholding Criteria

The best thresholding criterion was found by using RI and comparing manual versus automated segmentation, and by using the following notation: RI^{all} denotes the RI for all image data sets, RI^{clsm} denotes the RI for images with one threshold found manually, and RI^{op} corresponds to the RI for images with more than one threshold found manually. In Table 13.2, the resulting values for RI are shown for all image subsets. It is clear that ENTROPY is the best criterion for thresholding images with one threshold. On the other hand, OTSU is the best criterion for segmentation of images using more than one threshold. Overall, the ENTROPY criterion achieves

Table 13.2   RI for Different Image Subsets (Number of Thresholds Found Manually)

Index and Data Set	OTSU	ENTROPY	MINERROR
RI^{op}	0.7897	0.7300	0.7713
RI^{clsm}	0.7283	0.7767	0.6086
RI^{clsm} (global and threshold)	0.6542	0.7343	0.6079
RI^{all}	0.6184	0.7566	0.5846

the best performance for all images in the data set. Additionally, in the CLSM set of images, global thresholding is performed in order to compare the results with those of local thresholding. Local thresholding achieves better performance than global thresholding with all thresholding criteria.

For images obtained from CLSM, MSSRR and the correlation R are calculated to evaluate the differences between the threshold levels selected by automatic thresholding and the manual method, where $M = 616$ is the total number of images from CLSM. Figures 13.4a through c show correlation plots between manually set thresholds and thresholds obtained from ENTROPY, MINERROR, and OTSU, respectively. For an agreement between automatic and manual thresholding, we expect that the points fall in the diagonal line $y = x$. In Table 13.3, the resulting values for MSSRR and each thresholding criterion are depicted.

Figure 13.5 shows the plots of RI in increasing order of the performance value. It can be observed that the best performance is achieved by the ENTROPY criterion followed by the OTSU criterion.

Note the poor performance of the OTSU and MINERROR criteria for the segmentation of images with a single threshold. The reason that these methods perform poorly is that these criteria are based on clustering algorithms and, hence, the presence of a single peak in the histogram make these criteria assign the cluster to the center of the peak rather than allocating the threshold on one side of that peak.

The ENTROPY criterion attains better results for single-peaked histograms. This result is reasonable, since the formulation of this criterion leads to assigning lower entropy values when the class distribution is flat, whereas when the peaks are higher the entropy is higher too. This correlation implies that the ENTROPY criterion leads to results that are similar to those obtained by manual thresholding.

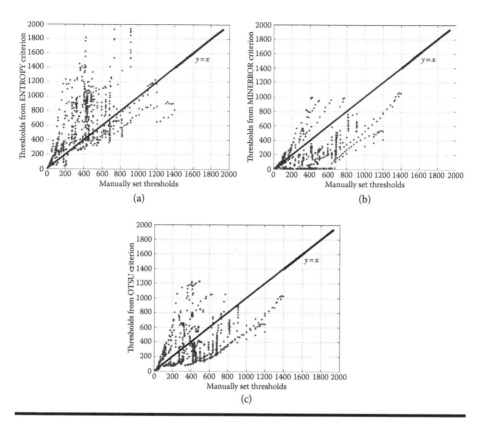

Figure 13.4 Correlation graphics between thresholding from automatic criteria and thresholds that are manually set: (a) correlation for the ENTROPY criterion, (b) correlation for the MINERROR criterion, and (c) correlation for the OTSU criterion.

Table 13.3 MSSRR and R^2 for Different Thresholding Criteria (CLSM Images)

	OTSU	ENTROPY	MINERROR
MSSRR	3.851	0.827	10.25
R^2	0.71	0.76	0.62

13.5.2 Determining the Best Number of Thresholds

The number of thresholds has a direct relation to the number of classes in which an image can be segmented. Therefore, a good way for solving this problem is to use a separate measurement on the error to estimate the number of thresholds

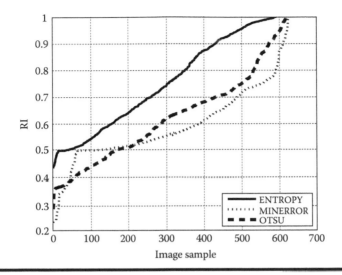

Figure 13.5 **The Rand index for confocal laser scanning microscopy images, for which the number of thresholds was found manually.**

Table 13.4 **MSE for the Estimation of the Best Number of Thresholds**

	Index/	*CH*	*DB*	*DN*
OTSU	7.44	221.97	96.63	212.76
ENTROPY	2.33	212.2	1.18	186.94
MINERROR	2.8	188.31	179.39	220.22

for the complete data set of images. Table 13.4 shows the mean-squared error (MSE) for each combination of thresholding criteria and clustering validity indices.

As can be observed, the DB index achieves the best performance with the ENTROPY criterion, which reaffirms that the combination DB attains a good performance in most of the cases for different data sets of images. Existing clustering validity indices have a direct relationship with each other in their formulation; however, each index has a different behavior depending on the number of thresholds selected. The behavior of each validity index can be observed in Figure 13.6. Although the plots are for only one of the images in the data set, they represent the behavior of the clustering validity indices for the entire data set.

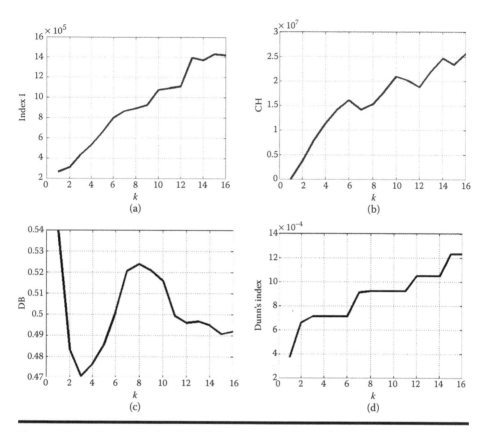

Figure 13.6 **General behavior of clustering validity indices: (a) Index I (IndexI), (b) Calinski Harabasz (CH), (c) Davies–Bouldin (DB), and (d) Dunn's (DN) indices.**

In Figure 13.6, one observes that the indices IndexI, CH, and DN are (for most values of k) monotonically increasing functions of k (Figures 13.6a, b, and d) that reach best performance when the function reaches the maximum point, that is, when $k = 61$, 64, and 64, respectively. This behavior, unfortunately, does not provide a clear answer regarding how to determine the optimal number of clusters with which an image should be segmented, and illustrates the high MSE values obtained by these indices to estimate the best number of thresholds. On the other hand, the DB index is the only index that shows a high independence in terms of the number of clusters. Thus, this index reaches its optimal performance when $k = 8$, which is a much more meaningful value than the values obtained by the other indices. Moreover, as k grows the DB index tends to give an almost constant rate, which reflects the fact that increasing the number of clusters does not cause the quality of the clustering to improve beyond a certain point (Figure 13.6c). In addition,

IndexI does not perform well because the term EK significantly reduces its value when the number of clusters grows.

This situation occurs because the dispersion of the data from the centers of each class is smaller when the values of k are larger, and the value of the distance between each point and the center becomes increasingly small since the classes have fewer members and these members are closer when dealing with one dimension. On the other hand, the DN index grows when k grows. This growth occurs because the diameter of the clusters is smaller when k is larger. In addition, the distance between the clusters is defined as the shortest distance between two members of each cluster. In a one-dimensional (1-D) histogram, this value corresponds to clusters of intensities located next to each other and, hence, this value has no impact on the value of the index and the index is almost entirely dominated by the diameter of each cluster, which decreases when k increases. The DB index is defined as the sum of the dispersion of pairs of clusters divided by the distance between them, implying that the index is lower when the dispersion of pairs of clusters is smaller than their distance, which does not have a direct influence on the number of clusters. This index has a behavior dominated by the nature of the clusters, favoring distant and compact clusters, which are equally desirable in the separation of peaks in a 1-D histogram.

13.5.3 Performance for Image Segmentation Techniques Combined with Cluster Validity Indices

Table 13.5 shows the values of RI for all biofilm images. From this test, the values included in Table 13.5 show that the best combination is ENTROPY + DB for RI^{all}. In addition, it is clear that the thresholding criterion with the best performance for this data set is based on the ENTROPY criterion. This result was predictable, because most of the images have one threshold and the best method of segmentation using one threshold is the ENTROPY criterion. The behaviors for different combinations of techniques for two separate cases, one and more than one threshold, are discussed in Sections 13.5.4 and 13.5.5.

Table 13.5 RI^{all} for All Automatically Segmented Biofilm Images

	IndexI	CH	DB	DN
OTSU	0.2163	0.2151	0.2969	0.2187
ENTROPY	0.2506	0.2351	0.7884	0.2385
MINERROR	0.2206	0.2332	0.2613	0.2085

13.5.4 One Threshold

All biofilm images obtained by confocal microscopy have a single optimal threshold (manually found by an expert). Table 13.6 shows the performance of different thresholding criteria and clustering validity indices for the image segmentation of biofilms with one threshold determined automatically. The combinations of ENTROPY and DB, corroborating the overall results, perform the best. In this case, the analysis shows the same pattern as that of overall performance, because ENTROPY is the best criterion for thresholding images with one threshold and DB is the best clustering validity index for an estimated value of k.

13.5.5 More Than One Threshold

Table 13.7 shows the performance of thresholding methods and cluster validity indices for segmentation of biofilms with more than one threshold. As can be seen, all methods perform well. The OTSU criterion combined with IndexI attains the best value of RI^{op}. Nevertheless, for this set of images, the performance with respect to the combination ENTROPY + DB does not differ significantly. However, it is clear that the number of clusters estimated by the clustering validity indices significantly influence the performances of the thresholding criteria.

Table 13.6 RI^{clsm} for Automatically Segmented Images of Biofilms with One Threshold Determined Automatically

	IndexI	*CH*	*DB*	*DN*
OTSU	0.6176	0.3901	0.5297	0.4002
ENTROPY	0.7573	0.4907	0.7634	0.5029
MINERROR	0.5844	0.3279	0.328	0.3075

Table 13.7 RI^{op} for Automatically Segmented Biofilm Images with More than One Threshold Determined Automatically

	IndexI	*CH*	*DB*	*DN*
OTSU	0.7739	0.6548	0.7070	0.6564
ENTROPY	0.6889	0.7046	0.7634	0.7077
MINERROR	0.7594	0.6657	0.7222	0.6302

13.5.6 Visual Validation

13.5.6.1 Image Segmentation over Isolated Images

Figure 13.7a shows the manual segmentation of a biofilm image with one threshold compared with the automatic segmentation using the ENTROPY criterion and the DB clustering validity index (Figure 13.8a). As can be seen from Figure 13.8b, automatic segmentation sets the thresholds with a value slightly lower than that for the manual segmentation (Figure 13.7b), implying that more pixels with high intensities are labeled with white color. Figures 13.9a and 13.10a show the manual segmentation of a biofilm with more than one threshold compared with the automatic segmentation that combines ENTROPY + DB. As can be seen from Figure 13.10a, the result of automatic segmentation is close to that of manual segmentation (Figure 13.9a), setting the thresholds to almost the same values when the segmentation is done by an expert.

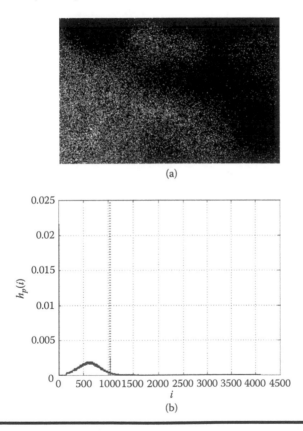

Figure 13.7 **Binary segmentation: (a) confocal microscopy image segmented manually and (b) histogram of confocal microscopy image segmented manually.**

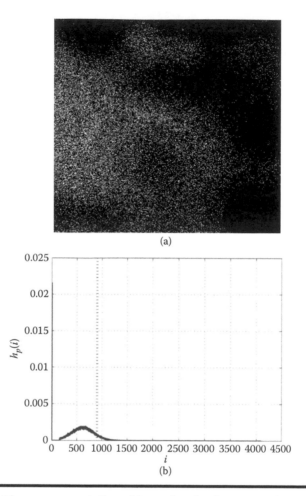

(a)

(b)

Figure 13.8 **Binary segmentation: (a) confocal microscopy image segmented automatically and (b) histogram of confocal microscopy image segmented automatically.**

13.5.6.2 Three-Dimensional Reconstruction of a Biofilm

Rebuilding the structure of a biofilm from a stack of images obtained by using confocal microscopy offers powerful visualization that allows one to observe the images from different angles. Figures 13.11a and 13.12a show the 3-D reconstruction of a biofilm through images segmented manually and automatic biofilm reconstruction by means of images segmented automatically through combinations of techniques, that is, ENTROPY + DB. This reconstruction is made with clouds of points; as can be seen from the figures, the image reconstructed automatically is similar to the one rebuilt manually by an expert. The existence of cross-sectional images in the biofilm

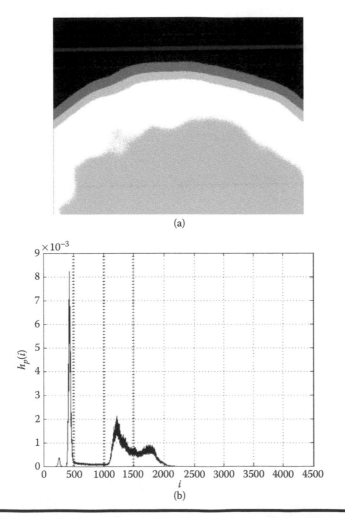

(a)

(b)

Figure 13.9   Multilevel thresholding segmentation: (a) optical image segmented manually and (b) histogram for the optical image segmented manually.

also makes it possible to use reconstruction techniques for 3-D images. One of the most common reconstruction algorithms is based on marching cubes (Lorensen and Cline 1987). This algorithm requires a cloud of points (ideally binary) as input. Then the algorithm connects these points between adjacent surfaces incrementally by means of a set of predetermined rules based on the type of distribution of the points in a cube. Figure 13.13 shows the 3-D reconstruction of a biofilm over 50 cross-sectional images. As can be seen from the figure, properties of luminosity of scenes and perspective of vectorial images enhance the visualization of the biofilm structure.

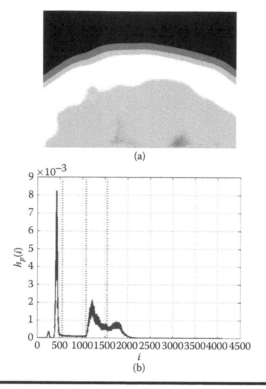

(a)

(b)

Figure 13.10 **Multilevel thresholding segmentation: (a) optical image segmented automatically and (b) histogram of the optical image segmented automatically.**

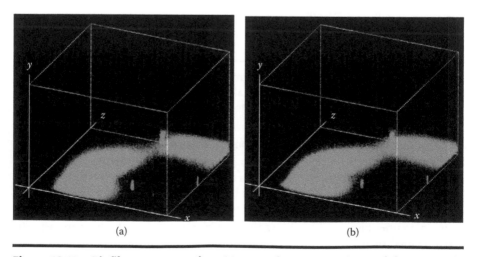

(a)   (b)

Figure 13.11 **Biofilm reconstruction: (a) manual reconstruction and (b) automatic reconstruction.**

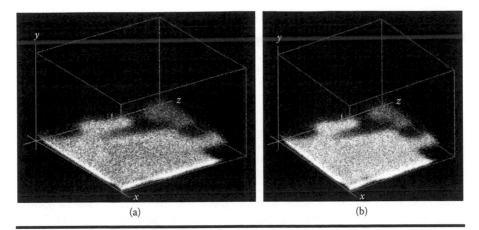

(a)   (b)

Figure 13.12 Biofilm reconstruction: (a) manual reconstruction and (b) automatic reconstruction.

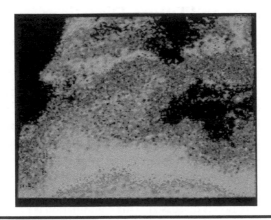

Figure 13.13 Biofilm reconstruction by marching cubes.

13.6 Quantification Validation

A quantification process was also done in a laboratory. In Table 13.8, we show the ratios of live and dead cells for 2 biofilms after 3 days of growth (3-D biofilms) and 4 days of growth (four-dimensional [4-D] biofilms). Ratios in the last row are obtained from laboratory measurements. The biofilm quantification techniques described by Beyenal et al. (2004) are applied on stacks of CLSM images (of these biofilms), which are automatically segmented by means of the ENTROPY, MINERROR, and OTSU criteria. Thus, the methods described by Beyenal et al. (2004) are used to estimate the ratios of live and dead cells of biofilm images after segmentation and 3-D reconstruction, and then compared with the ratios obtained

Table 13.8   Comparison between Cell Recounting and Biovolume Quantifications

	3-D Biofilm	4-D Biofilm
ENTROPY	89.3/10.7	92.4/7.6
MINERROR	46.1/53.9	50.2/49.9
OTSU	56.4/43.6	61.6/38.4
Laboratory	91.3/8.7	98.6/1.4

in the laboratory. As can be seen from the table, the ENTROPY criterion yields results closest to the values obtained in the laboratory. The MINERROR and OTSU criteria clearly underestimate the ratios obtained from the laboratory by a large margin, confirming again their poor performance in the RI results in Table 13.5.

13.7 Conclusions and Future Directions

Biofilms are complex microbiological structures that emerge from the behavior of bacteria, which are aggregated in a protective sticky matrix. Special types of microscopy, including CLSM and OM, are used to understand biofilms and obtain digital images. The study of these images, typically in the form of 3-D structures, is important for the microbiologist, and the most important of studies are automatic analyses that include segmentation and quantification. These steps involve an image segmentation process, which, if done correctly, does not propagate errors of appreciation in image quantification.

In this chapter, an extensive discussion on the automatic segmentation of biofilm images takes place and, in particular, a combination of these techniques for segmentation and validation is discussed. This combination is based on the efficient and optimal multilevel thresholding algorithm, which runs in polynomial-time complexity and allows for the estimation of the number of thresholds through the well-known DB clustering validity index. This index can find the best number of thresholds close to the criteria established by a human expert.

The thresholding criterion that performs as well as the manual segmentation is "maximum entropy." This discovery is assessed using the objective measure probabilistic RI, which compares segmentations done by the proposed method with segmentation done by an expert. Finally, automated reconstruction and visualization of the structures of 3-D biofilms is discussed, allowing us to view the structure from different angles and subsequently attaining a good representation in three dimensions. In addition, to validate the thresholding methods, we compare the ratio counts of live and dead cells in biofilms between laboratory measurements and image quantification (using the automatic segmentation methods); we obtained the best results with the combination ENTROPY + DB.

Segmentation of images allows the quantification processes to be performed and aids in studying other aspects of biofilms. Since multilevel thresholding is always optimal, it is possible to make a segmentation process free of subjectivity. Although only three main criteria are discussed in this chapter, other criteria can also be used for the segmentation of biofilm images whenever they satisfy the conditions stated in the work by Rueda (2008) and for which optimal thresholding can be achieved in polynomial time.

Quantification of biofilms enables researchers in the field to make an objective judgment about the relevant features of a biofilm and between different biofilms, by measuring coverage area, porous surface, spatial correlation, and entropy, along with other factors. On the other hand, vectorization of segmented images can facilitate visualization, allowing users to navigate the biofilm structure in three dimensions and helping them to observe the internal structure of the biofilm. These issues are worth investigating.

Acknowledgments

This work has been partially supported by the Natural Sciences and Engineering Research Council (NSERC) of Canada (Grant Nos. RGPIN261360 and RGPIN228117); the Canadian Foundation for Innovation, Canada (Grant No. 9263); the Ontario Innovation Trust, Canada; and the University of Atacama, Chile (University Grant for Research and Artistic Creativity, Grant No. 221172).

References

Adams, R., and L. Bischof. 1994. Seeded region growing. *IEEE Trans Pattern Anal Mach Intell* 16:641–647.

Beyenal, H., Z. Donovan, Z. Lewandowski, and G. Harkin. 2004. Three-dimensional biofilm structure quantification. *J Microbiol Methods* 59:395–413.

Beyenal, H., Z. Lewandowski, and G. Harkin. 2004. Quantifying biofilm structure: Facts and fiction. *Biofouling* 20:1–23.

Beyenal, H., A. Tanyolac, and Z. Lewandowski. 1998. Measurement of local effective diffusivity in heterogeneous biofilms. *Biotechnol Bioeng* 56:656–670.

Bow, S.-T. 2002. *Pattern Recognition and Image Preprocessing*. 2nd ed., New York: Marcel Dekker.

Branda, S., S. Vik, L. Friedman, and R. Kolter. 2005. Biofilms: The matrix revisited. *Trends Microbiol* 13:20–6.

Costerton, J., Z. Lewandowski, D. Caldwell, D. Korber, and M. Lappin-Scott. 1995. Microbial biofilms. *Annu Rev Microbiol* 49:711–745.

Gorur, A., D. Jaramillo, J. Costerton, P. Webster, C. Schaudinn, and G. Carr. 2009. Imaging of endodontic biofilms by combined microscopy (fish/clsm-sem). Technical Report.

Gotz, F. 2002. Stapylococcus and biofilms. *Mol Microbiol* 43:1367–1378.

Hall-Stoodley, L., J. Costerton, and P. Stoodley. 2004. Bacterial biofilms: From the natural environment to infectious disbases. *Nat Rev Microbiol* 2:95–108.

Hall-Stoodley, L., and P. Stoodley. 2009. Evolving concepts in biofilm infections. *Cell Microbiology* 11(7):1034–1043.

Heydorn, A., A. Nielsen, M. Hentzer, C. Sternberg, M. Givskov, B. Ersboll, and S. Molin. 2000. Quantification of biofilm structures by the novel computer program COMSTAT. *Microbiology* 146:2395–2407.

Johnson, L. 2008. Microcolony and biofilm formation as a survival strategy for bacteria. *J Theor Biol* 251:24–34.

Jorgensen, T., J. Haagensen, C. Sternberg, and S. Molin. 2003. Quantification of biofilm structures from confocal imaging. Technical report, Optics and Fluids Department, Riso National Laboratory.

Kaplan, J. 2010. Biofilm dispersal: Mechanisms, clinical implications, and potential therapeutic uses. *J Dent Res* 89(3):205–218.

Klapper, I. 2004. Effect of heterogeneous structure in mechanically unstressed biofilms on overall growth. *Bull Math Biol* 66:809–824.

Lewandowski, Z., S. Altobelli, P. Majors, and E. Fukushima. 1992. NMR imaging of hydrodynamics near microbially colonized surfaces. *Water Sci Technol* 26:577–584.

Lorensen, W., and H. Cline. 1987. Marching cubes: A high resolution 3D surface construction algorithm. *SIGGRAPH Comput Graph* 21(4):163–169.

Maulik, U., and S. Bandyopadhyay. 2002. Performance evaluation of some clustering algorithms and validity indices. *IEEE Trans Pattern Anal Mach Intell* 24:1650–1655.

McBain, A. 2009. In vitro biofilm models: An overview. *Adv Appl Microbiol* 69:99–132.

Merod, R., J. Warren, H. McCaslin, and S. Wuertz. 2007. Toward automated analysis of biofilm architecture: Bias caused by extraneous confocal laser scanning microscopy images. *Appl Environ Microbiol* 73:4922–4930.

Mueller, L., J. Almeida, J. de Brouwer, L. Stal, and J. Xavier. 2006. Analysis of a marine phototrophic biofilm by confocal laser scanning microscopy using the new image quantification software PHLIP. *BMC Ecol* 6:1–15.

Otsu, N. 1979. A threshold selection method from gray-level histograms. *IEEE Trans Syst Man Cybern* 9:62–66.

Rueda, L. 2008. An efficient algorithm for optimal multilevel thresholding of irregularly sampled histograms. In *7th International Workshop on Statistical Pattern Recognition (S+SSPR 2008)*, volume LNCS 5432, 612–621. Orlando, FL: Springer.

Russ, J. 2007. *The Image Processing Handbook*. 5th ed., CRC Press. Boca Raton, FL: Taylor & Francis Group.

Stewart, P., and J. Costerton. 2001. Antibiotic resistance of bacteria in biofilms. *Lancet* 358:135–138.

Stewart, P., and M. Franklin. 2008. Physiological heterogeneity in biofilms. *Nat Rev Microbiol* 6:199–210.

Sunner, J. A., I. B. Beech, and K. Hiraoka. 2005. Microbe-surface interactions in biofouling and biocorrosion processes. *Int Microbiol* 8:157–168.

Theodoridis, S., and K. Koutroumbas. 2006. *Pattern Recognition*. 2nd ed. Elsevier Academic Press.

Tolle, C., T. McJunkin, and D. Stoner. 2003. Mapper: A software program for quantitative biofilm characterization. Technical report, Idaho National Engineering and Environmental Laboratory.

Unnikrishnan, R., C. Pantofaru, and M. Hebert. 2007. Toward objective evaluation of image segmentation algorithms. *IEEE Trans Pattern Anal Mach Intell* 29:929–944.

Xu, C., D. Pham, and J. Prince. 1998. A survey of current methods in medical image segmentation. Technical report, Department of Electrical and Computer Engineering.

Yang, X., H. Beyenal, G. Harkin, and Z. Lewandowski. 2000. Quantifying biofilm structure using image analysis. *J Microbiol Methods* 39:109–119.

Yang, X., H. Beyenal, G. Harkin, and Z. Lewandowski. 2001. Evaluation of biofilm image thresholding methods. *Water Sci Technol* 35:1149–1158.

Yang, S., and Z. Lewandowsky. 1995. Measurement of local mass transfer coefficient in biofilms. *Biotechnol Bioeng* 48:737–744.

Zhang, Y. 1996. A survey on evaluation methods for image segmentation. *Pattern Recognit* 29:1335–1346.

Chapter 14

Discovering Association of Diseases in the Upper Gastrointestinal Tract Using Text Mining Techniques

S. S. Saraf, G. R. Udupi, and Santosh D. Hajare

Contents

14.1 Introduction

The medical field has seen a rapid rise (Lyman and Varian 2003) in the collection of data since the introduction of computers and information management systems. Data collection took place initially on paper media; it has now matured to electronic health record format. The need to collect data and maintain it in proper formats has led to specifications for electronic health records. Standards for the maintenance of electronic health records (Kwak 2005; Huser and Rocha 2007; Barretto, Warren, and Goodchild 2004) have been suggested and are being constantly upgraded (Lopez-Nores et al. 2009; De Potter et al. 2009; Arguello et al. 2009; Marinos, Marinos, and Koutsouris 2003; Hoerbst and Ammenwerth 2009; Becker and Sewell 2004).

The data collected is in various forms beginning with text-structured and -unstructured forms generated by the process of taking the medical history of a patient. Numerical values are generated by pathological results and physical observations. Grayscale and color images are generated by computed tomography (CT) scans, magnetic resonance imaging (MRI) scans (Wu-fan 2008; McAuliffe et al. 2001; Hui et al. 2005), and endoscopic observations (Liedlgruber and Uhl 2009). Video is generated by temporal observations of a scan and as a result of endoscopic observations (Cao et al. 2007). Therefore, the process of data collection has led to a multidisciplinary research that aids diagnosis.

Medical diagnosis is the art of determining the pathological status of a person based on the available set of findings. The efficacy of the process reflects the experience and the exposure of an expert. Data mining techniques are useful to discover patterns and knowledge for from the large amount of medical data (Wu 2004), initially, data mining work was concentrated on extracting information from databases (structured data). The process of discovering patterns in unstructured text led to text mining (Luo 2008).

In this chapter, we present the application of data/text mining (Qiu, Zhang, and Song 2009; Qiu and Ge 2007; Yu and Li 2009) to discover patterns in the observations made during an endoscopy of the upper gastrointestinal (GI) tract. The upper GI tract includes the following organs: esophagus, stomach, and duodenum. The observations that can be made and the diseases that can be discovered from a GI examination form a rich source of data for knowledge discovery.

The chapter is organized as follows: We briefly introduce the upper GI tract and the observations on the condition of its organs, which forms the basis for a controlled vocabulary for text mining, in Section 14.2. We introduce the process of text mining and knowledge discovery in Section 14.3. The application of knowledge discovery to the domain is discussed in Section 14.4, and we interpret results from Section 14.4 and similar implementations in Section 14.5. We conclude with the summary in Section 14.6.

14.2 Upper Gastrointestinal Tract

The upper GI tract includes the esophagus, stomach, and duodenum, as shown in Figure 14.1. The tract is called the "food pipe" in common language. The stomach is an organ in which the food is churned and prepared for proper absorption, which occurs in the later stages of digestion. The pH of the stomach is in the range of 1–3.

Under normal circumstances, food travels down the esophagus to the stomach because a valve-like action at the junction of the esophagus and the stomach called the "lower esophageal sphincter" (LES) allows the food to pass. If the LES opens spontaneously for varying periods of time allowing the contents of the stomach to enter the esophagus, then gastroesophageal reflux (GER) occurs. GER is often the cause of heartburn in patients. This manifests as *esophagitis*, a condition of erosion of esophageal mucosa due to exposure to the acidic contents of the stomach. Esophagitis is broadly classified as either erosive or nonerosive esophagitis. The specific classifications are the Savary–Miller classification (Savary and Miller 1978) and the Los Angeles classification (Wong et al. 2008). Savary–Miller classification involves observation of mucosal breaks in the esophagus and classification is done based on the extent of mucosal breaks, whereas Los Angeles classification deals with observation of mucosal breaks over multiple folds of the esophagus and the extent of the break over the circumference of the esophagus. Prolonged esophagitis may lead to *esophageal ulcers* (Allende and Yerian 2009) and bleeding (Laine and Peterson 1994). *Esophageal varices* is a condition in which submucosal veins in the lower esophagus are extremely dilated. This condition occurs because

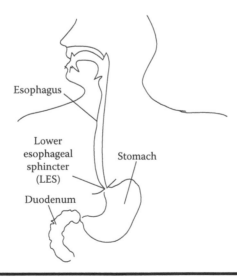

Figure 14.1 Upper gastrointestinal tract.

of *portal hypertension*, which is commonly due to cirrhosis. The patients are prone to bleeding. In cases where the LES is lax (loose), part of the stomach protrudes into the esophagus; this condition is called *hiatus hernia*.

The stomach is the churning machine in the food pipe. It is divided into four sections: (1) the cardia where the contents of the esophagus empty into the stomach; (2) the fundus, which is formed due to upper curvature of the organ; (3) the corpus body or the central region; and (4) the pylorus or antrum, which is the lower section of the stomach that connects it to the small intestine. The stomach lining is rich in secreting glands, and the food is subjected to various chemicals that enhance the absorption and motility of the stomach. Diseases of the stomach (Koch and Stern 1996) occur due to a process of loosening of the cardia, called *lax cardia*, which can cause esophagitis. *Gastritis* is an inflammation of the lining of the stomach, which manifests as pain in the upper abdomen; the condition is mainly caused by excessive alcohol consumption or the action of anti-inflammatory drugs. *Antral gastritis* is inflammation of the antrum. This inflammation affects the process of digestion. One of the major causes of infection in the stomach is *Helicobacter pylori* bacteria. In advanced stages, *H. pylori* leads to cancer of the stomach (CA stomach).

The duodenum is the first part of the small intestine; inside it, food is further broken down using enzymes. The disease conditions of the duodenum (Joffe, Lee, and Blumgart 1978) can be *duodenal ulcer*, usually caused by an infection of *H. pylori* bacteria. *Duodenal polyps* are lesions of gastric mucosa.

In this section, we have briefly highlighted the diseases of the upper GI tract. The terms indicated in italics are the terms used in text mining.

14.3 Text Mining and Knowledge Discovery

Due to computerization of data collection and the use of information management systems, algorithms have replaced manual data study for the analysis of data. As the amount of data collected has increased in size and form, data mining algorithms (Fernandez et al. 2002) have become important tools for sifting through huge amounts of data and identifying valid, potentially useful, and understandable patterns in data.

Text mining (Hearst 1999) attempts to glean meaningful information from natural language text. Since text is unstructured in nature, it does not easily yield to algorithmic approaches. Text mining algorithms are mainly designed assuming that a document is a collection of words with rules. This model is called the "bag of words" model.

Text mining is applied to perform the following tasks:

- Text summarization
- Document retrieval
- Text categorization

- Document clustering
- Language identification
- Identifying key phrases
- Entity extraction
- Information extraction

We will discuss text summarization, document retrieval, text categorization, and information extraction in Sections 14.3.1 through 14.3.4; these processes are relevant to the topic of our discussion. The other processes are briefly described at the end of Section 14.3.3.

14.3.1 Text Summarization

Text summarization is a process in which a summary of the text is produced. It also amounts to some sort of compression of text when a meaningful summary is extracted. The summaries are further classified as extract, abstract, indicative abstract, and critical abstract; each summary is an attempt to refine the presented text to generate a concise summary that justifiably covers the topic.

Summarization approaches are usually divided into two categories: (1) text extraction and (2) text abstraction. Text abstraction involves gleaning through the original text with all the linguistic and semantic constraints of the language applied to it and producing a document, which contains a similar theme as the text and conveys the intent with which the document has been written. Text extraction identifies relevant passages in one or more documents. The extraction process may be performed using statistical measures combined with linguistic constraints. The passages are combined to form the summary of the document.

The criterion for a good summary is that it must have a coherent and cohesive text, which effectively communicates the intent of the document. Coherence of the summary is constrained by the compression factor, that is, the number of words. It is usually seen that the summary tends to become incoherent as the compression ratio rises. We require a structure that focuses on identifying and maintaining the semantic chains in the text.

One of the examples of summarization is the SweSum summarization engine. Originally designed for Swedish language, SweSum has been adapted for English. The engine generates good summaries at 70% compression rates. SweSum works in three different phases (Figure 14.2): In the first phase, tokenization and keyword extraction is performed. In the second phase, ranking of sentences is performed, and, in the final phase, the summary is produced.

The aim of tokenization is to split the text into sentences. The tasks are constrained by the punctuation marks and abbreviations used in the text. The tokenization process refers to the language lexicon to prevent false detection of text. Topic detection is performed on sentences and based on the set criteria. The process involves searching for sentences that relate to the title of the document and the

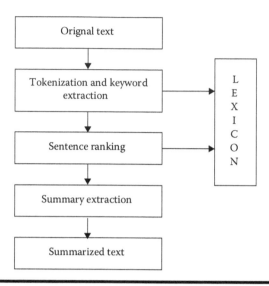

Figure 14.2   SweSum architecture.

keywords. The process refers to the dictionary to find the occurrences of words and topics. A score is assigned to a sentence based on user-defined criteria. The inclusion of that sentence in the summary is based on the score assigned to it.

Text summarization in the medical domain is a challenging task as the information is presented in various forms, like scientific articles, electronic medical records, Web documents, e-mailed reports, semistructured databases, X-ray images, and endoscopy video. Lexicon development for the medical domain is a task in itself. Text summarizers for the medical domain have to take into account the various forms of information and an authentic lexicon.

14.3.2 Document Retrieval

This process involves searching for documents containing some or all the words presented as a query for document search and retrieval. A query is expressed as a set or as a Boolean combination of words and phrases. The relevance-ranking process ensures that retrieval is performed efficiently.

Document retrieval systems are based on how matching and ranking are conducted. Some of the prevalent models are vector space, Boolean, probabilistic, and language modeling.

The document retrieval process is initiated by a user query to a search and fetch program. The process is illustrated in Figure 14.3.

The document retrieval process involves offline and online processes. The offline process comprises the collection and processing of documents for proper

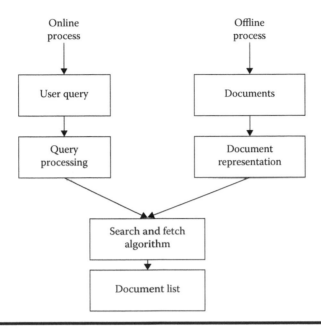

Figure 14.3   Document retrieval process.

representation so that they can be effectively retrieved. Document processing includes document normalization, which involves getting different types of documents to one format. For the next step, the document is divided into retrievable chunks. This process may help to build and index the document. The chunks in the document that can be indexed are identified in the next step. This step is important because it determines the retrieval efficiency of the system. An inverted file is produced, which contains a stored array of all units that can be indexed with information about the author and organization.

The query-processing step transforms the user query into a format acceptable to the search and fetch algorithm. The query is broken down into keywords, special symbols eliminated, and a query representation generated. Once the query representation is generated, it is presented to the search and fetch algorithm that examines the inverted files and calculates the score of the document. From this score, relevance is determined. The results are presented based on the relevance factors.

Document retrieval in the medical domain becomes relevant when a physician desires to seek answers from the medical literature. The documents extracted are in large numbers, making the task of referring tedious and time consuming. The relevance calculation determines the effectiveness of the retrieval. Systems with document retrieval are coupled with summarizers to present the summary of the retrieved documents reported in the literature.

14.3.3 Text Categorization

Text categorization is a process in which documents are classified to specific categories based on their content. The categories are predefined by a controlled vocabulary. The process involves extracting features from the documents with known categories, and machine learning techniques are used to assign a category to a document based on the extracted features. A controlled vocabulary that evolves, which could be standardized, results from this process.

A text categorization system (Figure 14.4) is an implementation of a classifier that assigns specific categories to documents. The process comprises two steps: (1) feature determination and (2) training a classifier for the classification process. The document is preprocessed by removing the stop words and transforming the document to an acceptable feature set, leading to an attribute-value representation of the text. Each word corresponds to a feature with the number of occurrences being the feature value, which leads to a large-dimensional feature space that hinders the classifier training process. The feature values are represented based on the information-gain approach, and features are represented with their inverse document frequency (idf), which is the most commonly used feature in the information retrieval field.

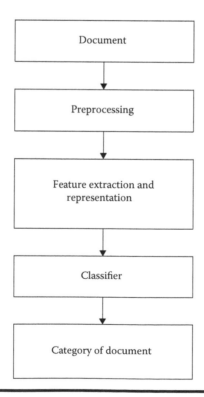

Figure 14.4 Steps in text categorization.

With the feature vectors in place, classifiers like k-nearest neighborhood (k-NN), neural networks, fuzzy k-NN, and support vector machines (SVMs) can be trained. The documents can be categorized based on the feature set. This sets the platform for "document clustering," wherein the feature set can be subjected to unsupervised clustering.

"Language identification" provides important information that can be stored as metadata for documents. Identification of key phrases helps to determine the possible class of a document. Key phrases serve as important metadata for the classification of documents. Key phrases can be from the controlled vocabulary. This process can be used to train classifiers to classify a document. Entity extraction is one of the important tasks of text mining. During entity extraction, the algorithm searches for structured information in the text. The entities can be names, e-mail addresses, etc.

14.3.4 Information Extraction

Once entity extraction is completed, relationships between the entities can be determined, which leads to knowledge. The knowledge could be a known fact or a new relationship, in which case, it is knowledge discovery. Various methods are used to determine entity relationships. One of them is a co-occurrence matrix that takes the occurrence frequencies of predefined phrases into account. The analysis of the matrix enables researchers to glean knowledge about the entities involved.

The process of relationship determination between entities, or knowledge discovery, as applied in our case study is described here. The document can be partitioned into m entities (E_1, E_2, E_3, ..., E_i, ..., E_m), and each entity will have a description or a set of n phrases ($P_{i,1}$, $P_{i,2}$, $P_{i,3}$, ..., $P_{i,j}$, ..., $P_{i,n}$) associated with it. The relationship can be expressed as a co-occurrence matrix that gives the occurrence frequency of the phrases. The co-occurrence matrix is shown in Table 14.1. Analyzing the matrix values (A_{ij}) will give the relationship between phrases and entities.

In the case of medical reports, the entities are organs and the phrases are observations of the organ status. The total frequency of the occurrence of a particular phrase is X, and the independent occurrence is Y. Then, the dependent occurrence Z is the

Table 14.1   Co-Occurrence Matrix

	$P_{i,1}$	$P_{i,2}$	$P_{i,3}$	...	$P_{i,n}$
$P_{i,1}$	A_{11}	A_{12}	A_{13}	...	A_{1n}
$P_{i,2}$	A_{21}	A_{22}	A_{23}	...	A_{2n}
$P_{i,3}$	A_{31}	A_{32}	A_{33}	...	A_{3n}
...	...	...	...	...	...
$P_{i,n}$	A_{i1}	A_{i2}	A_{i3}	...	A_{in}

difference of X and Y, indicating that if the independent value Y is lower the observations of one organ (entity) is dependent on the other organ (entity). We might get a case in which the dependent value is almost zero, indicating that the observation is not influenced by any other organ (entity).

The patterns could serve as knowledge for decision-making tasks. The process of knowledge discovery in a database (KDD; Kalavathy, Suresh, and Akhila 2007; Shadabi and Sharma 2008; Pazzani 2000 Gouda and Cheng 1998) is illustrated in Figure 14.5. We explain the various steps involved in the process of KDD here:

The first step is selection. During selection, a particular area is evaluated and information is segregated as per the requirements. In our case, we separated the reports of upper GI tract endoscopy.

The second step is preprocessing. In this step, unwanted data is removed; hence, this stage is also called the "data-cleaning" phase. It is important to check the validity of data, identify terms, and correct spelling mistakes in the reported diagnosis.

The third step is transformation. This process involves transforming the data into a form that is suitable for the data mining task. In our case, we created a database

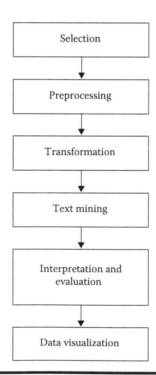

Figure 14.5 Stages of knowledge discovery in database.

with each field of memo data type capable of holding unstructured text. We created five fields: (1) esophagus, (2) stomach, (3) duodenum part I, (4) duodenum part II, and (4) final diagnosis.

The fourth step is data mining. This stage is concerned with extracting patterns from the data. The patterns can be found by unsupervised means or fine-tuned with a supervised process in which we provide the patterns for searching.

The fifth step is interpretation and evaluation. During this stage, rules that can be converted to knowledge to aid the decision-making process are determined.

The sixth step is data visualization. During data visualization, the data is made understandable for the analyst. Visual maps like dendograms can be constructed, and co-occurrence matrices may be displayed.

14.4 Application of Knowledge Discovery in a Database to the Domain

The application domain is used to find relationships between diseases occurring in the upper GI tract. The sources of information are reports of upper GI endoscopy. The process is highlighted in Figure 14.6.

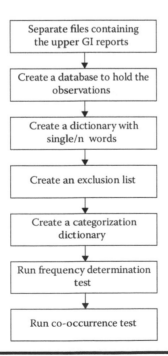

Figure 14.6   Flow chart of the domain application.

We explain the process in detail as follows:

Step 1: Separate files containing reports of upper GI tract endoscopy. The report format is shown in Figure 14.7.
The files containing the line Gatro-Duodenoscopy Report should be separated.

Step 2: Create a database, with five fields corresponding to four organs and the diagnosis, to hold the observations as shown in the format of the report and the endoscopic diagnosis. Each field should hold the description of the respective organ.

Step 3: Create a dictionary with single words and n ($n > 1$) words to make a meaningful combination of words.

Step 4: Create an exclusion list, the words like "the," "of," and "a" are included in this list. These words are not included in the content analysis.

Step 5: Create a category of words that indicate the diseases of the organs in the upper GI tract. This step is also called a categorization dictionary. The words included in the dictionary are listed in Table 14.2.

Step 6: The frequency test gives the number of occurrences of the words from the categorization dictionary in the upper GI tract endoscopy reports. The frequencies are listed in Table 14.3.

GASTRO-DUODENOSCOPY REPORT

OESOPHAGUS
E/O EROSIVE OESOPHAGITIS
NO ULCER/STRICTURE/GROWTH/VARICES
O.G.Jt AT 38 Cms.

STOMACH
E/O EROSIVE ANTRAL GASTRITIS
NO ULCER/STRICTURE/GROWTH/VARICES
FUNDAL VIEW-NO VARICE
E/O LAX CARDIA

DUODENUM PART I
NORMAL
NO ULCER/DEFORMITY/DUODENITIS

DUODENUM PART II
NORMAL

ENDOSCOPIC DIAGNOSIS
LAX CARDIA WITH EROSIVE OESOPHAGITIS
EROSIVE ANTRAL GASTRITIS

Figure 14.7   Format of the upper GI tract endoscopy report.

Table 14.2   List of Words in the Categorization Dictionary

Item No.	Words in the Categorization Dictionary
1	ANTRAL_GASTRITIS
2	CARCINOMA_STOMACH
3	DIFFUSE_EROSIVE_GASTRITIS
4	DIFFUSE_GASTRIC_EROSIONS
5	DUODENAL_EROSIONS
6	DUODENAL_POLYP
7	DUODENAL_ULCER
8	EROSIVE_ANTRAL_GASTRITIS
9	EROSIVE_GASTRITIS
10	EROSIVE_OESOPHAGITIS
11	GASTRIC_POLYP
12	GASTRIC_ULCER
13	HAEMORRHAGIC_GASTRITIS
14	HIATUS_HERNIA
15	LAX_CARDIA_WITH_EROSIVE_OESOPHAGITIS
16	LINEAR_EROSIVE_OESOPHAGITIS
17	MILDLY_LAX_CARDIA_WITH_NON_EROSIVE_OESOPHAGITIS
18	NON_EROSIVE_OESOPHAGITIS
19	OESOPHAGEAL_CANDIDIASIS
20	OESOPHAGEAL_VARICES
21	PORTAL_HYPERTENSION
22	ULCERATIVE_OESOPHAGITIS

Step 7: The co-occurrences test gives important information that relates the terms in the categorization dictionary. Since the words relate to the diseases of the organs involved in the upper GI tract, it gives an insight into the relationship and possible associations of diseases. The co-occurrence matrix is displayed in Table 14.4. The item number in Table 14.3 corresponds to the words shown in Table 14.2.

Table 14.3 Frequency of Occurrences of Words from the Categorization Dictionary

Item No.	Words from the Categorization Dictionary	Frequency
1	ANTRAL_GASTRITIS	54
2	CARCINOMA_STOMACH	13
3	DIFFUSE_EROSIVE_GASTRITIS	150
4	DIFFUSE_GASTRIC_EROSIONS	38
5	DUODENAL_EROSIONS	44
6	DUODENAL_POLYP	3
7	DUODENAL_ULCER	13
8	EROSIVE_ANTRAL_GASTRITIS	130
9	EROSIVE_GASTRITIS	7
10	EROSIVE_OESOPHAGITIS	69
11	GASTRIC_POLYP	3
12	GASTRIC_ULCER	14
13	HAEMORRHAGIC_GASTRITIS	14
14	HIATUS_HERNIA	12
15	LAX_CARDIA_WITH_EROSIVE_ OESOPHAGITIS	207
16	LINEAR_EROSIVE_OESOPHAGITIS	3
17	MILDLY_LAX_CARDIA_WITH_NON_ EROSIVE_OESOPHAGITIS	11
18	NON_EROSIVE_OESOPHAGITIS	48
19	OESOPHAGEAL_CANDIDIASIS	20
20	OESOPHAGEAL_VARICES	200
21	PORTAL_HYPERTENSION	32
22	ULCERATIVE_OESOPHAGITIS	10

Table 14.4 Co-ocurrence Matrix

Item No.	1	2	3	4	5	6	7	8	9	10	11	12	13	14	15	16	17	18	19	20	21	22
1	54	0	0	0	0	0	0	0	0	8	0	0	0	1	9	0	1	3	1	0	7	0
2	0	13	0	0	0	0	1	0	0	2	0	0	0	0	0	0	0	0	0	0	0	3
3	0	0	150	0	8	2	1	0	0	15	0	1	0	3	49	1	2	9	4	1	4	1
4	0	0	0	38	6	0	1	0	0	2	0	1	0	0	17	0	0	1	0	1	0	0
5	0	0	8	6	44	0	1	6	0	10	0	1	1	4	10	0	0	0	0	1	0	1
6	0	0	2	0	0	3	0	0	0	0	0	0	0	0	1	0	0	0	0	1	0	0
7	0	1	1	1	1	0	13	1	0	2	0	0	0	0	2	0	0	0	0	1	0	0
8	0	0	0	0	6	0	1	132	0	6	1	0	0	0	28	0	1	13	2	2	7	1
9	0	0	0	0	0	0	0	0	7	0	0	0	0	0	2	0	0	1	0	0	0	0
10	8	2	15	2	10	0	2	6	0	69	0	0	1	8	1	0	0	0	4	0	1	0
11	0	0	0	0	0	0	0	1	0	0	3	0	0	0	0	0	1	0	0	0	0	0

(Continued)

Table 14.4 Co-ocurrence Matrix (Continued)

Item No.	1	2	3	4	5	6	7	8	9	10	11	12	13	14	15	16	17	18	19	20	21	22
12	0	0	1	1	1	0	0	0	0	0	0	14	0	0	1	0	0	0	0	3	1	0
13	0	0	0	0	1	0	0	0	0	1	0	0	14	0	4	0	0	1	0	0	0	0
14	1	0	3	0	4	0	0	0	0	8	0	0	0	12	0	1	0	0	2	0	0	2
15	9	0	49	17	10	1	2	28	2	1	0	1	4	0	207	0	0	0	0	1	7	0
16	0	0	1	0	0	0	0	0	0	0	0	0	0	1	0	3	0	0	0	0	0	0
17	1	0	2	0	0	0	0	1	0	0	1	0	0	0	0	0	11	0	0	0	1	0
18	3	0	9	1	0	0	0	13	1	0	0	0	1	0	0	0	0	48	0	0	1	0
19	1	0	4	0	0	0	0	2	0	4	0	0	0	2	0	0	0	0	22	3	1	1
20	0	0	1	1	1	1	1	2	0	0	0	3	0	0	1	0	0	0	3	200	0	0
21	7	0	4	0	0	0	0	7	0	1	0	1	0	0	7	0	1	1	1	0	32	0
22	0	3	1	0	1	0	0	1	0	0	0	0	0	2	0	0	0	0	1	0	0	10

14.5 Discussion

The test was run on 900 cases. In this section, we focus on the relationships based on the co-occurrence matrix shown in Table 14.4. An important relationship, which can be derived, is between words occurring independently and words occurring in association (dependent) with other words. This is illustrated in Table 14.5.

Table 14.5   Dependent and Independent Occurrences of Words in the Categorization Dictionary

Item No.	Total	Dependent Occurrences	Independent Occurrences
1	54	30	24
2	13	6	7
3	150	101	49
4	38	29	9
5	44	35	9
6	3	2	1
7	13	10	3
8	132	68	64
9	7	3	4
10	69	60	9
11	3	2	1
12	14	8	6
13	14	7	7
14	12	10	2
15	207	132	75
16	3	2	1
17	11	6	5
18	48	29	19
19	22	18	4
20	200	14	186
21	32	30	2
22	10	9	1

It can be seen from row 20 of Table 14.5 that the independent occurrence is 93%. This number corresponds to esophageal varices, a condition caused by portal hypertension, due to cirrhosis. Cirrhosis is a chronic liver disease that results in loss of liver function.

Another example can be seen from row 3 (diffuse erosive gastritis). In this case, the dependent cases are 67.33%. In this 63.76%, 48% dependence is on item number 15 (lax cardia with erosive esophagitis) from the co-occurrence matrix.

The associations form the probability relationships between the diseases of organs in the upper GI tract. The limitation of this categorization model is the limitation of the vocabulary, which is generated out of the 900 cases.

Some implementations of text mining in the medical field are as follows:

The text mining of clinical records for cancer diagnosis (Lee, Chih-Hong, and Hsin-Chang 2007) describes text mining of reports to identify relationships between cancer and external factors from medical records in order to support a diagnosis. The implementation is in three phases: (1) The first phase is ontology-based keyword extraction. These keywords are used as index terms to encode the documents. (2) In the second phase, text clustering in unsupervised mode is performed to build self-organizing maps (SOMs). These maps lead to document cluster maps. (3) A SVM is trained to categorize the document in the third phase.

Another implementation (Claster, Shanmuganathan, and Ghotbi 2007) reads the reports of a CT scan and examines the usefulness of the CT scan advice to a patient. The method constructs a SOM based on the term frequency (tf) and idf. A neural network is trained to perform the categorization process.

In a study by McCowan et al. (2007), text classification techniques were used to train SVMs to extract elements of a stage listed in cancer-staging guidelines. The system uses staging data and pathological reports for diagnosing lung cancer patients.

14.6 Conclusion

With the medical field moving from the knowledge-intensive scenario to a data-intensive scenario, a shift that is possible due to the proliferation of data-collecting facilities to every bedside, the data mining and text mining techniques help to discover knowledge that aids decision-making structures. In this chapter, we report a case study based on the reports of upper GI tract endoscopy. The process of text mining, especially for information regarding diseases, shows the probability relationships between the diseases of organs. The efficacy of the results is limited by the size and quality of the categorization dictionary. The keywords extracted can be linked with the images or video features assisting efficient content-based retrieval of multimedia.

References

Allende, D. S., and L. M. Yerian. 2009. Diagnosing gastroesophageal reflux disease: The pathologist's perspective. *Adv Anat Pathol* 16:161–4.

Arguello, M., J. Des, R. Perez, M. Fernandez-Prieto, and H. Paniagua. 2009. Electronic Health Records (EHRs) standards and the semantic edge: A case study of visualising clinical information from EHRs. In eds. D. Al-Dabass, A. Orsoni, A. Brentnall, A. Abraham, and R. Zobel, 485–90. *UKSIM '09: Proceedings 11th International Conference on Computer Modeling and Simulation, March 25–27, 2009. Cambridge, UK,* IEEE Computer Society.

Barretto, S., J. Warren, and A. Goodchild. 2004. Designing guideline-based workflow-enabled electronic health records. In ed. Ralph H. Sprague, Jr. *Proceedings 37th Annual Hawaii International Conference on System Sciences, 5–8 January, 2004, Big Island, Hawaii,* IEEE Computer Society, Los Alamitos, California, 135–45.

Becker, M., and P. Sewell. 2004. Cassandra: Flexible trust management, applied to electronic health records. In IEEE Computer Society Technical Committee on Security and Privacy, *Proceedings 17th IEEE Computer Security Foundations Workshop, June 28–30, 2004, Pacific Grove, California,* IEEE Computer Society, Los Alamitos, California, 139–54.

Cao, Y., D. Liu, W. Tavanapong, J. Wong, J. Oh, and P. de Groen. 2007. Computer-aided detection of diagnostic and therapeutic operations in colonoscopy videos. *IEEE J Biomed Eng* 54:1268–79.

Claster, W., S. Shanmuganathan, and N. Ghotbi. 2007. Text mining in radiological data records: An unsupervised neural network approach. In eds. D. Al-Dabass, R. Zobel, A. Abraham, S. Turner. *Proceedings First Asia International Conference on Modelling and Simulation, 27–30 March, 2007, Phuket, Thailand,* IEEE Computer Society, Los Alamitos, California 329–33.

De Potter, P., P. Debevere, E. Mannens, and R. Van de Walle. 2009. Next generation assisting clinical applications by using semantic-aware electronic health records. In Institute of Electrical and Electronics Engineers, *CBMS 2009 Proceedings 22nd IEEE International Symposium on Computer-Based Medical Systems CBMS 2009, 2–5 Aug. 2009, Albuquerque, NM,* Institute of Electrical and Electronics Engineers, 1–5.

Fernandez, C., J. Martinez, A. Wasilewska, M. Hadjimichael, and E. Menasalvas. 2002. Data mining - a semantic model. In Institute of Electrical and Electronics Engineers, *Proceedings of the 2002 IEEE International Conference on Fuzzy Systems, 2002. FUZZ-IEEE '02, May 12–17, 2002, Honolulu, Hawaii,* vol. 2, 938–43.

Gouda, K., and J. Cheng. 1998. Using relevant reasoning to solve the relevancy problem in knowledge discovery in databases. In IEEE Systems, Man, and Cybernetics Society, *Proceedings IEEE International Conference on Systems, Man, and Cybernetics, August 8–12, 1988, Beijing and Shenyang,* International Academic Publishers, vol. 2, 1473–8.

Hearst, M. A. 1999. Untangling text mining. In *Proceedings Annual Meeting of the Association for Computational Linguistics ACL99, 20–26 June 1999,* University of Maryland, Maryland, USA, 3–10.

Hoerbst, A., and E. A. Ammenwerth. 2009. Structural model for quality requirements regarding electronic health records - state of the art and first concepts. In Institute of Electrical and Electronics Engineers, *Proceedings ICSE Workshop on Software Engineering in Health Care SEHC '09, 18–19 May 2009, Vancouver, British Columbia, Canada,* Institute of Electrical and Electronics Engineers, 34–41.

Hui, D., W. Guangzhi, H. Bo, Z. Yiyi, Y. Zhi, M. Meng, and G. ShangKai. 2005. Construction of a knowledge center for medical image processing. In Institute of Electrical and Electronics Engineers, *Proceedings 27th Annual International Conference of the Engineering in Medicine and Biology Society IEEE-EMBS, 1–4-September 2005, Shanghai, China*, Institute of Electrical and Electronics Engineers, 82–5.

Huser, V., and R. Rocha. 2007. Retrospective analysis of the electronic health record of patients enrolled in a computerized glucose management protocol. In Institute of Electrical and Electronics Engineers, *Proceedings Twentieth IEEE International Symposium on Computer-Based Medical Systems CBMS '07, 20–22 June 2007, Maribor*, Institute of Electrical and Electronics Engineers, 503–08.

McCowan, I. A., D. C. Moore, A. N. Nguyen, R. V. Bowman, B. E. Clarke, E. E. Duhig, and M.-J. Fry. 2007. Collection of cancer stage data by classifying free-text medical reports. *J Am Med Inform Assoc* 14:736–45.

Joffe, S. N., F. D. Lee, and L. H. Blumgart. 1978. Duodenitis. *Clin Gastroenterol* 7:635–50.

Kalavathy, R., R. Suresh, and R. Akhila. 2007. KDD and data mining. In Institution of Engineering and Technology, *Proceedings International Conference on Information and Communication Technology in Electrical Sciences (ICTES 2007) IET-UK, 20–22 December 2007, Chennai, India*, Curran Associates, Inc., 1105–10.

Koch, K. L., and R. M. Stern. 1996. Functional disorders of the stomach. *Semin Gastrointest Dis Oct* 7:185–95.

Kwak, Y. S. 2005. International standards for building Electronic Health Record (EHR). In ed. Heung Kook Choi, *Proceedings 7th International Workshop on Enterprise networking and Computing in Healthcare Industry HEALTHCOM 2005, 23–25 June 2005, Busan, Korea*, Institute of Electrical and Electronics Engineers, 18–23.

Laine, L., and W. L. Peterson. 1994. Bleeding peptic ulcer. *N Engl J Med* 331:717–27.

Lee, C.-H., W. Chih-Hong, and Y. Hsin-Chang. 2007. Text mining of clinical records for cancer diagnosis. In ed. J-S. Pan, *Proceedings Second International Conference on Innovative Computing, Information and Control (ICICIC), 5–7 September 2007, Kumamoto, Japan*, Institute of Electrical and Electronics Engineers, 172–5.

Liedlgruber, M., and A. Uhl. 2009. Endoscopic image processing - an overview. In ed. P. Zinterhof, *Proeedings 6th International Symposium on Image and Signal Processing and Analysis ISPA 2009, 16–18 September 2009, Salzburg, Austria*, Institute of Electrical and Electronics Engineers, 707–12.

Lopez-Nores, M., J. Pazos-Arias, J. Garcia-Duque, Y. Blanco-Fernandez, and M. Ramos-Cabrer. 2009. Entering information about medication intake in standard electronic health records from the networked home. In Institute of Electrical and Electronics Engineers, *Digest of Technical Papers International Conference on Consumer Electronics ICCE '09, 10–14 January 2009, Las Vegas, Nevada*, Institute of Electrical and Electronics Engineers, 1–2.

Luo, Q. 2008. Advancing knowledge discovery and data mining. In ed. Q. Luo, *Proceedings First International Workshop on Knowledge Discovery and Data Mining WKDD 2008, 23–24 January 2008, Adelaide, Australia*, Institute of Electrical and Electronics Engineers, 3–5.

Lyman, P., and H. R. Varian. 2003. How much information. Study Report. http://www.sims.berkeley.edu/how-much-info/ (accessed February 25, 2010).

Marinos, G., S. Marinos, and D. Koutsouris. 2003. Towards an XML-based user interface for electronic health record. In Institute of Electrical and Electronics Engineers, *Proc. 25th Annual International Conference of the IEEE Engineering in Medicine and Biology Society, 17–21 Sept. 2003, Cancun, Mexico*, Institute of Electrical and Electronics Engineers, vol. 2, 1402–05.

McAuliffe, M., F. Lalonde, D. McGarry, W. Gandler, K. Csaky, and B. Trus. 2001. Medical image processing, analysis and visualization in clinical research. In Institute of Electrical and Electronics Engineers, *Proceedings 14th IEEE Symposium on Computer-Based Medical Systems CBMS 2001, 26–27 July 2001, Bethesda, MD, USA*, IEEE Computer Society, 381–6.

Pazzani, M. 2000. Knowledge discovery from data? *IEEE Intell Syst Appl* 15:10–12.

Qiu, Y., and J. Ge. 2007. Research and realization of text mining algorithm on web. In ed. Y. Wang, *Proceedings International Conference on Computational Intelligence and Security Workshops CISW 2007, 15–19 December 2007, Harbin, Heilongjiang, China*, Institute of Electrical and Electronics Engineers, 413–6.

Qiu, Y., Y. Zhang, and M. Song. 2009. Text mining for bioinformatics: State of the art review. In ed. W. Li, *Proceedings 2nd IEEE International Conference on Computer Science and Information Technology ICCSIT 2009, 8–11 August 2009, Beijing, China*, Institute of Electrical and Electronics Engineers, 398–401.

Savary, M., and G. Miller. 1978. *Handbook and Atlas of Endoscopy*. Solothurn: Gassmann.

Shadabi, F., and D. Sharma. 2008. Artificial intelligence and data mining techniques in medicine—success stories. In Institute of Electrical and Electronics Engineers, *Proceedings International Conference on BioMedical Engineering and Informatics BMEI 2008, 27–30 May 2008, Sanya, Hainan, China*, Institute of Electrical and Electronics Engineers, 235–39.

Wong, R.-K.M., Y. Khay-Guan, G. Kok-Ann, T. Hui-Wen, and H. Khek-Yu. 2008. Validation of structured scoring using the LA classification or esophagitis and endoscopically suspected Barrett's esophagus in a tertiary Asian endoscopy center. *J Gastroenterol Hepatol* 24:103–6.

Wu, X. 2004. Data mining: Artificial intelligence in data analysis. In Institute of Electrical and Electronics Engineers, *Proceedings IEEE/WIC/ACM International Conference on Intelligent Agent Technology (IAT 2004), September 20–24, 2004, Beijing, China*, Institute of Electrical and Electronics Engineers, 7–14.

Wu-fan, C. 2008. Hotspots on modern medical imaging and image analysis. In Institute of Electrical and Electronics Engineers, *Proceedings 5th International Summer School and Symposium on Medical Devices and Biosensors ISSS-MDBS 2008, 1–3 June 2008, Hong Kong, China*, Institute of Electrical and Electronics Engineers, 8–9.

Yu, L., and Q. Li. 2009. A novel web text mining method based on semantic polarity analysis. In Institute of Electrical and Electronics Engineers, *Proceedings 5th International Conference on Wireless Communications, Networking and Mobile Computing WiCom '09, 24–26 September 2009, Beijing, China*, Institute of Electrical and Electronics Engineers, 1–4.

Chapter 15

Mental Health Informatics: Scopes and Challenges

Subhagata Chattopadhyay

Contents

15.1 Introduction

Effective applications of "information communication technology" (ICT) have largely changed the structure and functions of present-day health-care systems globally (Haux et al. 2001; Suleiman 2001; Westbrook et al. 2009; Klein and Kajbjer 2009). Hence, there are attempts to replace traditional paper-based health care with electronic or digital health care, popularly termed "e-health" by organizations such as the World Health Organization (2005). Most of the developed world uses e-health as the basis of health care because of its obvious benefits. These benefits include secure and effective data sharing over distances, reduction of referral costs, and provision of faster care with improvement in overall quality (Gerkin 2009; Akematsu and Tsuji 2009), amid issues, debates, and arguments (Khoja 2008; Shoaib et al. 2009; Ernstmann et al. 2009; Durani and Khoja 2009). Despite various isolated attempts of implementing e-health (WHO 2005; Kapoor et al. 2004; Chattopadhyay 2010), the developing world is still limping behind due to various social, political, and economic reasons (Chattopadhyay 2009). To understand e-health better, we need to understand its principal structural and respective functional components, which are discussed as follows:

For electronic storage of patient data (e.g., both image and nonimage data), computers with embedded data security and effective data management are important. For facilitation of effective communication through local area network (LAN) or wide area network (WAN), Internet access is necessary. This type of communication introduces fast, secure, and effective data sharing across hospitals and thereby helps in the electronic referral of patients within the network with continuous feedback facility. Patients (especially those with chronic disease conditions) need not travel physically for specialist consultation; therefore, travel time and cost are reduced. Other benefits are that health-care professionals can communicate with their specialist colleagues, discuss critical/complicated cases, and learn from them (indirect learning) apart from direct learning by accessing electronic knowledge materials such as journal databases available on the Web. The Internet helps to enhance and upgrade the physician knowledge base, a process that is popularly called "electronic learning" or "e-learning."

Another important component of e-health is human resources. Trained labor that is able to manage health care through e-health by junior staff training, educating colleagues in handling technology, and building local capacity at organizational levels are a vital part of e-health.

"Medical informatics" (also called health-care informatics) is defined as the application of information technology (IT) in the field of health care (Hersh 2002; Chen et al. 2005); it comprises both the computer and human resources components of e-health, which are described earlier in this section. Under the first component, computer, this term refers to mathematical modeling of medical concepts and data, applications of computational algorithms for data analysis, preserving data with desired security, facilitating effective data sharing through communication technologies across stakeholders, and so forth. On the other hand, it requires human experts to interface the health-care requirements and technology and facilitate user acceptance; hence, human resources are significant components of e-health (Venkatesh et al. 2003). Today, for efficient management of health data, medical informatics has become a necessity in handling raw data, which is grossly unstructured, is subjective in nature, and is growing at exponential rates.

"Knowledge engineering," popularly known as "data mining and knowledge discovery," is defined as applications of mathematical logic and computer algorithms to extract hidden patterns or knowledge from raw medical data (Chen et al. 2005). It is considered to be the heart of medical informatics (Chen et al. 2005). The knowledge thus extracted is incorporated into a control system for medical decision making; this step is the key objective of knowledge engineering, and such a system is popularly known as a "knowledge-based system" (KBS). Here, a knowledge base is nothing but the combination of databases and rule sets. A KBS can also learn from experiments and practice using a learning algorithm. It mimics the way in which human experts make decisions using their state of learning.

Medical doctors learn throughout their career by interacting with patients. Every time they examine a case, they use the IF-THEN rules for diagnosing it. Each new case is then appended to their patient database. The diagnostic rules (correct and incorrect) are similarly added to their rule sets. This is a continuous process and with time leads to diagnostic precision by reducing the error rate. In a KBS, the same procedure is mimicked. However, completely mimicking this task is difficult and is, hence, a cutting-edge research challenge in medical informatics, especially in those domains where the data are highly subjective, for example, mental health. Figure 15.1 shows the interrelations among e-health, medical informatics, knowledge engineering, and medical decision support (i.e., KBS).

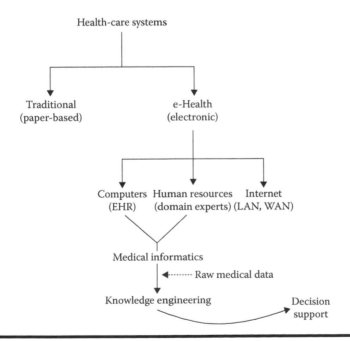

Figure 15.1 Interrelations among e-health, medical informatics, knowledge engineering, and medical decision support.

Section 15.2 gives an overview of various challenges in mental health or psychiatry to determine how the concepts of medical informatics can be beneficial in understanding and analyzing psychiatry data.

15.2 Mental Health: Challenges

Mental health is a specialized domain under neuromedicine, which is again a large field under health sciences. From the medical informatics perspective, the challenges encountered in psychiatry are as follows:

- Highly subjective clinical data (e.g., history, signs, and symptoms)
- Questionable data reliability
- Nonavailability of any direct measuring tool (as found in other domains of medical science, e.g., nephrology, neurology, cardiology, and gastroenterology)
- Bizarre presentations in patients (e.g., history of onset and progression of illness, signs and symptoms reflecting the severity of the illness)
- Varied interpretations of these presentations by clinicians based on their individual medical logic
- Underdiagnosis or overdiagnosis that tends to increase patients' sufferings in terms of loss of time and money

Psychiatry data are subjective and have questionable reliability (Kaplan et al. 1994). Often, the information given by patients (i.e., the primary data) cannot be considered, as they are mentally ill. On the other hand, their relatives sometimes provide inadequate or exaggerated information (i.e., the secondary data). Unfortunately, there is no direct measuring tool available in mental health diagnosis, and the assessment scales are interpreted manually and are, therefore, susceptible to human error. The presentation (e.g., history, signs, and symptoms) of any illness is also ambiguous as patients suffering from similar diseases often present differently (e.g., schizophrenia and schizophreniform disorders under psychotic disorders). On the other hand, there are other possibilities such as the existence of look-alike diseases (e.g., depression and psychosis) and combinations of various diseases behind a group of presentations (e.g., psychotic depressions). Another important issue in psychiatry is the prevailing diagnostic ambiguities. Doctors often fail to perceive the actual disease amid multiple look-alike signs and symptoms. Because of their individualized knowledge base (also called medical logic), they often arrive at a wrong diagnosis, an underdiagnosis, or an overdiagnosis of the actual disease. Let us summarize this discussion to understand the issue more clearly:

- A particular disease presents differently in a patient.
- Patients suffering from two or more different illnesses may present similarly.

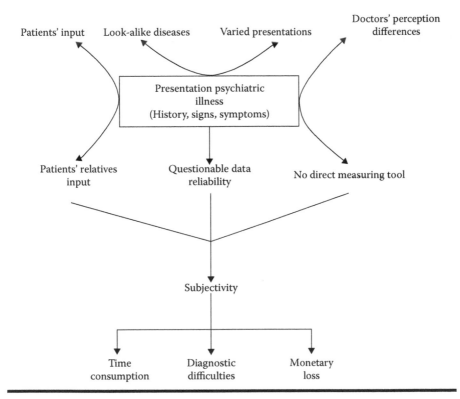

Figure 15.2 The onset and distribution of subjectivities in mental health.

- There may be multiple illnesses behind a set of symptoms.
- The perception of cases based on clinical presentation differs among doctors.

Figure 15.2 shows the hurdles faced in psychiatry.

Given this scenario, this chapter focuses on the critical analysis of various knowledge engineering techniques in psychiatry and attempts to throw light on its current state and future prospects, especially the issues related to realistic implementations. Before doing so, we investigate the basics of medical knowledge engineering in Section 15.3.

15.3 Medical Knowledge Engineering: Core Concepts

Medical knowledge engineering is defined as the application of mathematical logic and computer algorithms to mine interesting hidden patterns from a given raw medical data set. The term raw data indicates unstructured, unformatted, non-standardized, and nonlinear data, which may be extremely complex in nature. Hence, these data formats are difficult to process manually for mining hidden patterns/information/knowledge. It is also important to remember that the patterns extracted from raw data may not necessarily be the desired knowledge until these are conceptualized by the domain experts. In other words, whatever the pattern or information retrieved by the computer methods and algorithms, it should be conceptualized and processed by human experts in an understandable format to produce sharable knowledge. Hence, domain experts play important and rather critical roles in processing the extracted information into understandable and sharable knowledge (Chattopadhyay, 2002). A lack of domain expertise inevitably leads to extraction and sharing of erratic knowledge, which is undesirable in real-world situations.

In health care, raw data is produced in millisecond time stamps and from all sectors, for example, clinics, accounts, pharmacies, and laboratories. Raw data are extracted from clinical data, that is, history, signs, symptoms, laboratory data, pharmacy data, accounts and billing data, data related to hospital occupancy, human resources data, and so forth, under a hospital information system. The focus of this chapter is on clinical data, for example, history, signs, and symptoms, which are highly subjective as found in mental health practice. Therefore, the remaining types of data are not discussed here. Figure 15.3 shows the basic principle of medical knowledge engineering.

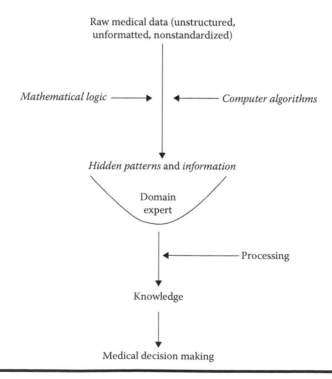

Figure 15.3 A flowchart of medical knowledge engineering.

15.4 Medical Knowledge Management in Psychiatry: Scope

Now, we understand the challenges in mental health that may be addressed by e-health approaches. Section 15.4.1 contains a systematic discussion of various e-health applications in psychiatry for knowledge acquisition, sharing, and extraction, which together may be called "knowledge management."

15.4.1 Knowledge Acquisition and Sharing by e-Mental Health

"Telemedicine" is one of the most important arms of ICT applications in e-health care. Telemedicine is, essentially, health care delivered to remotely situated patient populations, especially those in rural sectors in large countries, using multiway interactive audio–video telecommunications, computers, and telemetry. Telemedicine mitigates the sociocultural barrier of the face-to-face interview of patients (Lee, 2009). The key objective of developing and implementing telemedicine is to establish an effective communication between two hospitals and their health-care staff for day-to-day health-care management, such as rural hospitals or primary health centers (that often lack specialists and specialized infrastructure) with city-based tertiary hospitals equipped with specialists and higher-level infrastructure. By establishing seamless interactive communication, the use of ICT telemedicine aims to cater effective and quality health care to the rural community (Hilty et al. 2009; Alexender and Lattanzio 2009). In other words, telemedicine necessarily renders an interface by which specialists can be reached by virtually sitting in rural health-care units, thereby facilitating the diagnosis process (Marks et al. 2009) and reducing travel time and cost (Whitacre et al. 2009). Psychiatry is a specialized domain in health science, and there are several socioeconomic and legal issues related to it, making it one of the most important medical domains. Psychiatric patients are screened and diagnosed by analyzing the history, signs, and symptoms of patient behavioral pattern (Marks et al. 2009; Hilty et al. 2009). Most of the diseases are chronic and can be managed locally. Therefore, telemedicine is a useful way to screen the diseases and continue treatment at local levels without much physical traveling.

Knowledge engineering in this circumstance is quite broad, manual, and concept oriented. General physicians (i.e., doctors without specialist qualifications and required experience) are available in the rural hospitals, and they use the interactive approach of telemedicine to discuss a case (which they find complicated) with specialists at rural health centers. Patient information is sent to specialists using ICT (i.e., Internet). This mechanism is popularly termed an "e-referral" where the patients are virtually referred to a specialist, and there is no need for the patient to travel from villages to cities for consultation. Such a method curbs travel costs. Moreover, there is no major loss of time from the patients' side for a specialist

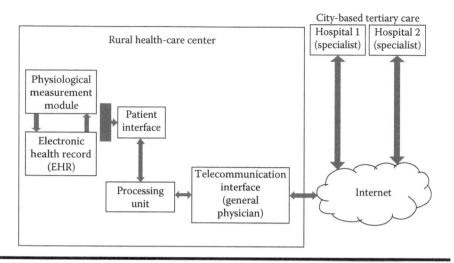

Figure 15.4 The workflow diagram of telepsychiatry.

consultation. From the medical knowledge engineering point, general physicians gather knowledge from the specialists during teleconsultation and use the derived knowledge in future cases, reducing the future consultation load to the specialists and, thereby, enhancing the process of local capacity building, reducing overall ICT consumption, and the costs involved in its use (Scaturo and Huszonek 2009). This process may also be termed as electronic learning or e-learning. The learning is seamless, shared among other staff members of rural health centers, and effectively used when the situation demands. In psychiatry practice, telemedicine (called telepsychiatry) is one effective method for patient consultation, and it is gaining popularity day by day. Figure 15.4 shows the workflow of the telepsychiatry system.

In this figure, specialized knowledge is transferred from remotely situated specialists through telemedicine's interactive interfaces to general physicians in a seamless manner during audio–video interactive patient consultations. Such a mode of knowledge gathering is fast, secure, and effective by virtue of ICT applications and is a challenge in e-mental health research. However, the true challenge lies in measuring its benefits (e.g., cost reduction, quality enhancement, knowledge implementation, and the availability and accessibility of the technology) across patient populations, which comprises the future work of this author.

15.4.2 *Knowledge Engineering in e-Mental Health: Current Status*

Medical knowledge engineering techniques endeavor toward extracting class patterns by interesting feature selection after analyzing raw clinical data. It essentially mimics the way doctors arrive at a diagnosis by manually extracting meaningful information from patient input. Doctors use their medical logic, which is a combination of supervised and unsupervised learning processes. On the other hand, in knowledge engineering, computer logic is applied and learning algorithms are utilized for supervised or unsupervised learning. If the learning is supervised, it is called "classification." Otherwise, it is called "clustering."

In psychiatry, knowledge engineering methods have been used in various spheres, which are discussed in Sections 15.4.2.1 and 15.4.2.2.

15.4.2.1 Knowledge Engineering at the Cellular Level

Advances in genetics and cell biology research have enabled us to explore and understand the significant molecular and genetic factors behind the occurrences of mental illnesses. All these factors apparently remain undermined unless explored. Laboratory-based experimental research reveals detail mechanisms and functional (e.g., biochemical) pathways at the molecular level about the occurrence of mental illnesses (Pauling 1968; Harris 1974; Wermter et al. 2009).

Most psychiatric diseases are genetically predisposed, for example, schizophrenia (Srivastava et al. 2009), manic-depressive psychosis (Okumura et al. 2009), and obsessive-compulsive disorder (Kwon et al. 2009). Hence, there are strong familial tendencies of occurrences. There are studies where various techniques of knowledge engineering are applied to analyze and correlate psychiatric diseases. These are described as follows:

Bioinformaticians commonly use the convergent functional genomics (CFG) approach to engineer first-generation genome-wide association studies (GWAS) data sets, which are raw in nature, that is, unformatted, unstructured, and unprocessed. The key reason for doing so is to identify the significant genetic signals that often remain underexplored while using the genetics-only approach. Supportive studies by Le-Niculescu et al. (2009) conclude that CFG is a useful method for integrating a genetics-only approach and functional genomics in bipolar disorder (also called manic-depressive psychosis). The authors then concluded that CFG could identify a series of new candidate genes and blood biomarkers and, thus, help physicians and researchers understand the mechanism of actions of these genes in the occurrence of bipolar disorders in human beings.

There is another method called "prioritizing disease genes by analysis of common elements" (PDG-ACE) that has been applied by McEachin et al. (2008) to corroborate the correlations of two important genes: (1) the major depressive disorder gene and (2) the alcohol use disorder gene. Using the aforementioned technique, the authors were able to prove a strong correlation between these two sets of genes, and their research provides an integrated bioinformatics approach toward knowledge engineering of etiopathology and comorbidities in psychiatric diseases in the vulnerable population, that is, occurrences of major depression in alcohol abusers and vice versa.

Pattern array (PA) is another useful knowledge engineering technique in bioinformatics. Kafkafi et al. (2008) used this technique to classify the psychopharmacological drugs for measuring the in vivo carcinogenicity (i.e., generation of cancer or cancer-like features in the vulnerable cells) of such drugs. The aim of their study was to discover how such adverse effects can be predicted at an early stage by variations in gene expressions. Three classes of drugs were chosen for this study: (1) opoids, (2) psychotomimetic drugs, and (3) psychomotor stimulants. The authors observed that PA successfully classifies these drugs based on their dosages, dependencies, and many other pharmacokinetic and pharmacodynamic parameters when tested in vivo.

15.4.2.2 Knowledge Engineering at the Clinical Data Level

Clinical data, as mentioned in Section 15.3, indicates history, signs, and symptoms of patients suffering from diseases. The exciting features that entice researchers to use knowledge engineering techniques in mental health data are associated extreme subjectivities, data unavailability due to various ethical issues, and the presence of nonlinear relations among parameters. Hence, there is not as much research in the mental health domain as in other domains of health sciences. Knowledge engineering techniques are used in this domain especially for disease screening and diagnosis, risk prediction, prognosis determination, etc. The techniques used are traditional statistical, hard computing, soft computing, and hybrid techniques. Some recent studies are described as follows:

In one study, Baca-Garcia et al. (2006) used a traditional multivariate statistical technique to observe the level of significance of various psychiatry parameters that may assist psychiatrists in assessing whether suicidal risks for patients require admission. The authors found that traditional statistical data mining techniques are still the reliable predicting tools with 99% sensitivity and 100% specificity. In another study, Baca-Garcia et al. (2007) also used random forest and forward selection techniques to explore the psychiatry variables that might be associated with familial suicide attempts in a group of suicide attempters. Random forest is a special technique in data mining that can develop multiple classification trees where each tree votes for classes or even assigns new classes and counts the maximum number of votes to make the final selection. The tree can be grown as large as possible without much pruning. In this study, using random forest and forward selection techniques Baca-Garcia et al. (2007) were able to classify patients who had attempted suicide at some point in their lives. In addition to good classification, the authors could also obtain a new class called "alcohol abuse" as the comorbidity of suicide enhancing the clinical hints.

Random forest has also been used by a study conducted by Marinic et al. (2007). In this study, the authors used the random forest classifier on 120 psychiatric patients, 60 of whom suffered from posttraumatic stress disorder (PTSD). The aim of the study was to explore the salient diagnostic features of PTSD in the study sample. The authors developed two predictive models: One was based on structured psychiatric interview and the other one used clinician-administered PTSD scale (i.e., CAPS), positive and negative syndrome scale (i.e., PANSS), Hamilton anxiety scale (or HAMA), and Hamilton depression scale (or HAMD). The authors observed that the first predictive model distinguished PTSD from other neurotic, stress-related, or somatoform disorders.

A multidimensional classifier was also used in a study to correlate alcoholism and homicidal tendencies (Reulbach et al. 2007). Multidimensional classification system (MCS)-based results show that there is a high chance of committing murders by those who abuse alcohol, especially people having cyclothymias and anxiety disorders as associated illnesses.

Several researchers (Chattopadhyay (2007); Chattopadhyay et al. (2007a); Chattopadhyay et al. (2008a); Chattopadhyay and Daneshgar 2009) studied suicidal risks in a set of real-world psychiatric adult patients. The central objective of these series of studies was to explore behavioral patterns among patients who either attempted or committed suicide. In these studies, the Pierce's suicide intent scale (PSIS; Pierce 1977) was used to capture the behavioral evidence as symptoms in the backdrop of suicide in adult Indian population that may be at risk. In the study by Chattopadhyay (2007), linear regressions were performed to note the effect of individual symptoms on the degree of suicidal risk. The experiment evaluated whether properly modeled regressions are still effective measures to assess the relationships among process variables. However, the author failed to appropriately distinguish the risk boundaries, that is, mild-moderate, moderate-severe, and severe-to-very severe risks. In the study by Chattopadhyay et al. (2007a), the authors used a support vector machine (SVM) to redefine the risk boundaries. The critical patient population as support vectors could be identified as the threshold for mild-moderate, moderate-severe, and severe-very severe risk categories. The shortcoming of this study is that it failed to assess the level of similarity among patients within the same risk group. Another study (Chattopadhyay, Pratihar, and De Sarkar 2008a) addressed this issue, and the authors used the similarity-based approach to assess the likeliness among patients in a particular group and to note whether similarity-based classification could be improved if more parameters were added to the matrix model. The results showed that distance calculations (by Euclidean method) were effective in measuring similarities among the patients, and addition of multiple parameters improved the performance of classifiers. Chattopadhyay et al. (2008b) developed a classifier on a JAVA-based platform using the Euclidean distance as the similarity measure, that was able to predict the suicidal risk levels in young adults. The authors also observed that increased number of suicide-intent parameters/attributes would be useful in begetting better classifications.

In one study, Chattopadhyay and Daneshgar (2009) attempted to model suicidal risks with comorbid psychiatric illnesses. The authors first used regressions for assessing the risk levels among patients and then used a single-linkage hierarchical clustering algorithm (SLHCA) for correlating the risk levels with psychiatric morbidities. In SLHCA, the authors used Euclidean distance–based measures to draw the dendrogram of the illnesses depending on the severity level. Based on the observations, the authors generated rules and measured their quality to ascertain that the rules were genuine and not coincidental. The results were verified by human experts, and it was found that SLHCA could also be useful in correlating suicidal risk levels and psychiatric morbidities. The study might be a breakthrough for clinicians in predicting suicidal risks based on the associated psychiatric illness in patients.

Nebekar et al. (2007) used hierarchical optimal classification tree analysis (HOCTA) as the knowledge engineering method for rapidly developing clinically meaningful surveillance rules for a set of 3987 administrative patient data. The authors explored and validated surveillance rules for drug-induced bleeding, delirium, and psychosis as the adverse effects of drugs using HOCTA.

Chattopadhyay et al. (2010) used statistical data mining techniques to predict seven types of psychosis in adults: (1) schizophrenia, (2) mania, (3) psychotic depression, (4) delusional disorder, (5) schizoaffective disorder, (6) organic psychosis, and (7) catatonia. The authors used the brief psychiatric rating scale (BPRS-F2; Overall and Gorham, 1962) to capture the symptoms of these disorders and performed statistical modeling using multiple linear regressions with significance testing. The aim was to extract the salient diagnostic features. The authors applied the Plackett–Burman design of experiments, popularly known as PBD (Plackett and Burman 1946), which is a fractional factorial statistical design of experiments to measure the effects of various factors (derived based on BPRS-F2) on the seven adult psychoses. The results were consistent with the real-world clinical picture. Similarly, many other significant factors were extracted from the given set of patient data for screening the remaining illnesses. This study claims that the wise use of Plackett-Burman Design of Experiments (PBD) is still recommended in knowledge engineering to extract hidden patterns from a set of psychiatry data that are highly subjective in nature.

In psychiatry, patients often present with symptoms that are similar in many psychiatric as well as nonpsychiatric illnesses, and there are possibilities for a mixture of disorders. As diseases have no clear-cut boundaries, it is very difficult to differentiate one illness from another. Given this scenario, Chattopadhyay et al. (2007b) applied various fuzzy clustering techniques to differentiate look-alike illnesses for a given set of symptoms. The authors used fuzzy c-means (Bezdek 1981) and entropy-based fuzzy clustering (Yao et al. 2000) algorithms. Both algorithms work by iteratively measuring similarities (neighborhood functions) of data points (i.e., patients) from the respective centroids (i.e., cluster centers), given a threshold value to decide the extent to which similarity data points will be included in a particular cluster. The added advantage of these techniques is that they can identify outliers, that is, those data points that cannot be clustered. The algorithms were then used on a set of real-world clinical psychiatry data, and they were able to cluster the diseases (schizophrenia, mania, depression mixed with psychosis, delusional disorder, schizoaffective disorder, organic psychosis, and catatonia). Multidimensional data (representing number of patients) are then mapped in two dimensions for visualization using the self-organizing map (Kohonen 1995).

Apart from traditional statistical and fuzzy logic–based techniques (Zadeh 1965), neural network (NN) also poses a significant method in disease classification by virtue of its reasoning ability using synapses among neurons, similar to the human brain. Attempts have also been made to use NN-based psychiatry disease classifications. Lowell and Davis (1994) classified and then ranked psychiatric diseases based on the period of hospital stay using a multilayer feedforward NN. Yana et al. (1994) used a three-layered perceptron for differentiating psychotic disorders (e.g., schizophrenia and depression) from neurotic disorders (e.g., anxiety disorders). The authors found that a backpropagation NN could differentiate these groups of disorders much more precisely.

The NN has also been used in psychiatry several times with different aims. Zou et al. (1996) also classified psychiatric illnesses using the NN technique. Aruna, Puviarasan, and Palaniappan (2005) proposed a neuro-fuzzy hybrid approach to diagnose psychosomatic disorders. The authors defined fuzzy membership values based on linguistic (i.e., interview-based) inputs, which were fed, in turn, to the multilayer feedforward NN as its inputs. The outputs thus obtained were further tuned by backpropagation to obtain the final outputs. The authors then compared this hybrid approach with the traditional Bayesian belief network and linear discriminant analysis, and found superior results. There are many other studies that use NN as a classification tool. For example, Buscema et al. (1998) classified various eating disorders. Jefferson et al. (1998) utilized NN for predicting the chance of depression after occurrence of manic episodes in a large group of patients.

Chattopadhyay et al. (2009) used a genetic fuzzy system for the screening and prediction of seven adult psychotic illnesses, namely schizophrenia, mania, schizophreniform disorder, catatonia, depression with psychosis, schizoaffective disorder, and delusional disorder. The authors adopted the technique described by Takagi and Sugeno (1983) to develop a fuzzy controller for making the inference/diagnosis, which has been optimized by a binary-coded genetic algorithm (GA; Holland 1975). Results have shown that such a hybrid technique can diagnose these diseases more accurately than can traditional statistical techniques, such as multiple regressions and significance tests.

All these studies display various attempts to explore and understand highly subjective psychiatry data with the help of several data mining techniques to reduce the chances of human-induced diagnostic biases. However, it is worth mentioning that none of these studies were able to bypass the research boundaries and propagate to real-world psychiatry practice. Therefore, brining these tools into clinical practice is a challenging task. In Section 15.5, we detail the issues related to the practical implementation of these tools and discuss how these hurdles can be minimized.

15.5 Discussions and Future Scopes

It is important to remember that effective applications of ICT have been important in getting a smarter health-care system. By virtue of effective ICT applications, the diagnoses are now quicker, more accurate, and more precise. The human anatomy, physiology, biochemistry, and disease patterns and their representations have not changed remarkably over these years. In addition, human expertise for capturing, processing, and understanding patient information for the diagnosis of disease has also not been drastically improved. Nevertheless, immense improvements in the application of health sciences have been observed over the past three to four decades, owing to the continuous evolution of computer theories, algorithms, and techniques that are more targeted toward meeting real-life needs.

The key objective of this chapter is to display various applications of knowledge acquisition methods in relation to the mental health domain by virtue of ICT. The concept of e-mental health has been proposed as a mode of mental health practice through telepsychiatry and medical knowledge engineering, leading to medical decision support systems. The knowledge acquisition by general practitioners through telepsychiatry (also called indirect e-learning) needs to be evaluated using large multicenter trials.

Research activities using computer algorithms and methods are mostly focused on extracting patterns from raw mental health data using the techniques of knowledge engineering and propagating the science toward predicting illnesses with the help of classifiers and controllers. However, most research activities are confounded in pen and paper and are not used much in real-world psychiatry practice because of problems such as lack of interdisciplinary attitude and approach among the medical fraternity and poor adoption of technology by clinicians. While examining these two problems, we need to look into some important issues: Doctors are uncomfortable when using computers (Briggs 2001; Carroll 2008). They are completely apathetic toward the use of decision support tools in practice. One reason could be complete detachment from mathematical logic during their medical training, which only teaches manual pattern recognition through literature and practice, and the other reason is the apprehension that they might become more dependent on computer-aided decision support and lose their natural ability to diagnose a case.

There is a strong tendency among doctors to cling to traditional medical practices.

There are also other related issues, such as inadequacy of traditional policy decisions in health care at the government and organization levels, economic issues, and personal attitudes toward the adoption of informatics in health care by the medical fraternity. Considering all these issues, we argue in favor of the design, development, and applications of the concept of informatics that is much needed in today's mental health care. We hope to influence researchers to put their research into practice. We also show that a balance should be adopted between clinical practice and computer-aided practice in the mental health domain. However, determining which of these methods to use for a particular medical situation should be the choice of the doctor.

References

Akematsu, Y., and M. Tsuji. 2009. An empirical analysis of the reduction in medical expenditure by e-health users. *J Telemed Telecare* 15:109–11.

Alexander, J., and A. Lattanzio. 2009. Utility of telepsychiatry in aboriginal Australians. *Aust N Z J Psychiatry* 43:1185.

Aruna, P., N. Puviarasan, and B. Palaniappan. 2005. An investigation of neurofuzzy systems in psychosomatic disorders. *Expert Syst Appl* 28:673–9.

Baca-Garcia, E., M. M. Perez-Rodriguez, I. Basurte-Villamor, J. Saiz-Ruiz, J. M. Leiva-Murillo, P. de, R. Santiago-Mozos, A. Artes-Rodriguez, and L. J. de. 2006. Using data mining to explore complex clinical decision: A study of hospitalization after a suicide attempt. *J Clin Psychiatry* 67:1124–32.

Baca-Garcia, E., M. M. Perez-Rodriguez, D. Saiz-Gonzalez, I. Basurte-Villamor, J. Saiz-Ruiz, J. M. Leiva-Murillo, P. de, R. Santiago-Mozos, A. Artes-Rodriguez, and L. J. de. 2007. Variable associated with familial suicide attempts in a sample of suicide attempters. *Prog Neuropsychopharmacol Biol Psychiatry* 31:312–6.

Bezdek, J. C. 1981. *Pattern Recognition with Fuzzy Objective Function Algorithms*. Norwell, MA: Kluwer Academic Publishers.

Briggs, B. 2001. Doctors sound off on IT concerns. *Health Data Manag* 9:38–40.

Buscema, M., M. Mazzetti de Pietralata, V. Salvemini, M. Intraligi, and M. Indrimi. 1998. Application of artificial neural network in eating disorders. *Subst Use Misuse* 33:765–91.

Carroll, J. 2008. Doctors say getting wired too costly, time-consuming. *Manag Care* 17:5–6.

Chattopadhyay, S. 2002. Can information technology be another useful tool to diagnose and understand schizophrenia? *Curr Sci* 83:1057–8.

Chattopadhyay, S. 2007. A study on suicidal risk analysis. In *Proceedings of 9th IEEE International Conference on e-Health Networking, Applications and Service (HEALTHCOM 2007)*, 74–9. Taipei, Taiwan: IEEE Xplore.

Chattopadhyay, S. 2010. A framework for studying perceptions of rural healthcare staff and basic ICT support for e-health use: An Indian experience. *Telemed J E Health* 16(1):80–8.

Chattopadhyay S., F. Daneshgar. 2009. A Study on Suicidal Risks in Psychiatric Adults. *Int J Biomed Eng Technol*; accepted, in press.

Chattopadhyay, S., D. K. Pratihar, and S. C. De Sarkar. 2007b. Fuzzy clustering of psychosis data. *Int J Bus Intell Data Min* 2:143–59.

Chattopadhyay, S., D. K. Pratihar, and S. C. De Sarkar. 2008a. Developing fuzzy classifiers for predicting the chance of occurrence of adult psychoses. *Knowl Based Syst* 20:479–97.

Chattopadhyay, S., D. K. Pratihar, and S. C. De Sarkar. 2009. Fuzzy logic-based screening and prediction of adult psychoses: A novel approach. *IEEE Trans Syst Manage Cybern* 39:381–7.

Chattopadhyay S., D. K, Pratihar S. C. De Sarkar. 2010. Statistical Modelling of Psychoses Data. *Computer Methods and Programs in Biomedicine* 100:222–36.

Chattopadhyay, S., P. Ray, H. S. Chen, M. B. Lee, and H. C. Chiang. 2008b. Suicidal risk evaluation using a similarity-based classifier. In *Proceedings of International Conference on Advanced Data Mining and Applications (ADMA 2008)*, eds., Changjie Tang, Charles X. Ling, Xiaofang Zhou, Nick J. Cercone, Xue Li 51–61. Chengdu, China: Springer.

Chattopadhyay, S., P. Ray, M. B. Lee, and H. S. Chen. 2007a. Towards the design of an e-health system for suicide prevention. In *Proceedings of IASTED Conference on Artificial Intelligence and Soft Computing (ASC)*, ed., A. P. del Pobil 191–6. Palma de Malorca, Spain: ACTA Press.

Chen, H., S. S. Fuller, C. Friedman, and W. Hersh. 2005. Knowledge management, data mining, and text mining in medical informatics. In *Medical Informatics Knowledge Management and Data Mining in Biomedicine,* ed. H. Chen, S. S. Fuller, C. Friedman, and W. Hersh, 4–30. New York: Springer's Integrated Series in Information Systems.

Durrani, H., and S. Khoja. 2009. A systematic review of the use of telehealth in Asian countries. *J Telemed and Telecare* 15:175–81.

Ernstmann N., O. Ommen, M. Neumann, A. Hammer, R. Voltz, and H. Pfaff. 2009. Primary care physician's attitude towards the German e-health card project: Determinants and implications. *J Med Syst* 33:181–8.

Gerkin, D. G. 2009. E-health can be a two-edged sword for the medical doctor. *Tenn Med* 102:7–8.

Harris, H. 1974. The development of Penrose's ideas in genetics and psychiatry. *Br J Psychol* 125:529–36.

Haux, R., P. Knaup, A. W. Bauer, W. Herzog, E. Reinhardt, K. Uberla, W. van Eimeren, and W. Wahlster. 2001. Information processing in healthcare at the start of the third Millennium: Potential and limitations. *Methods Inf Med* 40:156–62.

Hersh, W. R. 2002. Medical informatics: Improving health care through information. *JAMA* 288:1955–8.

Hilty, D. M., P. M. Yellowlees, P. Sonik, M. Derlet, and R. L. Hendren. 2009. Rural child and adolescent telepsychiatry: Success and struggles. *Pediatr Ann* 38:228–32.

Holland, J. H. 1975. *Adaptation in Natural and Applied Systems.* Ann Arbor, MI: Univ Michigan Press.

Jefferson, M. F., N. Pendleton, C. P. Lucas, et al. 1998. Evolution of artificial neural network architecture: Prediction of depression after mania. *Methods Inf Med* 37:220–5.

Kafkafi, N., D. Yekutieli, P. Yarowsky, and G. I. Elmer. 2008. Data mining in behavioral test detects early symptoms in a model of amyotrophic lateral sclerosis. *Behav Neurosci* 122:777–87.

Kaplan, H. I., B. J. Saddock, and J. A. Greb. 1994. *Synopsis of Psychiatry, Behavioural Sciences and Clinical Psychiatry,* 803–23. New Delhi, India: B.I. Waverly Pvt. Ltd.

Kapoor, L., S. K. Mishra, and K. Singh. 2005. Telemedicine: Experience at SGPGIMS, Lucknow. *J Postgrad Med* 51:312–5.

Khoja, S., R. Scott, and S. Gilani. 2008. E-health readiness assessment: Promoting "hope" in the health-care institutions of Pakistan. *World Hospitals and Health Services* 44:36–38.

Klein, G. O., and K. Kajbjer. 2009. eHealth tools for patients and professionals in a multicultural world. *Stud Health Technol Inform* 150:297–301.

Kohonen, T. 1995. *Self-Organizing Maps.* Heidelberg, Germany: Springer-Verlag.

Kwon, J. S., Y. H. Joo, H. J. Nam, M. Lim, E. Y. Cho, M. H. Jung, J. S. Choi, et al. 2009. Association of the glutamate transporter gene SLC1A1 with atypical antipsychotics-induced obsessive-compulsive symptoms. *Arch Gen Psychiatry* 66:1233–41.

Le-Niculescu, H., S. D. Patel, M. Bhat, R. Kuczenski, S. V. Faraone, M. T. Tsuang, F. McMahon, et al. 2009. Convergent functional genomics of genome-wide association data for bipolar disorder: Comprehensive identification of candidate genes, pathways, and mechanisms. *Am J Med Genet B Neuropsychiatr Genet* 150:155–81.

Lee, O. 2009. Telepsychiatry and cultural barriers in Korea. *Stud Health Technol Inform* 144:145–8.

Lowell W. E., and G. E. Davis. 1994. Predicting length of stay for psychiatric diagnosis groups using neural networks. *J Am Med Inform Assoc* 1(6): 459–66.

Marinic, I., F. Supek, Z. Kovacic, L. Rukavina, T. Jendricko, and D. Kozaric-*Kovacic*. 2007. Posttraumatic stress disorder: Diagnostic data analysis by data mining methodology. *Croat Med J* 48:185–97.

Marks, S., U. Shaikh, D. M. Hilty, and S. Cole. 2009. Weight status of children and adolescents in a telepsychiatry clinic. *Telemed J E Health* 15:970–4.

McEachin, R. C., B. J. Keller, E. F. Saunders, and M. G. McInnis. 2008. Modeling gene-by-environment interaction in comorbid depression with alcohol use disorders via an integrated bioinformatics approach. *BioData Min* 1:2.

Nebekar, J. R., P. R. Yarnold, R. C. Soltysik, B. C. Sauer, S. A. Sims, M. H. Samore, R. W. Rupper, et al. 2007. Developing indicators of inpatient adverse drug events through nonlinear analysis using administrative data. *Med Care* 45:81–8.

Okumura, T., T. Kishi, T. Okochi, M. Ikeda, T. Kitajima, Y. Yamanouchi, Y. Kinoshita, et al. 2009. Genetic association analysis of functional polymorphisms in neuronal nitric oxide synthase 1 gene (NOS1) and mood disorders and fluvoxamine response in major depressive disorder in the Japanese population. *Neuropsychobiology* 61:57–63.

Overall, J. E., and D. R. Gorham. 1962. The brief psychiatric rating scale. *Psychol Rep* 10:779–812.

Pauling, L. 1968. Orthomolecular psychiatry. Varying the concentrations of substances normally present in the human body may control mental disease. *Science* 160:265–71.

Pierce, D. W. 1977. Suicidal intent in self-injuries. *Bri J Psychiatry* 130: 377–85.

Plackett, R. L., and J. P. Burman. 1946. The design of optimum multi-factorial experiments. *Biometrika* 33:305–25.

Reulbach, U., T. Biermann, S. Bleich, et al. 2007. Alcoholism and homicide with respect to the classification systems of Lesch and Clininger. *Alcohol Alcohol* 42:102–7.

Scaturo, D. J., and J. J. Huszonek. 2009. Collaborative academic training of psychiatrists and psychologists in VA and medical school settings. *Acad Psychiatry* 33:4–12.

Shoaib, S. F., S. Mirza, F. Murad, A. Z. Malik. 2009. Current status of e-health awareness among healthcare professionals in teaching hospitals in Rawalpindi. *Telemed J E Health* 15:347–52.

Srivastava, V., S. N. Deshpande, and B. K. Thelma. 2009. Dopaminergic pathway gene polymorphisms and genetic susceptibility to schizophrenia among north Indians. *Neuropsychobiology* 61:64–70.

Suleiman, A. B. 2001. The untapped potential of Telehealth. *Int J Med Inform* 61:103–12.

Takagi, T., and M. Sugeno. 1983. Derivation of fuzzy control rules from human operator's control action. In *Proceedings of IFAC Symposium on Fuzzy Information, Knowledge Representation and Decision Analysis,* ed. E. Sanchez, 55–60, Marseilles, France.

Venkatesh, V., M. G. Morris, G. B. Davis, and F. D. Davis. 2003. User acceptance of information technology: Toward a unified view. *MIS Q* 27:425–78.

Wermter, A. K., M. Laucht, B. G. Schimmelmann, T. Banaschweski, E. J. Sonuga-Barke, M. Rietschel, and K. Becker. 2010. From nature versus nurture, via nature and nurture, to gene x environment interaction in mental disorders. *Eur Child Adolesc Psychiatry* 19:199–210

Westbrook, J. I., J. Braithwaite, K. Gibson, R. Paoloni, J. Callen, A. Georgiou, N. Creswick, and L. Robertson. 2009. Use of information and communication technologies to support effective work practice innovation in the health sector: A multi-site study. *BMC Health Serv Res* 9:201.

Whitacre, B. E., P. S. Hartman, S. A. Boggs, and V. Scott. 2009. A community perspective on quantifying the economic impact of teleradiology and telepsychiatry. *J Rural Health* 25:194–7.

World Health Organization (WHO). 2005. *e-Health Report by Secretariat, 58th World Health Assembly.* A58/21, Geneva, Switzerland: WHO.

Yana, K., K. Kawachi, K. Iida, Y. Okubo, M. Tohru, and F. Okuyania. 1994. A neural set screening of psychiatric patients. In *Proceedings of 16th Annual International Conference of the IEEE Engineering in Medicine and Biology Society* 16:1366–7.

Yao, J., M. Dash, and S. T. Tan, et al. 2000. Entropy-based fuzzy clustering and fuzzy modeling. *Fuzzy Sets Syst* 113:381–8.

Zadeh, L. A. 1965. Fuzzy sets. *Inf Control* 8:338–53.

Zou, Y., Y. Shen, L. Leu, et al. 1996. Artificial neural network to assist psychiatric diagnosis. *Br J Psychiatry* 169:64–7.

Chapter 16

Systems Engineering for Medical Informatics

Oliver Faust, Rajendra Acharya U., Chong Wee Seong, Teik-Cheng Lim, and Subhagata Chattopadhyay

Contents

16.1 Introduction

As humans get older, they become ever more dependent on biomedical systems for their well-being. In recent years, this realization has sparked rapid development and widespread deployment of biomedical systems, which have progressed from single-purpose island systems, such as traditional X-ray machines, to massively networked health-care systems in which medical imaging, such as positron emission tomography–magnetic resonance imaging (PET-MRI), is only a small part of the overall system. These networks distribute an ever-increasing amount of biomedical data. Therefore, our society is becoming more and more dependent on a technology that is getting more and more complex each day. This growing complexity is a problem, because the combination of high dependence and high complexity represents an aggravating factor that heavily impacts the system requirements.

Over time, this impact has caused an almost exponential rise in systems requirements. Unfortunately, projects that follow an evolutionary design methodology cannot cope with an exponential increase in requirements, because evolution relies on the assumption that progress is only made from one system generation to another. For biomedical systems, the life cycle is still counted in years instead of months or weeks; therefore, evolutionary progress takes time. But as outlined earlier in this section, the requirements (or demand) outpace the rate of progress achievable with evolutionary design methodologies. Therefore, all solutions to this fundamental problem require a paradigm shift.

Having established the need for a paradigm shift, we have to look for solutions from other areas; otherwise, we are in danger of reinventing the wheel. Engineers deal with systems all the time; therefore, it is likely that there are engineering solutions for this problem. In biomedical engineering, the problem is that we have to build highly complex systems, which are very reliable, fast, and within budget.

This scenario is frequently confronted by aerospace engineers. These engineers build flying machines that carry humans and goods through the third dimension from point A to point B. In 1961, point A was Cape Canaveral here on Earth and point B was the southern Sea of Tranquility on Moon. The Apollo program was set up to build a physical solution for this aerospace problem. The engineers while designing this space rocket experienced a large amount of growth in aerospace requirements (NASA 2010). This growth was fuelled by the same factors that currently drive the requirements for biomedical systems: technical complexity and reliability. Therefore, it is natural to investigate how engineers, during that time, overcame such massive obstacles. First, they had some of the brightest minds working on this project. However, there is a limit of what even the brightest minds can do without appropriate support structures, in this case, design methodologies. More specifically, the Apollo program used an engineering method called "systems engineering" (Sadeh 2006). Many experts believe that the application of systems engineering principles leads to better management, which paves the way to ultimate success.

In this chapter, we make the case for biomedical systems engineering by exploring the systems engineering methodology and by outlining an example project. We show that systems engineering can be applied to biomedical problems and can solve a variety of problems posed by the exponential growth in requirements. The problem solutions come from holistic management, that is, the fruitful interaction between engineers and management. Therefore, systems engineering enables project teams, engineers, and managers to build reliable and trustworthy biomedical systems through a systematic development process.

16.2 Systems Engineering

Systems engineering is an interdisciplinary design methodology for complex engineering projects. It gives guidelines on how such projects should be managed. Such guidelines are very important for biomedical systems, because their requirements increase exponentially with project size. Systems engineering outlines work processes and guidelines to handle large projects. These guidelines can be mechanized, which leads to systems engineering tools. The design methodology draws from both technical and human-centered disciplines such as control engineering and project management.

16.2.1 History

The origin of systems engineering can be traced back to the 1940s (Schlager 1956). Researchers at the Bell Telephone Laboratories felt the need to identify and manipulate the properties of a system as a whole. This need became more pressing, because of an exponential increase in systems complexity (Chestnut 1965). For example, more people became connected to the telephone network. Initially, a telephone network was a point-to-point connection. However, such point-to-point connections became impractical as the number of possible communication partners increased. To overcome this problem, engineers determined that because not everybody was using the phone at all times only one line to an end user was necessary. These assumptions led to the invention of human- and, later, machine-operated switchboards.

However, introduction of switchboards meant a dramatic increase in system complexity. For the system to work the caller must provide routing information. Whereas in the early days routing information was verbally delivered to a human operator, today we use identification in the form of telephone numbers. Opportunities for cost saving that arise because not everyone talks at the same time make the system more practical. However, these cost savings could only be realized when the communication system followed complex statistical models (Angus 2001).

Furthermore, the communication companies had to create sophisticated and manpower-intensive management and support structures to deliver a constant or a continuously improving quality of service to their customers. The discussion of this early communication system supports the statement that in complex engineering projects the requirements increase roughly exponentially with project size. This increase in requirements goes against the common perception that overall system complexity can be established by adding the complexities of individual system parts. Because only complex projects are affected by this exponential rise in requirements, the systems engineering methodology is mainly driven by large organizations that require huge engineering systems, such as the U.S. Department of Defense and National Aeronautics and Space Administration (NASA; Hall 1962). Besides defense and aerospace, many information- and technology-based companies, software development firms, and industries in the field of electronics and communications require systems engineers to be part of their team.

These organizations saw that it was no longer possible to rely on design evolution for their systems to improve. Furthermore, existing tools were not sophisticated enough to meet the growing demands that were posed by the accelerating system complexity. Therefore, within the systems engineering framework, new methods could be developed that addressed system complexity directly (Sage 1992). Currently, the systems engineering methodology is evolving, in areas of both the development and the identification of new methods as well as modeling techniques. Current methods provide a better comprehension of engineering systems, because they use the latest research results for management and engineering to address the problems resulting from increases in system complexity.

The National Council on Systems Engineering (NCOSE) was founded in 1990 by representatives from a number of U.S. corporations and organizations as a professional society for systems engineering. In 1995, the name of the organization was changed to the International Council on Systems Engineering (INCOSE) as a result of the growing involvement of systems engineers from outside the United States (INCOSE Group 2004). The INCOSE developed the aim to address the need for improvements in systems engineering in both practice and education. Now, universities in several countries offer graduate programs in systems engineering, and continuing education options are also available for practicing engineers (INCOSE Education and Research Technical Committee 2006).

16.2.2 Concept

Systems engineering represents both an approach and a discipline in engineering. It is used to formalize the design approach and, in doing so, to identify new methods and research opportunities. In this respect, systems engineering is not different from other fields of engineering. However, it is different from other disciplines in that systems engineering favors holistic and interdisciplinary thinking and design approaches (Adcock 2001).

Traditional engineering embraces areas like design, development, production, and operation of physical systems. From this perspective, systems engineering is just engineering. However, in traditional engineering, there is an almost automatic tendency to segment the engineering process into individual processes and thereby fragment it. This fragmentation is not something bad per se; it is necessary in design and development. Engineers learn early the well-known fragmentation technique of "divide and conquer," which helps them to process complex technical problems. Unfortunately, this principle is not an overall guiding principle, because it reflects neither the system goals nor the often complex organizational structures. Systems engineering aims to be such an overall guiding principle that enables the combination of management and engineering teams to deliver the system goals in complex organizational structures. The use of the term systems engineering has evolved over time to embrace a wider and a more holistic concept of systems and of engineering processes (Goode and Machol 1957).

16.2.3 Holistic View

Early in the development cycle, systems engineering focuses on management techniques to capture requirements. It addresses documenting requirements, before proceeding with design synthesis and system validation. All these steps are done with an awareness of the complete problem, the system life cycle. On an abstract level, the systems engineering process can be decomposed into two parts:

1. Systems engineering technical process: The technical process includes assessing available information and defining effectiveness measures.
2. Systems engineering management process: The goal of the management process is to organize and coordinate the technical efforts throughout the product life cycle.

The motivation for using these processes is to create a number of models, such as behavioral and structural models. Furthermore, these processes must deliver a trade-off analysis and create sequential build and test plans (Oliver et al. 1997). Both engineering and management processes are interlinked. These links provide feedback and cross-fertilization; they help to identify and understand the aforementioned stages. Examples of such models include the waterfall model and the VEE model (SEOR 2007).

16.2.4 Interdisciplinary Field

Most of the time, building a physical problem solution requires contributions from diverse technical disciplines (Ramo and St.Clair 1998). By providing a systems (holistic) view of the development effort, systems engineering helps us to put all technical contributors into a unified team effort, forming a structured development process that proceeds from concept to design and implementation as well as from production to operation and, in some cases, to termination and disposal.

This perspective is often replicated in educational programs in which faculty teach systems engineering courses at departments other than their home departments to create an interdisciplinary environment (Cornell University 2007a; MIT 2007).

16.2.5 Managing Complexity

The need for systems engineering arose with the increase in complexity of systems and projects. In this context, complexity refers to engineering systems as well as logical and human organization of data (Shishko 1995). At the same time, a system can become more complex due to an increase in size, that is, an increase in the number of interacting entities; due to an increase in the amount of data or variables; or by combining previously independent systems. Biomedical health-care systems show increases in all three hallmarks of complexity. For example, in recent years we have seen an exponential rise in the amount of data produced by biomedical imaging systems. This growth is fuelled by the invention of new methods, the combination of existing methods, as well as improvements in sensor and processor technologies. At the same time, general health administration systems continue to grow, and the amount of database entries and the number of fields continue to increase. To deliver a higher standard of health care, biomedical imaging data must be combined with administrative data to form comprehensive health records. Once such systems are in operation nationwide, they become truly large.

The development of smarter control and routing algorithms, microprocessor design, as well as analysis of environmental systems also come within the purview of systems engineering. Systems engineering encourages the use of tools and methods to better comprehend and manage complexity in systems. The following list outlines some areas where tools are applied (Cornell University 2007b):

- Modeling and simulation
- Optimization
- System dynamics
- Systems analysis
- Statistical analysis
- Reliability analysis
- Decision making

Following an interdisciplinary approach for engineering systems is inherently complex, because the behavior of system components and the interaction among them are not always immediately clear or well defined. The systems engineering methodology aims to guide both management and engineering teams in defining and characterizing systems and subsystems, as well as documenting the interaction between these systems. In doing so, the gap that exists between informal requirements from users, operators, marketing organizations, and technical specifications can be bridged (DoD 2001).

The results of a study conducted by the INCOSE Systems Engineering Center of Excellence indicate that an optimal level of systems engineering effort is about 15%–20% of the total project effort (Honour 2004). At the same time, other studies have shown that systems engineering essentially leads to cost reduction among other benefits. However, no quantitative survey at a larger scale encompassing a wide variety of industries has been conducted until recently. Such studies are underway to determine the effectiveness and to quantify the benefits of systems engineering (Elm 2005; Valerdi et al. 2004).

Systems engineering encourages the use of modeling and simulation techniques to validate assumptions or theories on systems and the interactions within them (Sage and Olson 2001; Smith 1962). Safety engineering uses a specific flavor of systems engineering, which emphasizes early detection of possible stable failures. Early detection is so important because decisions made at the beginning of a project, which have consequences that are not clearly understood, can have enormous implications later in the life cycle of a system. In safety engineering, it is the task of the systems engineer to explore these issues and make critical decisions. There is no method that guarantees decisions made today will still be valid when a system goes into service years or decades after it is first conceived. Therefore, systems engineering enforces reevaluation or at least strong incentives for constant reevaluations of project-relevant decisions. As more and more lives become directly dependant on the correct functioning of biomedical systems, the importance of safety engineering and fault tolerance continues to grow.

16.2.6 Systems Engineering Process

Systems engineering methodology refers to the process that applies systems engineering techniques to turn ideas into physical problem solutions. Systems engineering processes are related to the stages in a system life cycle. The systems engineering process should begin with the design stage of the system life cycle at the very beginning of a project; however, systems engineering can also start at the middle of the life cycle. A variety of organizations propose different systems engineering processes. Figure 16.1 shows a coarse-grain systems engineering process diagram.

Several systems engineering process guidelines have been developed since 1969. These guidelines describe all the activities and deliverables of a systems engineering project.

As outlined in Section 16.2.2, even systems engineering cannot escape the principle of divide and conquer; there are different systems engineering processes for different stages of the system life cycle (Boehm 2005).

Systems engineering provides a so-called metamodel, which serves as the blueprint of the methodology. The metamodel is an abstraction of models, which highlights model properties. Therefore, it can be applied to model processes for a wide range

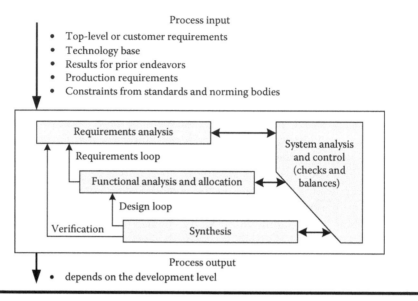

Figure 16.1 **Overview of systems engineering process diagram.**

of projects. The model that leads from an initial idea to a new product is project specific; however, all project models share the same properties. The systems engineering metamodel captures these common properties. Although most of them are derived from existing standards, ANSI/EIA 632 (G47 1999) deserves special mention because many current systems engineering methods are based on this standard. Further, ISO/IEC 15288 (Arnold 2004) and the INCOSE Systems Engineering Handbook (INCOSE group 2000), which are more recent, serve as sources for this method. The final source is an older standard, Mil-Std-499B (DoD 1992). The activities and concepts in these models can be explained as follows. The model is separated into four main processes: (1) agreement, (2) project, (3) technical process, and (4) evaluation. Figure 16.2 shows the systems engineering metamodel. The left side of the figure depicts the process network, that is, the entities that translate the need for a new system into a physical problem solution. The right side shows the deliverables of individual processes, as defined by the systems engineering methodology. Sections 16.2.7 through 16.2.10 discuss the main processes that make up the metamodel.

16.2.7 Agreement Processes

The first step in the systems engineering process is to establish an agreement with the customer on whether or not to build a new system. In general, a new system is built when the outcome of the feasibility study is positive: There is a need for a new system and there is no other system that can be used, or it will be more cost effective to create a new system. When the outcome is negative, the project will end here.

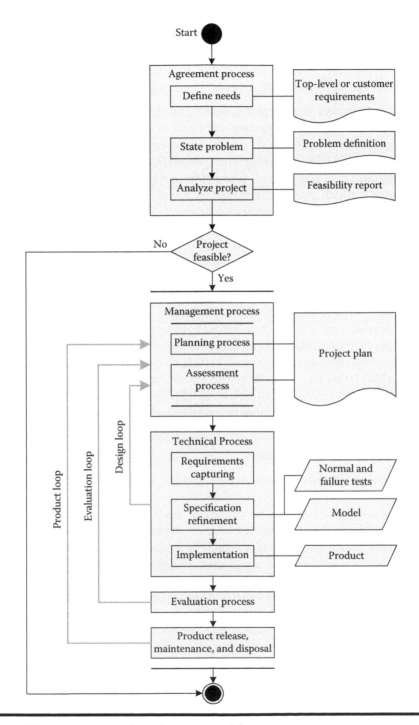

Figure 16.2   Systems engineering metamodel.

The input of the feasibility study comes from an acquisition process and a problem definition process.

16.2.8 Project Processes

After the stakeholders agree that it is feasible to build a new system, the project planning process begins. This process will result in an initial project plan, which can be modified as the technical processes are executed. The technical processes are executed in parallel with the management processes, because a project needs planning, assessment, and control at all time points in the life cycle. The result is a process network in which individual processes exchange information with one another. The guiding principle for process network design is that a management process will assess the output of each individual process. The resulting assessment will determine whether changes have to be made or the output can be passed to the next technical process. An example for this regulatory process can be found in the management process, where changing or updating the schedule is a common task. Furthermore, the project management ensures that objectives are met on both time and budget. To aid project management, a work breakdown structure is made. This is required for the general success of the project. The configuration management process records the changes that are made in either design or requirements. These records enable stakeholders to comment on or amend certain changes in the proposal. Another process, which is very important for project success, is risk control. Identifying risks and thinking about solutions at an early stage will reduce cost, because tackling risks in the later stages of a project is labor intensive and, therefore, costly.

16.2.9 Technical Processes

Technical processes cover the design, development, and implementation phases of the system life cycle. The technical process is executed in parallel with the management process after top-level (or customer) requirements are established in the agreement process. The set of top-level requirements is translated into either software or hardware requirements that define the product functionalities. These software or hardware requirements can lead to a range of alternate product designs. Each requirement is periodically examined for validity, consistency, desirability, and attainability by the management process. Design decisions are made based on these examinations or evaluations. With the chosen design, a requirements analysis is performed and a decision is made on whether or not to do a functional design. Upon making a positive decision, the technical process progresses to the specification stage. The specification stage yields a functional design. In general, the functional design is a product description in the form of a model, which is called "functional architecture." This model describes what the product does and what items the product consists of; this description is known as "allocation and synthesis." The model must be as precise as possible in order to avoid ambiguities in later stages of development. Such precise models are known as "formal" models; with appropriate tools, it is possible to translate these models into implementations without human intervention. The specification step must also define appropriate tests for the implementation. These tests must evaluate both normal and faulty behaviors of the device. After specification refinement, the product can be developed, integrated, and implemented in the user environment. This is done in the so-called implementation phase of the technical process.

16.2.10 Evaluation Processes

The evaluation loop is another feedback path in the systems engineering metamodel. During and after product creation, the following questions must be answered: Does the product do what it is intended to do? Are the requirements met? Are the requirements valid and consistent? How are the requirements prioritized? Therefore, the requirements step in the technical process must outline specific implementation tests. For example, these tests must determine whether or not the implementation of a biomedical system complies with health and safety standards. The specification step must outline how the requirements are achieved. For example, what specific measurements must be performed and what are the acceptable measurement results for complying with health and safety standards. Another scenario that calls for test specification comes from the functionality requirement. In the specification step, mathematical models yield a specific classification rate. A valid implementation test is constructed as follows: The product is provided with the same input vectors as the mathematical model, and the product is checked to determine whether or not it produces the same output as the mathematical model. The product passes the test if the outputs for the model and the implementation are the same.

This section describes how systems engineering can be used to build biomedical systems. To make this description accurate, we follow an example from the area of data mining and knowledge management. To be specific, the example is an automatic mental state detection system. Section 16.2.11 introduces the project, and it is structured such that it reflects the systems engineering metamodel described in Section 16.2.6.

16.2.11 Need Definition: The Case for Automated Mental State Detection

Before we can assess the need to build an automated mental state detection system, we need a clear problem description. The mental state of a human being can be used to diagnose a wide range of diseases. Unfortunately, with current technology the scope of this justification is too wide. We have to focus on specific mental states in order to present a realistic need definition. In the proposed project, we focus on three mental states: (1) normal, (2) epileptic, and (3) alcoholic. The normal mental state serves as a reference against which the two abnormal states are measured. The need to detect the epileptic state arises because roughly 1% of the human population suffers from epilepsy and approximately 30% of epileptic cases cannot be treated (ILAE 1993).

Epileptic state detection systems are needed to soften the impact of this disease on both patients and the environment of patients. A strong support for detecting the alcoholic state comes from the fact that neuronal death is one of the most serious consequences of alcohol exposure during development (West et al. 1990). Both alcoholism and epilepsy are brain disorders. Therefore, it is not surprising that there is a link between them. Seizures occur in relation to alcohol withdrawal, following a period of prolonged intoxication in serious alcoholics. Such seizures constitute a special withdrawal syndrome with important prognostic and therapeutic implications (Morris and Victor 1987). Devetag et al. (1983) conducted a review of convulsive seizures in 153 alcoholic patients. After this review, the authors proposed a classification method. They analyzed the pattern of so-called alcoholic epilepsy and distinguished it from other alcohol-related seizures (Devetag et al. 1983).

We continue the assessment by reviewing state-of-the-art procedures. The currently used technique for detecting different mental states involves either clinical observation or electroencephalogram (EEG) analysis by senior clinicians. Both methods are labor intensive, and they require specialized personal information to conduct the tests and analyze the results.

At the moment, there is no commercial system available that automates mental state detection, although there is a great need for such systems. Therefore, we propose to use signal processing and data mining techniques to automate the detection of mental states. These automated systems lower the cost by freeing specialized personnel and by speeding up the diagnosis process.

16.2.12 Management Process

The management process for the automatic mental state detection system project was executed as weekly meetings between the project members and supervisor/manager. The discussions that took place during these meetings were used for both project planning and assessment. The most tangible result of the management process is a project plan.

In this particular project, we used a Gantt chart (Herrmann 2005) for project planning. From the Gantt chart, shown in Figure 16.3, it is clear that the agreement process took about 1 month. Based on the discussion in Section 16.2.11, we decided to proceed with the project. The duration of the project under discussion was 18 months. Therefore, the time line in the Gantt chart also spans 18 months. The individual work packets follow the systems engineering metamodel. These work packets are customized by filling them with project-relevant tasks. After the agreement process, both technical and management processes were executed. The management process was active for the remainder of the project, whereas the

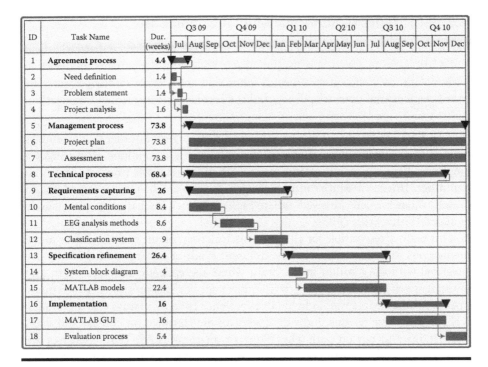

ID	Task Name	Dur. (weeks)	Q3 09			Q4 09			Q1 10			Q2 10			Q3 10			Q4 10		
			Jul	Aug	Sep	Oct	Nov	Dec	Jan	Feb	Mar	Apr	May	Jun	Jul	Aug	Sep	Oct	Nov	Dec
1	**Agreement process**	4.4																		
2	Need definition	1.4																		
3	Problem statement	1.4																		
4	Project analysis	1.6																		
5	**Management process**	73.8																		
6	Project plan	73.8																		
7	Assessment	73.8																		
8	**Technical process**	68.4																		
9	**Requirements capturing**	26																		
10	Mental conditions	8.4																		
11	EEG analysis methods	8.6																		
12	Classification system	9																		
13	**Specification refinement**	26.4																		
14	System block diagram	4																		
15	MATLAB models	22.4																		
16	**Implementation**	16																		
17	MATLAB GUI	16																		
18	Evaluation process	5.4																		

Figure 16.3   Gantt chart of the automated mental state detection system.

technical process gave way to the evaluation process at the end of month 17. The product release maintenance and disposal process is not part of this example.

16.2.13 Technical Process

Management and technical processes are executed in parallel. The technical process follows the systems engineering metamodel: requirements capturing, specification refinement, and implementation. Sections 16.2.13.1 through 16.2.13.4 detail these sequential processes.

16.2.13.1 Requirements Capturing

The requirements capturing process starts with reviewing the current knowledge of automated mental state detection. We define types of abnormal mental conditions and show how these conditions can be detected based on EEG recordings. The detection methods are formalized such that they lead to a set of features that capture information about mental state. This feature set is used for automatic classification. The last topic of requirements capturing for this project is machine classification algorithms. The way in which these individual parts make up the automated mental state classification system is discussed in Section 16.2.13.2.

16.2.13.1.1 Types of Abnormal Mental Conditions

The domain of abnormal mental conditions is a wide and often controversial field studied by neuroscientists and psychologists. Following the systems engineering methodology, we focus on two mental states that are relevant to our project. The first of these mental states is epilepsy, and the second is severe alcohol intoxication.

16.2.13.1.1.1 Epilepsy—Epilepsy refers to a neurological disorder that assembles neurons in the brain in an abnormal condition. It is a condition that creates a temporary disturbance in the paths over which messages travel from one brain cell to another. This disturbance causes messages to come to a halt or get mixed up. As a result, the nerve cells involved radiate high-frequency discharges that are barely regulated (Fisher et al. 2005).

Our brain is responsible for all conscious functions of our body as well as many unconscious functions; therefore, the different kinds of seizures we experience are correlated with the location in the brain where the epileptic activities begin and how extensively and swiftly the activities develop. For this reason, there are many types of seizures. Moreover, individuals experience epilepsy in ways that are unique to them. For these reasons, individual seizures are classified on clinical and EEG observations and not on pathophysiology or anatomy (LAE 1981). Seizures are defined as the characteristics of epilepsy and are used to describe the abnormal fits that are the characteristic symptom of the disease. Epilepsy can be categorized into two main classes, depending on the brain region that is affected during seizures. These two main classes are (1) partial seizure and (2) generalized seizure. Partial seizure refers to abnormal bursts of energy that occur in one region of the brain, whereas generalized seizure is abnormal bursts of energy that affect the nerve cells throughout the brain. Epilepsy affects people regardless of their age, nations of origin, and race.

The cause of epilepsy is clear only in some instances (Frucht et al. 2000). Epilepsy can be due to brain damage caused by a difficult birth, a stern blow to the head, a stroke caused by the lack of oxygen in the brain, or a virus infecting the brain such as in the condition of meningitis. Very rarely is the cause of epilepsy a brain tumor.

Epilepsy with an identified cause is known as "symptomatic" epilepsy. Six out of 10 patients do not have any known cause of epilepsy. These cases come under "idiopathic" epilepsy.

When individuals suffer from epilepsy, their normal EEG pattern is likely to be interrupted by irregular bursts of electrical energy, which are more powerful than signals caused by normal brain activity (Figure 16.4). For patients with seizure disorders or epilepsy, the spikes or bursts of electrical activities in their brains can be observed from the EEG readings recorded.

Two factors that can classify the exact type of the seizure are the (1) patterns of EEG and (2) location of these waves. The brain wave patterns of other disorders

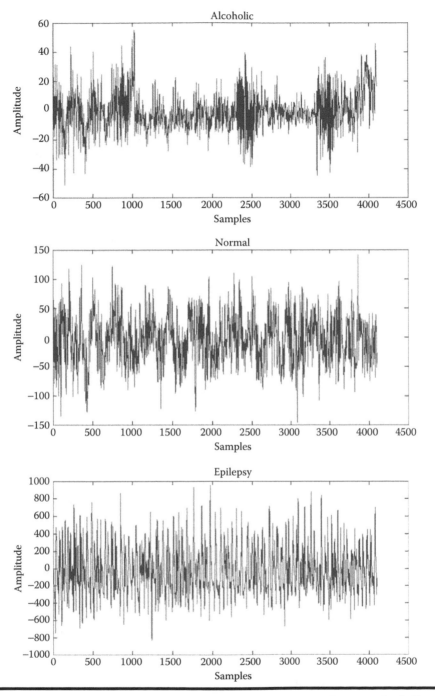

Figure 16.4 Electroencephalogram signals: (a) brain wave showing alcoholism, (b) normal brain wave, and (c) brain wave showing epilepsy.

differ broadly. Generally, brain wave patterns in patients with brain disease, mental retardation, and brain injury show overall slowing. Several types of childhood epilepsy have characteristic epileptic EEG patterns that lead to a specific diagnosis and treatment (Brorson and Wranne 1987).

16.2.13.1.1.2 Alcohol and the Brain

16.2.13.1.1.2 Alcohol and the Brain—All brain functions involve communications among nerve cells in the brain. The communication signals travel over small gaps (synapses) from one neuron to another. When alcohol reaches the brain, it causes physical effects like intoxication and sleepiness, by hindering the signals traveling over the synapses. Prolonged exposure to alcohol causes alterations in the brain that can lead to alcoholism.

Alcoholism is a chronic disorder characterized by a dependence on alcoholic beverages. This treatable disease is linked to an individual's uncontrollable need for alcohol. Both the addictive need for alcohol and the damage to health caused by the disease are sources of huge economic and social costs (NIH 2009).

Alcohol affects the central nervous system. These effects can be analyzed with EEG readings, because EEG measures the electrical activity in the brain, which is governed partly by external sensory stimuli. It has been shown that EEG patterns in alcoholics are different from those of subjects with a normal mental state (Faust et al. 2007). Alcoholic subjects can be differentiated from subjects with a normal mental state based on their EEG alpha activities. A person who consumes alcohol will have a greater increase in slow alpha activity as well as a decrease of rapid alpha activity when compared with a person with a normal mental state. Figure 16.4 shows the typical EEG signals of alcoholic, normal, and epileptic subjects.

16.2.13.1.2 Power Spectral Density Analysis

Power spectral density (PSD) shows the strength of the energy (power) of a time-domain series as a function of frequency. It is defined as the Fourier transform of time-series autocorrelation sequences or the square of the Fourier transform of time series scaled by a proper constant period. The definition of PSD requires the existence of a Fourier transform signal, meaning that the signals are square summable or square integrable. The unit of PSD is power per frequency, and power can be acquired within a specific frequency band by integrating PSD within that frequency range. The PSD is useful for signal analysis, as it identifies oscillatory signals and provides the amplitudes of the oscillations, that is, it provides a measure of both oscillation frequency and power.

In the project, PSD was calculated using Burg's method, a form of autoregressive (AR) modeling. Burg's method for AR spectral estimation is based on minimizing both forward and backward prediction errors at the same time by executing the Levinson–Durbin recursion. In contrast with other AR estimation techniques, Burg's method refrains from calculating the autocorrelation function; the reflection coefficients are estimated directly. The primary advantages of Burg's method lies in its ability to resolve closely spaced sinusoids in signals with low noise levels, and to estimate short data records, in which case the AR PSD estimates are very close to the true values. In addition, Burg's method ensures stable AR models and the algorithm is computationally efficient (Übeyli and Güler 2004).

The accuracy of Burg's method is lower for high-order models, long data records, and high signal-to-noise ratios (which can cause line splitting, or the generation of extraneous peaks in the spectrum estimate). Burg's method is also susceptible to frequency shifts (relative to the true frequency) resulting from the initial phase of noisy sinusoidal signals. This effect is magnified when analyzing short data sequences. Burg's method is capable of resolving closely spaced sinusoids in signals with low noise levels.

16.2.13.1.3 Burg's Algorithm

We define a stationary time series $x_1, x_2, \ldots, x_N$. The autocorrelation function $R_x(k)$ of this series can be computed as follows:

$$R_x(k) = \frac{1}{N} \sum_{i=1}^{N-|k|} x_i x_{x_i+k}^* \tag{16.1}$$

where "*" is the conjugate complex. The PSD $S_x(f)$ is defined as the Fourier transform of $R_x(k)$. We advance the discussion by defining maximum entropy as a spectral estimate that maximizes the entropy rate:

$$H = \frac{1}{2B} \int_{-B}^{B} \log[S_x(f)] df \tag{16.2}$$

with respect to the unknown $R_x(k)$ values. B indicates the signal bandwidth, which corresponds to a sampling period of $T = 1/2B$. With that, the maximum entropy spectrum is calculated as follows:

$$\hat{S}_x(f) = \frac{P_m}{B\left|1 + \sum a_k \exp(-j2\pi kkT)\right|^2} \tag{16.3}$$

where P_m is the final prediction error. From this equation it follows that the maximum entropy spectrum has the same spectrum as an AR model with coefficients a_i. Burg's algorithm constitutes a method to estimate these coefficients.

Let the linear prediction estimate of x_i be given by $\hat{x}_i$:

$$\hat{x}_i = -\sum_{k=1}^{P} a_k x_{i-k} \tag{16.4}$$

for an mth-order AR model. With the prediction error defined as $e_i = x_i - \hat{x}_i$, the prediction error power is as follows:

$$P_p = \sum_{k=0}^{P} a_k R_x(-k), \ 1 \le p \le m \tag{16.5}$$

Equation 16.5 can be expressed as a recursive relation:

$$P_p = P_{p-1}\left(1 - |r_p|^2\right) \tag{16.6}$$

where r_m is the so-called reflection coefficient. These coefficients can be obtained from the time series with the well-known Levinson–Durbin recursion (Levinson 1947; Durbin 1960).

Burg has shown that the absolute value of all refection coefficients must be smaller than 1 for the filter to be of minimum phase. This coefficient is also sufficient for the filter to be stable. To achieve both minimum phase and stability, Burg's method minimizes the average prediction error power:

$$P_p = \frac{1}{2}\left(P_{f,p} + P_{b,p}\right) \tag{16.7}$$

where $P_{f,p}$ is the forward error power and $P_{b,p}$ is the backward error power.

$$P_{f,p} = (N - p)^{-1} \sum_{i=1}^{N-p} e^p_{f,i} e^{p*}_{f,i} \ P_{b,p} = (N - p)^{-1} \sum_{i=1}^{N-p} e^p_{b,i} e^{p*}_{b,i}, \tag{16.8}$$

where $e_{f,i}$ and $e_{b,i}$ are forward and backward errors, respectively. The starting point for Burg's algorithm is $p = 0$; thus, it follows that

$$P_0 = R_x(0) = \frac{1}{N} \sum_{i=1}^{N} x_i x_i^* \tag{16.9}$$

Having established $P(0)$, the subsequent average prediction error powers follow from Equation 16.6. With $\partial P_1/\partial r_1 = 0$ we find r_1 for which P_1 is minimum. The general expression for reflection coefficients is as follows:

$$r_p = \frac{-2 \sum\limits_{i=p+1}^{N} e_{f,i}^{p-1} e_{f,i-1}^{(p-1)*}}{\sum\limits_{i=p+1}^{N} \left(\left| e_{f,i}^{p-1} \right|^2 + \left| e_{f,i-1}^{(p-1)} \right|^2 \right)}, \quad p = 1, 2, \ldots \qquad (16.10)$$

It can be shown that this equation yields a stable minimum-phase filter.

16.2.13.1.4 Peak Amplitude Detection

Figure 16.5 shows the results of Burg's method of PSD estimation used on a normal EEG signal. The first parameter, **f1 Max**, represents the mean and variance of the peak power. The second parameter, **a1 Max**, indicates the mean and variance of the frequency where **f1 Max** is located. The third parameter, **f1 Min**, corresponds to the peak power divided by its corresponding frequency.

16.2.13.1.5 Machine Classification

After surveying feature extraction methods, we now focus on machine classification. Classification of data is a common procedure for many biomedical processing systems, especially automated diagnosis systems. In machine classification, objects are placed in groups with quantitative details of one or more features in objects, such as traits, variables, and characters, using a training set of previously classified items. Based on both previous studies (Faust et al. 2010) and experience, we selected Gaussian mixture model (GMM) as the algorithm for automated classification.

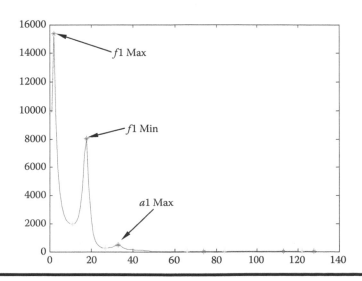

Figure 16.5   Local maxima and minima in Burg's power spectral density.

The GMM is commonly used to solve clustering problems in areas such as statistical data analysis and density estimation. One of the important characteristics of GMM is its ability to form smooth approximations for any arbitrarily shaped densities. As the actual data has a multimodel distribution, GMM provides a useful tool to model the characteristics of the data. Another useful characteristic of GMM is the possibility of employing a diagonal covariance matrix instead of a full covariance matrix. Hence, the amount of computational time and complexity is reduced significantly. The GMM has been extensively used in many areas of pattern recognition and classification, and brings about great success in the area of identification and verification.

In general, this technique has been especially successful in speaker identification and verification (Reynolds 2000; Seo 2001). Density estimation uses a sample of collected data to obtain a population probability density function.

The probability mixture model is a probability distribution that results from a convex computation of other probability distributions. The probability mass functions of a discrete random variable X is defined as the weighted sum of its component distributions $f_{Y_i}(x)$:

$$f_X(x) = \sum_{i=1}^{n} a_i f_{Y_i}(x) \tag{16.11}$$

for some mixture proportions $0 \leq a_i \leq 1$ where $a_1 + \ldots + a_n = 1$. Using this equation information, we can establish a parametric mixture model with unknown parameters Θ_i:

$$f_X(x) = \sum_{i=1}^{n} a_i f_I(x; \Theta_i) \tag{16.12}$$

In this project, the model parameters Θ_i are estimated through training so that we can maximize the likelihood of the observations using the expectation-maximization algorithm (Bilmes 1997).

16.2.13.2 Specification Refinement

This section describes specification refinement, which defines how to build the mental state detection system. Following the systems engineering metamodel, specification refinement is part of the technical process. The discussion starts with the block diagram shown in Figure 16.6.

In accordance with systems engineering methodology, we outline implementation tests here in the specification section. The first of these tests is based on the statistical method called "analysis of variance" (ANOVA) between groups (Addelman 1969). This method uses variances to decide whether or not the means, which are evaluated independently for each class of input parameters, are different. The test results are documented with a so-called p value. A high p value indicates that the means of

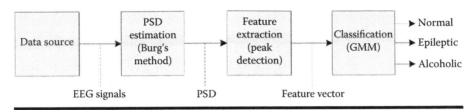

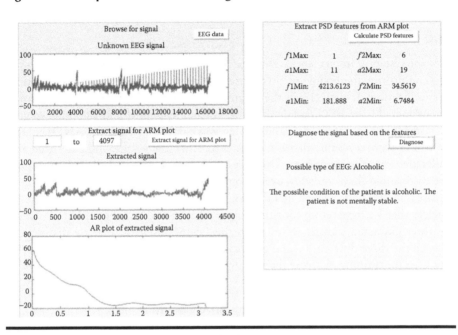

Figure 16.6 Implementation block diagram.

Figure 16.7 Screenshot showing an alcoholic case.

individual classes are similar. Therefore, for classification, problems with low p values are desired, because low values indicate that the features form independent classes.

16.2.13.3 Implementation

Following the systems engineering metamodel, we next implement our system. As indicated in the project plan, shown in Figure 16.3, this step takes longer than all the other steps, except for running MATLAB® models. The fact that modeling requires more time than implementation indicates a very modern workflow, because most of the time is spend on modeling. We focus on the systems engineering aspects of this work. To provide some evidence that the implementation was successful, we present a screenshot of the graphical user interface (GUI) for the automated mental state classification system for correctly classified normal, alcoholic, and epileptic mental states in Figure 16.7. The GUI consists of input EEG data and its waveform image, the waveform image of extracted EEG data for AR modeling (Burg's method), the waveform of an AR plot on the extracted signal, a display textbox on the values of the extracted signal, and a classification textbox.

In the first part of this GUI, we browse for the EEG signal so that we can analyze and display it. In the second part, we window the relevant section of the EEG signal and compute the PSD with Burg's method. The last two parts of the GUI are feature extraction and automated diagnosis.

16.3 Results

After the system is implemented, the next logical step is testing. Following the systems engineering methodology, this section describes the tests that are outlined in the specification. Table 16.1 lists the mean and variance for individual features that were extracted from EEG signals showing normal, epileptic, and alcoholic stages. The last column contains the p value, which was calculated using the ANOVA test. The results show a p value that is almost zero. This low p value indicates that the results are clinically significant. This test helps to determine whether data groups have discernible characteristics, even though they show identical mean values.

The statistical data, shown in Table 16.1, reflects the classification rate for the 10 matrices that we have set up for this project to train and test the GMM classifier. The three extracted features were fed to the GMM inputs to make a distinction between the three different mental conditions. Out of the 330 data obtained, 90% were used for training, whereas the remaining 10% were used to test the classifier.

Table 16.1 ANOVA Test Results Using Burg's Method

	Normal	*Epileptic*	*Alcoholic*	p *Value*
f_1 Max	1.705 ± 0.457	6.8700 ± 2.98	1.3333 ± 0.479	0.0001
a_1 Max	16.685 ± 3.44	33.580 ± 31.3	16.533 ± 9.80	0.0001
f_1 Min	80907 ± 2.163E + 05	7.41485E + 05 ± 1.046E + 06	1532.0 ± 1.621E + 03	0.0001
a_1 Min	16350 ± 3.070E + 04	2.04595E + 05 ± 2.811E + 05	92.660 ± 141	0.0001
f_2 Max	10.040 ± 3.02	25.760 ± 24.5	12.167 ± 8.73	0.0001
a_2 Max	29.690 ± 13.1	63.720 ± 30.0	13.033 ± 11.1	0.0001
f_2 Min	1836.1 ± 890	80142 ± 1.065E + 05	49.483 ± 64.1	0.0001
a_2 Min	751.10 ± 1.001E + 03	5172.3 ± 1.735E + 04	10.273 ± 19.6	0.0005

For this project, an average classification rate of 96.9697% with sensitivity and specificity of 100% and 95.2%, respectively, was obtained.

16.4 Conclusions

This chapter introduces the use of systems engineering in biomedical projects involving EEG scans of epileptics and alcoholics. We argue that biomedical systems are becoming more complex each day, because advanced diagnosis support systems require a high level of complexity to relieve medical personnel from even simple routine tasks. To cope with this increasing complexity, it is necessary to move away from evolutionary system designs and move toward systems-based development. Evolutionary systems designs cope with the complexity problem only from one revision to the next. In contrast, systems-based development incorporates feedback loops within the (systems engineering) design methodology. Therefore, it can cope with the complexity problem much better. To support this argument, Section 16.2 of this chapter contains a case study in which the systems engineering methodology was applied to a biomedical project. We used the systems engineering methodology to design and develop an automated mental state decision system. The design methodology places extra emphasis on structure; for this reason, the presentation of this example project follows the structure outlined by the systems engineering metamodel. However, it is not possible to capture the feedback process adequately; therefore, the discussion includes only the final, agreed-upon version of the individual steps.

Systems engineering is applicable to a wide range of biomedical projects. When it is employed in these projects, as shown by our examples, it guides both technical and management teams such that all members see the bigger picture, thereby providing a holistic view and more technical detail, and saving time. Therefore, each management decision, which impacts the work of the teams, is seen as an intrusion and usually causes resistance. Systems engineering aims to minimize this resistance by providing adequate structures for information exchange between the teams. This makes the development process flexible and ultimately enables the project to cope with the inherent complexity problem of modern biomedical systems.

References

Adcock, R. 2001. *Principles and Practices of Systems Engineering.* Fort Belvoir, VA: Defense Acquisition University Press.

Addelman, S. 1969. The generalized randomized block design. *Am Stat* 23 (4):35–6. http://www.jstor.org/stable/2681737 (accessed December 2010).

Angus, I. 2001. An introduction to erlang B and erlang C. *Telemanagement* 187:6–8.

Arnold, S., and ISO/IEC 15288. 2004. Systems and software engineering—System life cycle processes.

Bilmes, J. 1997. *A Gentle Tutorial on the EM Algorithm and its Application to Parameter Estimation for Gaussian Mixture and Hidden Markov Models.* Technical report ICSI-TR-97-021, University of Berkeley.

Boehm, B. 2005. Some future trends and implications for systems and software engineering processes. In *Systems Engineering.* Vol. 9, No. 1 (2006).

Brorson, L. O., and L. Wranne. 1987. Long-term prognosis in childhood epilepsy: Survival and seizure prognosis. *Epilepsia* 28(4):324–30.

Chestnut, H. 1965. *Systems Engineering Tools.* Hoboken, NJ: Wiley.

Cornell University. 2007a. *Systems Engineering Program at Cornell University.* Cornell University. http://systemseng.cornell.edu/people.html (accessed May 25, 2007).

Cornell University. 2007b. *Core Courses, Systems Analysis–Architecture, Behavior and Optimization.* University. http://systemseng.cornell.edu/CourseList.html (accessed May 25, 2007).

Devetag, F., G. Mandich, G. Zaiotti, and G. Toffolo. 1983. Alcoholic epilepsy: Review of a series and proposed classification and etiopathogenesis. *Ital J Neurol Sci* 4(3):275–84.

DoD. 1992. *Department of Defense, Military Standard-499B Systems Engineering,* (Draft). US Department of Defense.

DoD. 2001. *Systems Engineering Fundamentals.* Fort Belvoir, VA: Defense Acquisition University Press.

Durbin, J. 1960. The fitting of time series models. *Rev Int Statist Inst* 28:233–43.

Elm, J. P. 2005. Surveying Systems Engineering Effectiveness. In *NDIA SE Conference,* 24–7. San Diego, CA.

Faust, O., R. U. Acharya, R. A. Allen, and C. M. Lim. 2007. Analysis of EEG signals during epileptic and alcoholic states using AR modeling techniques. *IRBM,* San Diego, CA, 1(29):44–52.

Faust, O., R. U. Acharya, C. M. Min, and B. H. C. Sputh. 2010. Automatic identification of epileptic and background EEG signals using frequency domain parameters. *Int J Neural Syst* 20(2):159–76.

Fisher, R., W. van Emde Boas, W. Blume, C. Elger, P. Genton, P. Lee, and J. Engel. 2005. Epileptic seizures and epilepsy: Definitions proposed by the International League Against Epilepsy (ILAE) and the International Bureau for Epilepsy (IBE). *Epilepsia* 46(4):470–2.

Frucht, M. M., M. Quigg, C. Schwaner, and N. B. Fountain. 2000. Distribution of seizure precipitants among epilepsy syndromes. *Epilepsia* 41(12):1534–9.

G47- Systems Engineering Committee. 1999. *Processes for Engineering a System.* ANSI/EIA-632-1999, ANSI/EIA.

Goode, H. H., and R. E. Machol. 1957. *System Engineering: An Introduction to the Design of Large-scale Systems.* New York: McGraw-Hill. 8. LCCN 56-11714.

Hall, A. D. 1962. *A Methodology for Systems Engineering.* Van Nostrand Reinhold.

Herrmann, J. W. 2005. History of decision-making tools for production scheduling. In *Proceedings of the 2005 Multidisciplinary Conference on Scheduling: Theory and Applications,* New York, July 18–21.

Honour, E. C. 2004. Understanding the Value of Systems Engineering. In *Proceedings of the INCOSE International Symposium,* Toulouse, France.

ILAE (Commission on Epidemiology and Prognosis, International League Against Epilepsy). 1993. Guidelines for epidemiologic studies on epilepsy. Commission on Epidemiology and Prognosis, International League Against Epilepsy. *Epilepsia* 34(4):592–6.

INCOSE Group (International Council on Systems Engineering Group). 2004. Genesis of INCOSE. http://www.incose.org/about/genesis.aspx (accessed July 11, 2006).

INCOSE Education & Research Technical Committee. 2006. *Directory of Systems Engineering Academic Programs*. http://www.incose.org/educationcareers/academic-programdirectory.aspx (accessed July 11, 2006).

INCOSE group (Volunteer group of contributors within the International Council on Systems Engineering). 2000. In *Systems Engineering Handbook*, ed. J. Whalen, R. Wray, and D. McKinney, version 2a. INCOSE, San Diego, CA.

LAE. 1981. Proposal for revised clinical and electroencephalographic classification of epileptic seizures. From the commission on classification and terminology of the international league against epilepsy. *Epilepsia* 22(4):489–501.

Levinson, N. 1947. The Wiener RMS error criterion in filter design and prediction. *J Math Phys* 25:261–78.

MIT. 2007. *ESD Faculty and Teaching Staff*. Engineering Systems Division, MIT. http://esd.mit.edu/people/faculty.html. (accessed May 25, 2007).

Morris, J. C., and M. Victor. 1987. Alcohol withdrawal seizures. *Emerg Med Clin North Am* 5(4):827–39.

NASA. 2010. *NASA Langley Research Center's Contributions to the Apollo Program*. http://www.nasa.gov/centers/langley/news/factsheets/Apollo.html (accessed May, 2010).

NIH (National Institute of Health). 2009. Alcoholism. http://www.nlm.nih.gov/medlineplus/ency/article/000944.htm.; National Library of Medicine (accessed January 15, 2009).

Oliver, D. W., P. Timothy, J. G. K. Kelliher Jr. 1997. *Engineering Complex Systems with Models and Objects*, 85–94. McGraw-Hill.

Ramo, S., and R. K. St. Clair. 1998. *The Systems Approach: Fresh Solutions to Complex Problems Through Combining Science and Practical Common Sense*. Anaheim, CA: KNI, Inc. http://www.incose.org/ProductsPubs/DOC/SystemsApproach.pdf.

Reynolds, D., T. Quatieri, and R. Dunn. 2000. Speaker Verification Using Adapted Gaussian Mixture Models. *Digital Signal Processing* 10(23):19–41.

Sage, A. P. 1992. *Systems Engineering*. Hoboken, NJ: Wiley IEEE.

Sage, A. P., and S. R. Olson. 2001. *Modeling and Simulation in Systems Engineering*. New York: SAGE Publications. http://intl-sim.sagepub.com/cgi/content/abstract/76/2/90. (accessed June 2, 2007).

Schlager, J. 1956. Systems engineering: Key to modern development. *IRE Trans EM*-3:64–6.

Sadeh, E. 2006. Societal impacts of the Apollo program. In *AIAA Space 2006 conference*, San Jose, CA USA.

Seo, C., K. Y. Lee, and J. Lee. 2001. GMM based on local PCA for speaker identification. *Electron Lett* 37(24):1486–8.

SEOR. 2007. *The SE VEE. SEOR*. Fairfax, VA: George Mason University. http://www.gmu.edu/departments/seor/insert/robot/robot2.html (accessed May 26, 2007).

Shishko, R. 1995. *NASA Systems Engineering Handbook*. Kennedy Space Center, FL: Diane Pub Co 1995.

Smith, E. C. Jr. 1962. Simulation in systems engineering. *IBM Syst J* 1(1):33–50.

Übeyli, E. D., and I. Güler. 2004. Spectral analysis of internal carotid arterial Doppler signals using FFT, AR, MA, and ARMA methods. *Comput Biol Med* 34(4):293–306.

Valerdi, R., C. Miller, and C. Miller. 2004. Systems engineering cost estimation by consensus. In *17th International Conference on Systems Engineering*, September 2004, Las Vegas, NV.

West, J. R., C. R. Goodlett, D. J. Bonthius, K. M. Hamre, and B. L. Marcussen. 1990. Cell population depletion associated with fetal alcohol brain damage: Mechanisms of BAC-dependent cell loss alcoholism. *Clin Exp Res* 14(6):813–8.

www.nasa.gov/centers/langley/news/factsheets/Apollo.html (accessed December, 2010).

Index

Milton Keynes UK
Ingram Content Group UK Ltd.
UKHW031125141024
449569UK00006B/434